DIFFUSE MATTER IN THE SOLAR SYSTEM: COMET HALLEY AND OTHER STUDIES

DIFFUSE MATTER IN THE SOLAR SYSTEM: COMET HALLEY AND OTHER STUDIES

PROCEEDINGS OF A
ROYAL SOCIETY DISCUSSION MEETING
HELD ON 21 AND 22 MAY 1986

ORGANIZED BY SIR WILLIAM McCREA, F.R.S.,
G. TURNER, F.R.S., R. HUTCHISON, A. J. MEADOWS
AND C. T. PILLINGER

AND EDITED BY
G. TURNER, F.R.S., AND C. T. PILLINGER

LONDON
THE ROYAL SOCIETY
1987

Printed in Great Britain for the Royal Society
by the
University Press, Cambridge

ISBN 0 85403 325 4

First published in *Philosophical Transactions of the Royal Society of London*,
series A, volume 323 (no. 1572), pages 247–449

Copyright

Published by the Royal Society
6 Carlton House Terrace, London SW1Y 5AG

CONTENTS

[Five plates]

CONTENTS

Phil. Trans. R. Soc. Lond. A **323**, 249–250 (1987)
Printed in Great Britain

Introductory remarks

BY SIR WILLIAM McCREA, F.R.S.
Astronomy Centre, University of Sussex, Brighton BN1 9QH U.K.

Edmund Halley was one of the world's greatest scientists. The Royal Society is proud of Halley's almost lifelong intimate involvement in its affairs. He himself belonged to the first generation of astronomers to make telescopic observations of the comet that bears his name. We belong to the first generation to send missions to make effective material contact with that comet. Here we have the privilege of the presence among us today of some of the pioneering participants in that enterprise.

I shall not say, 'Little did Halley think...' On the contrary, everything we know about his vision and foresight should convince us that he would not be totally surprised at the sort of observations about which we shall hear. And it is meet that as nearly as possible the first public discussion of the results should be in Halley's native London at a meeting of the Society to which he contributed so much.

This means, however, that the organizers had to arrange a programme before they knew what material would be available. Indeed, when they started to arrange it the comet must have been still about as far away as Jupiter, and it was to be a year or more before any of the space missions to observe it had left the ground. Also, no matter how successful the missions have been, it is still too soon to expect to be told their detailed results. This may be all to the good because it has, we hope, resulted in a programme such that what we shall learn about Comet Halley will be seen in a wider astronomical context.

Astronomers must naturally emphasize the extent to which their knowledge of anything outside the Earth comes from the electromagnetic radiation they are able to receive from it. But that is not quite the whole story. Bits of stone and iron do fall upon Earth from outside; smaller fragments enter the atmosphere and burn up; other interplanetary dust and gas is detected in other ways; material particles from the Sun produce aurorae; other more energetic (cosmic ray) particles may come from the Sun but mostly come from far outside the Solar System, some probably from outside the Galaxy. As regards the dust and gas, astronomers still do not know how much is orginal Solar-System material and how much is picked up as the System endlessly threads its way through its galactic environs.

One feature plays a special role. A 'new' comet (in the accepted sense) comes to the vicinity of Earth from a distance comparable to that of the Sun's nearest stellar neighbours, and it may return to such a distance. Thus we recognize that we are in some sense materially linked to our stellar surroundings by portions of matter that travel to and fro between us and them. So we may have to regard a comet as a bit of the astronomical Universe that is somehow shared between us and the rest of that Universe. On the other hand, the Solar System may possess a huge store of comets forming a vast surrounding cloud from which, from time to time, one gets deflected into a motion that brings it sufficiently close to the Sun for us to have sight of it; indeed, as a 'new' comet. Were this so, new comets would be specimens of material of the Solar System that has been in deep freeze ever since the System was formed, presumably some four-

and-one-half billion years ago. Comets would then be significant as a link between the Solar System and its own past, rather than a link with its present surroundings.

This last is a main reason why astronomers are so eager to have a close inspection of a cometary nucleus. They cannot hope to obtain this in the case of a 'brand new' comet, because there would never be time between the discovery of the comet and its subsequent disappearance sufficient for the mounting of a *Giotto*-type mission to observe it. Comet Halley provides the best known compromise. It does spend most of its time in a state of deep freeze, but it comes back in a well-predictable fashion.

The first half of the meeting will begin with the general study of comets, and of sorts of material that may have gone into their formation. Then we are to learn about the physical state and behaviour of small bodies and diffuse matter now in the Solar System, and the significance of all this for knowledge of the past evolution of the System.

The second half will be very much Halley's day. Special attention is naturally to be given to observations made during the present apparition of Comet Halley, with some review of the significance of what has already been learned from these observations. But there is to be some attention to other cometary topics as well.

Anyone acquainted with the field is likely to wonder why some particular topic does not figure explicitly in the programme. Some such will be found simply to be subsumed under a more comprehensive heading; others will be dealt with in brief contributions, or in the discussion sections.

Phil. Trans. R. Soc. Lond. A **323**, 251–267 (1987)

Printed in Great Britain

Nucleosynthesis contributions to the solar nebula

By M. Arnould

*Institut d'Astronomie, d'Astrophysique et de Géophysique, CP 165,
Université Libre de Bruxelles, avenue F. D. Roosevelt 50, B-1050 Bruxelles, Belgium*

The discovery of isotopic anomalies in meteorites suggests that the Solar System is made of material from compositionally different and imperfectly mixed reservoirs. One of them, which comprises the bulk Solar System material, is considered to be made of the well-homogenized ashes of many nucleosynthesis events. Its composition can be studied through models of the chemical evolution of the Galaxy. The main nucleosynthetic agents responsible for that evolution are very briefly reviewed, as well as the level of reliability of the model predictions.

The remaining reservoir(s) contain(s) isotopically anomalous material, which probably represents only a very minute fraction of the total Solar System material. The great astrophysical importance of the existence of such reservoirs is emphasized. Some selected examples are given to illustrate the rich diversity of potential nucleosynthetic mechanisms that possibly produced the isotopically anomalous material. The difficulties encountered and uncertainties involved in trying to interpret the array of anomalies within nucleosynthesis models are stressed, as well as the key importance of correlated anomalies.

1. Introduction

The knowledge of the elemental and isotopic composition of the fraction of galactic material out of which the Solar System formed (such a composition being referred to as 'primordial' is the following) has always played a prominent role in studies of the physicochemical evolution of the Solar System. It has also greatly influenced the science of nucleosynthesis, which tries in particular to unravel the astrophysical origin of the primordial Solar System abundances.

Meteoritic analyses have always played a pivotal role in attempts to determine such a composition. In addition to the wealth of information that they provide on the physical and chemical conditions at the time of formation of the Solar System, as well as on its further history, meteorites have indeed been shown to provide high-quality information on the bulk primordial Solar-System composition. In particular, the CI1 carbonaceous chondrites, considered as the least-altered samples of primitive solar matter presently accessible to Man, are now generally recognized as the best, most complete and coherent source of data when estimating the primordial Solar System abundances of non-volatile elements (see, for example, Anders & Ebihara 1982). Solar information, which now comes in quite good agreement with CI1 data for a large variety of elements (see Grevesse 1984), has to be used for the volatile elements H, He, C, N, O, Ne (some other astronomical data may also be useful in this respect), whereas interpolations guided by theoretical considerations are still required in some specific cases (Ar, Kr, Xe, Hg). Evaluating the primordial abundances derived on such grounds, Anders & Ebihara (1982) conclude that 'Though a few weak spots remain, most abundances now seem to be accurate to $\pm 10\%$ or better'.

On the other hand, and apart from a few exceptions (concerning largely volatile elements), terrestrial isotopic abundances are taken as representative of the Solar System as a whole. This common practice is justified in particular by the realization that the bulk terrestrial isotropic composition closely follows the meteoritic pattern, and by the canonical model of a uniform primordial composition. Also note the inability of the present spectroscopic techniques to provide highly reliable solar isotopic compositions for most of the elements.

In the past few years, it has been discovered that primitive meteorites contain a small amount of isotopically 'anomalous' matter (see, for example, Begemann 1980; Wasserburg & Papanastassiou 1982; Papanastassiou 1985; Shima 1986 for reviews). Operationally, such anomalies can be defined as deviations from the bulk Solar System isotopic composition that cannot be understood on the grounds of processes that are known to affect the meteoritic isotopic patterns (see, for example, Shima 1986).

Now there are more than 20 elements that are known to exhibit isotopic anomalies in meteorites. Even if their characterization is not free from ambiguities and difficulties (e.g. normalization problems), their very existence is generally regarded as providing new clues to many important astrophysical problems, like the nucleosynthetic yields of various stars, the physics and chemistry of interstellar dust grains, the formation and growth of grains in the vicinity of nucleosynthetically active objects, the circumstances under which stars (and especially Solar System-type structures) can form, as well as the early history of the Sun (in its T-Tauri phase†) and of Solar System solid bodies. In particular, such anomalies contradict the canonical model of a homogeneous and gaseous protosolar nebula, and suggest instead that the Solar System solid bodies started to condense out of compositionally different and not perfectly mixed reservoirs made of gas and/or dust.

One of those reservoirs, which accounts for the bulk Solar System material, is generally considered as made of the ashes of many nucleosynthetic events, which could mix in the interstellar medium (or at least in that portion of it to be incorporated in the Solar System) before isolation of the protosolar nebula from the galactic material. The understanding of the composition of that (by definition) isotopically normal reservoir (made of gas and possibly of dust, with a possible gas–dust fractionation) is the formidable problem addressed by most works in the field of nucleosynthesis, and by models of the chemical evolution of the Galaxy. This question is briefly examined in §2.

The remaining reservoir(s) contain(s) the isotopically anomalous material, which probably comprises only a very minute fraction of the total Solar System material. The very origin of such reservoir(s) is still debated and uncertain. In fact, several models have been proposed for its (their) origin, namely the following.

1. *Model 1*. Portions of the pristine Solar System matter have been differentially irradiated by an intense flux of energetic particles from the Sun. This model thus proposes a local origin for the isotopic anomalies.

2. *Model 2*. The Solar System contamination is due to different interstellar-medium (ISM) grain components, the composition of which is more or less markedly different from the solar

---

† T-Tauri stars are variable stars with spectral types G to M, which are located above the main sequence band in the Hertzsprung–Russell diagram. They are interpreted as young (age not more than about 10 Ma) pre-main sequence low-mass stars ($M \lesssim 3\,M_\odot$, where $M_\odot$ is the solar mass). They are evolving by gravitational contraction towards the main-sequence phase, which corresponds to hydrogen-burning in the central stellar regions (for a review of T-Tauri star properties, see, for example, Cohen 1984).

one, and the production of which is not tightly related spatially and temporally to the Solar System formation. Those grains might be created around red giants, novae, supernovae (SNe), or massive mass losing stars (MMLS), especially of the Wolf–Rayet (WR) type.† These various grains carry (at least part of) the nucleosynthesis signature of their production sites, and may lead to isotopic anomalies in meteorites through a variety of mechanisms (incomplete homogenization of, or fractionation between grain components and gas phase). The fate of such grains (referred to in the following as fossil grains) in the ISM has to be followed with an 'extended' model for the chemical evolution of the Galaxy.

3. *Model 3.* The exotic material comes in the form of gas and/or grains from (i) a single or several MMLS (of the Of‡ or WR type) or SN(e), or (ii) a highly evolved low-mass star (*ca.* 1 $M_{\odot}$) during its Asymptotic-Giant-Branch (AGB) or post-AGB phase§, and its production is tightly related spatially and temporally to the Solar System formation. In fact, through radiation pressure or steady and/or catastrophic mass losses, one or another of the contaminating objects could also have caused the compression of the proto-solar cloud to the point where gravitational instability led to the formation of the Solar System.

The merits and difficulties of the various models sketched above, and referred to as models 1, 2 and 3 in the following, have been discussed in many places (e.g. Clayton 1979, 1982; Arnould & Nørgaard 1981; Wasserburg & Papanastassiou 1982; Papanastassiou 1985). Without entering such a debate here, let us simply state that all scenarios probably have their difficulties and virtues. In fact, it is quite plausible that a composite scenario combining at least some of the above-mentioned models is more appropriate for the interpretation of the whole array of anomalies.

Whatever the precise model or combination of models, the anomalous Solar System material offers the very exciting perspective of testing theoretical nucleosynthesis predictions for a very limited number of events, even possibly a single one. This is particularly true in model 3. As discussed previously, such a situation is in marked contrast with the one encountered when trying to confront nucleosynthesis models with the bulk primordial Solar System composition. No wonder then that many new nucleosynthesis studies have been stimulated by the discovery of isotopic anomalies in meteorites.

At this point, it has to be emphasized, however, that the data do not strictly require a purely nuclear origin for the isotopic anomalies if suitable non-nuclear effects different from those already subtracted out to extract an anomaly (see above) can be found. In fact, several such mechanisms have been proposed, calling for mass-independent fractionation processes (see, for example, Consolmagno & Cameron 1980; Thiemens & Heidenreich 1983; Heidenreich & Thiemens 1985; Thiemens & Jackson 1985; Robert 1985; Reisse & Mullie 1985). Let us also note that Manuccia & Clark (1976) and Arrhenius *et al.* (1978) have reported large nitrogen isotopic fractionations (which are not explicitly stated to be mass-independent) in glow-discharge

---

† WR stars are massive (not less than about 10 $M_{\odot}$) stars characterized by huge mass-loss rates (typically of the order of 3×10^{-5} $M_{\odot}$ per year). Various WR subtypes are defined, following the identity of the elements that dominate their emission spectra: WN, WC, and WO stars have especially intense lines of He and N, He and C, and He and O, respectively. The WN and WC or WO stars are thought to correspond to early and late central He-burning stages, respectively (see, for example, Prantzos *et al.* 1986 *c, d* and references therein).

‡ Of stars are defined as emission-line O stars. They are considered to be in a central H-burning phase intermediate between the O and WR phases.

§ AGB stars are interpreted as low- and intermediate-mass (not more than about 8 $M_{\odot}$) stars composed of a double (H- and He-) burning shell surrounding a degenerate C–O core (see, for example, the review by Iben & Renzini 1983).

and plasma reaction systems. Such conditions would not normally be considered by meteoriticists, but none the less the data might be relevant to interpretations of isotope anomalies.

In this work we do not intend to present an exhaustive review of the various nucleosynthesis scenarios that have been proposed to account for all the discovered isotopic anomalies. We will instead concentrate on some selected cases that could be the signature of some specific nucleosynthetic events associated with different classes of stars and stellar evolution phases. This is the subject of §§3–6.

2. THE BULK SOLAR-SYSTEM COMPOSITION AND THE CHEMICAL EVOLUTION OF THE GALAXY

As previously stated, the bulk Solar System composition is thought to be made of the well-homogenized ashes of many nucleosynthesis events that occurred at a pregalactic era (during the Big Bang epoch of nuclide production, or after it, but before galaxy formation), or within the Galaxy before isolation of the protosolar nebula. In such views, the bulk composition has to be interpreted through models for the chemical evolution of the galactic disc in the solar neighbourhood.

Such a problem has been the subject of many investigations (see, for example, Tinsley 1980), and it goes largely beyond the scope of this work to explore its many intricacies in detail. Let us just stress a few points which may be of interest here.

A typical chemical evolution model involves several major ingredients, namely (i) the frequency at which stars of a given mass are formed in the galactic disc, (ii) the rate at which a given nuclide is withdrawn from the ISM and locked-up in forming stars ('astration'), or, on the contrary, is ejected into the ISM by a given star (through mere survival or 'fresh' nucleosynthesis), and (iii) the amount and composition of material possibly infalling on the galactic disc. It is generally assumed that such a material has mainly a Big-Bang composition, dominated by D, ^{3}He, ^{4}He, and possibly ^{7}Li (see, for example, Boesgaard & Steigman 1985).

Thus, the mass balance of a given nuclide in the ISM appears to be very difficult to establish. Apart from the stellar creation rate, it requires the knowledge of the evolution of a large variety of stars (say with masses M in the $1 \lesssim M \lesssim 100\ M_\odot$ range), and of the mass of each species a given star is able to eject in the ISM through non-explosive or explosive mass losses.

Roughly speaking, the most active nucleosynthetic agents in the Galaxy are the following.

(i) Low- and intermediate-mass ($1 \lesssim M \lesssim 8\ M_\odot$) red giants, which suffer steady mass losses and planetary nebula ejection before becoming white dwarfs. The white-dwarf material cannot be recycled into the ISM, except perhaps if such a configuration is involved in a SN explosion, as mentioned in (iv). The nucleosynthetic red-giant yields have been discussed in many places (e.g. Iben & Renzini 1983). Let us simply say that the ejected material is especially enriched with various H- and He-burning products, and possibly also with products of the s-process†.

---

† In nuclear astrophysics, the stable isotopes of the elements heavier than Fe are classically divided into the so-called s-, r- and p-nuclei (see, for example, Cameron 1982). The s-nuclei are the most tightly bound isotopes of an element, whereas the r- and p-nuclei are somewhat less bound neutron-rich and neutron-deficient isotopes, respectively. Many stable heavy nuclides have a mixed (sr- or sp-) character. That nomenclature relates directly to the principal mode(s) of production of the heavy nuclides, referred to as the s-, r- and p-processes. Schematically, the s- and r-processes are made of chains of neutron captures, which take place at rates that are typically smaller and larger than β-decay rates, respectively (of course, intermediate cases may exist; for the sake of simplicity, they are not identified here as a separate class) (see Mathews & Ward 1985, for a review). The very nature of the p-process is still debated. It might consist of proton captures or photodisintegrations in high-temperature (not less than about 10^9 K) non-explosive or explosive environments (see Arnould 1980, for a review).

However, many uncertainties of various origins still remain in the predicted composition of the ejected material. Let us also note that the central He flash which may affect certain of the considered stars ($M \lesssim 2.2\ M_\odot$) has sometimes been held responsible for the production of some r-nuclei. However, such a model faces severe difficulties (Deupree 1984a, b; Truran 1984; Mathews & Ward 1985).

(ii) Massive ($M \gtrsim 8\ M_\odot$) stars, partly through stellar-wind mass losses (especially in the MMLS case), and through SN explosions. In particular, some of the considered stars are thought to be the progenitors of type-II SNe (which are estimated to occur every 44 years in the Galaxy now (Tammann 1982)), or of some special SN events. Those explosions might be accompanied with the formation of a neutron star. However, at least in some cases, a black-hole formation or, in contrast, the ejection of the entire stellar material in the ISM cannot be excluded. On the other hand, some of those massive stars might also just collapse into black holes without SN explosions. Many aspects of the modelling of the final fate of those massive stars are still highly uncertain and debated (see, for example, Hillebrandt 1986; Woosley & Weaver 1986).

In spite of that, massive star explosions have classically been considered as particularly important agents for the chemical evolution of the Galaxy. Many observations have attempted to shed some light on the composition of the SN remnants, and many nucleosynthesis calculations have been devoted to SN yield calculations (see, for example, Hillebrandt 1986; Woosley & Weaver 1986). Here we just want to comment briefly on some aspects of that vast and difficult question.

From the use of a typical stellar birthrate (see, for example, Miller & Scalo 1979) and typical massive star SN yields, it has been often concluded that the galactic enrichment in nuclei heavier than He is due to the evolution and explosion of $ca.$ 20–30 $M_\odot$ stars (see, for example, Woosley 1986). A comparison between the yields from an $M = 25\ M_\odot$ star and the bulk Solar System composition for nuclides with mass number $A \leqslant 64$ is given in figure 1. The general agreement appears to be quite good, the most underproduced species being possibly made elsewhere (e.g. ^{12}C, ^{13}C and ^{14}N in lower-mass stars – see (i); ^{15}N is novae – see (iii)). However, it has to be stressed that those results are still quite uncertain, the situation arising from astrophysical (pre-SN and SN modelling) or nuclear physics uncertainties (a striking example being provided by the influence of the revision of the ^{12}C$(\alpha, \gamma)^{16}$O rate, as can be seen in Thielemann & Arnett (1985)). There are other reasons for interpreting the data of figure 1 with great care. On the one hand, it may well be that a $M \approx 25\ M_\odot$ star is not a good representation of the dominant contributor to the galactic heavy elements as a whole (see, for example, Arnould 1986a). On the other hand, even if it is, it has to be realized that the relative heavy element yields depend strongly on the stellar mass. Thus, it remains to be seen if the general trends of figure 1 will remain unchanged, or be drastically modified, when the required integration over the whole relevant mass range is performed. As far as the $A > 64$ nuclides are concerned, massive stars might contribute to some s-nuclei enrichment at a galactic level through steady mass losses or SN ejection of s-nuclei made in pre-SN stages, such as He- or C-burning (e.g. Prantzos $et\ al.$ 1986a, b; Arnett & Thielemann 1985). On the other hand, SNe are also very often held responsible for the synthesis of, and ISM enrichment with the r- and p-nuclei. However, many uncertainties remain in that respect (see, for example, Arnould 1980; Truran 1984; Mathews & Ward 1985), and no quantitative SN yield predictions have been made yet.

(iii) Novae, the ejecta of which are known to exhibit special compositions, and which are also predicted possibly to contribute to the galactic enrichment of several nuclei (see, for

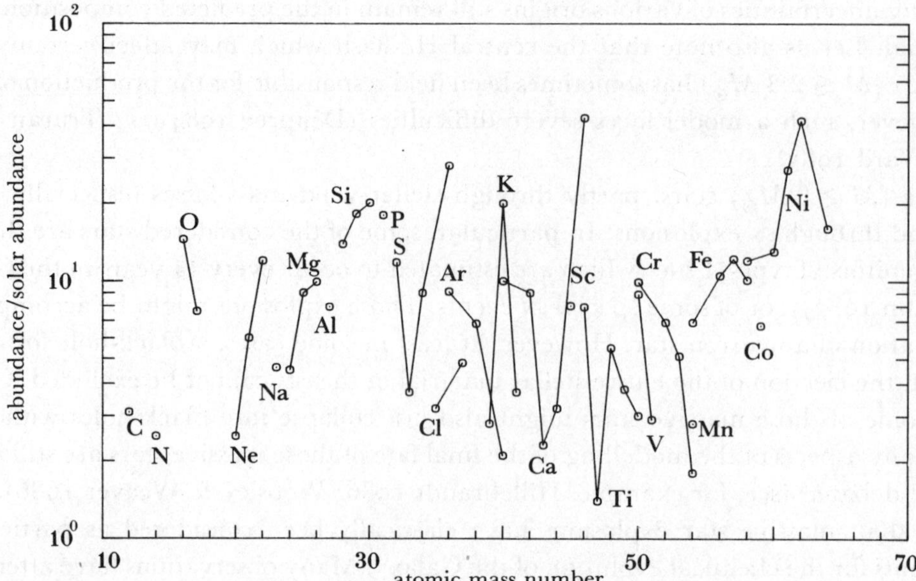

FIGURE 1. Comparison between the bulk primordial Solar System composition of the nuclides with mass number $A \leqslant 64$ and the composition of the material ejected from a 25 $M_\odot$ type-II supernova explosion. An average overproduction factor (with respect to solar, and before any dilution with other galactic material) of about 9 characterizes the calculated yield (adapted from Woosley & Weaver 1986).

example, Truran 1986; Wiescher *et al.* 1986); however, their yields are still not very reliably predictable.

(iv) Type-I sne (estimated to occur every 36 years in the Galaxy now (Tammann 1982)), typical events of that kind being interpreted in a 'standard' model calling for an exploding (because of C-deflagration) white dwarf accreting matter in a binary system (other models also exist, which cannot be entirely ruled out). The concomitant nucleosynthesis has been reviewed by, for example, Hillebrandt (1986), Woosley & Weaver (1986) and Thielemann *et al.* (1986), a typical example of which is provided in figure 2. It appears that such explosions eject predominantly iron-group nuclei, with some additional contributions from incomplete Si-, and explosive O-, Ne- and C-burnings. It has to be stressed that those type-I sn modellings and yields are still very uncertain.

(v) Spallation of ism atoms by galactic cosmic rays, which is able to account for the bulk Solar-System ^{6}Li, ^{9}Be, ^{10}B, most of ^{11}B, and at least some ^{7}Li (see, for example, Audouze & Reeves 1982; Arnould 1986*b*).

In conclusion, many agents possibly responsible for the chemical evolution of the Galaxy have been identified. On such grounds, a very preliminary explanation of the bulk Solar System composition can be provided. However, many aspects of the question of the chemical evolution of the Galaxy have been simplified almost beyond recognition in the models developed so far. In addition, most nucleosynthetic yields are still highly uncertain. Thus, much remains to be done before being able to provide a reasonably reliable and complete explanation for the bulk Solar System composition.

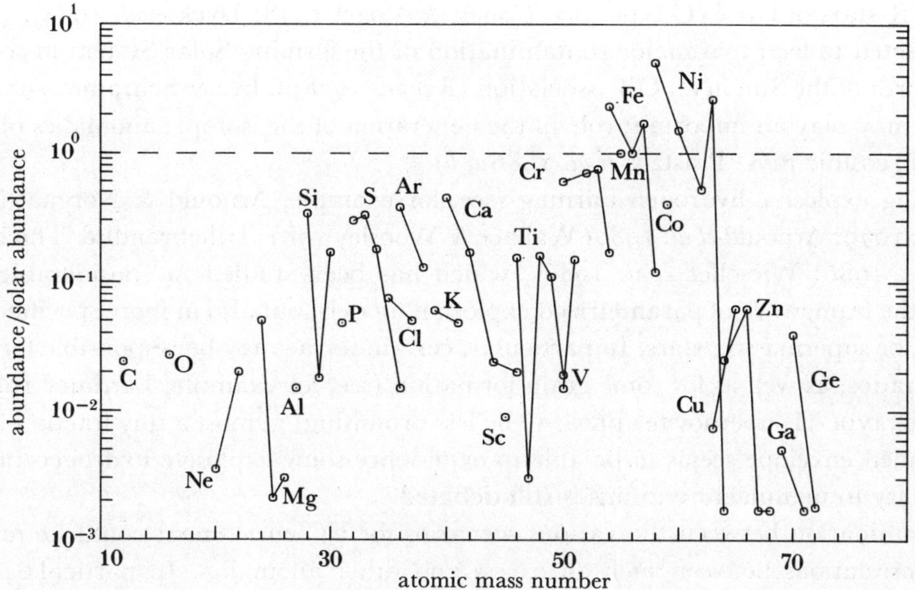

FIGURE 2. Comparison between the bulk primordial Solar System composition of the nuclides with mass number $A \leqslant 72$ and the composition of the material ejected from a 'standard' (C-deflagration) type-I supernova explosion. The calculated yields are *normalized to the solar* ^{56}Fe *abundance* (adapted from Thielemann *et al.* 1986).

3. WHAT CAN WE LEARN FROM ANOMALOUS CARBON AND RELATED ANOMALIES?

Various anomalous carbon components are now known to exist in primitive meteorites (Swart *et al.* 1983; Carr *et al.* 1983; Yang & Epstein 1984; Ming *et al.* 1985; Niederer *et al.* 1985), which exhibit a $^{12}C/^{13}C$ abundance ratio that is lower than the solar value of *ca.* 90. Although mass-independent fractionation processes might affect the C isotopic composition (e.g. Reisse & Mullie 1985; Wright & Pillinger 1985), it is quite possible that those components have been shaped by nucleosynthesis processes, and are of extrasolar origin. As reviewed by Arnould (1985*a*), the ^{13}C enrichment might be due to a 'conventional' or 'hot' CN(O) hydrogen-burning occurring during the 'normal' or 'explosive' evolution of certain stars (see, for example, Truran 1984), and, more specifically, in the following.

(I) At various stages during the evolution of intermediate-mass stars ($1 \lesssim M/M_\odot \lesssim 8$), and namely during the Red-Giant Branch (RGB), the AGB, or even the postplanetary nebula phase (Iben *et al.* 1983). In relation to some scenarios for the origin of isotopically anomalous reservoirs (§1), it is also worth noting that carbon-rich grains may form around certain of the mentioned stars at various stages of their evolution, namely if high enough surface carbon enrichments can be reached (see, for example, Lattimer 1982). Arnould (1985*a*) has reviewed the uncertainties still affecting the precise evaluation of the ^{13}C-enrichment provided by the considered stars as a function of their mass, stellar wind, initial stellar composition, or evolutionary stage.

(II) In more massive stars, as a result of the combination of hydrogen burning and strong stellar winds. This situation is best exemplified by $M \gtrsim 40\ M_\odot$ stars in their Of stage, or during the WN stage of their WR phase. In such stages, $^{12}C/^{13}C$ abundance ratios much lower than solar can indeed be achieved (Prantzos *et al.* 1986*c, d*). Note also that (*a*) grains are observed to form in WR stars of late WC type (e.g. Cohen & Vogel 1978; Dyck *et al.* 1984), (*b*) WR stars are expected to lead to a major contamination of the forming Solar System in scenarios putting the birth of the Sun in an OB association (Reeves 1978; Olive & Schramm 1982), and (*c*) WR stars may play an important role in the generation of the isotopic anomalies observed in the galactic cosmic rays (Prastzos *et al.* 1986*a, b*).

(III) During explosive hydrogen burning (see, for example, Arnould & Nørgaard 1978; Lazareff *et al.* 1979; Arnould *et al.* 1980; Wallace & Woosley 1981; Hillebrandt & Thielemann 1982; Truran 1986; Wiescher *et al.* 1986), which has been studied in considerable detail especially in the framework of parametrized explosion models, but also in more specific models of novae, sne, or supermassive stars. In particular, certain novae may be responsible for rather low $^{12}C/^{13}C$ ratios, as well as for some grain formation (see, for example, Lattimer 1982). In these instances, type-II supernovae appear to be less promising; at most a tiny fraction (if any) of their extended envelope seems to be able to experience some explosive hydrogen burning, and their ability to manufacture grains is still debated.

Some discrimination between the various scenarios for ^{13}C enrichment could be made on grounds of correlations between such an excess and other anomalies. In particular, recent experiments indicate that an isotopically heavy C component, referred to as C-β, is associated with an anomalous Xe (and possibly Kr) component considered to be synthesized in the s-process, and referred to as s-Xe (and s-Kr). On the other hand, another heavy C component, dubbed C-α, is accompanied with a special neon component called Ne-E(L) (see §4). These correlations are discussed in detail by Swart *et al.* (1983), Lewis & Anders (1983) and Carr *et al.* (1983).

The association of heavy C with s-Xe might be explained in the framework of scenario (I), as at least certain of the stars under consideration are perhaps responsible for the development of the s-process (see, for example, Mathews & Ward 1985). In contrast, the Of and WR stars implicated in scenario (II) might have difficulties in explaining such a correlation. Indeed, the s-process material that could be brought to the stellar surface and into the interstellar medium during the WC stage is predicted to be correlated with a ^{12}C excess. In contrast, ^{13}C is overabundant just during the preceding Of and WN phases, which cannot produce any s-process enrichment (Prantzos *et al.* 1986*a, b*). Nor does the association of heavy C with s-Xe seem to be in favour of the explosive origin of scenario (III). It is indeed far from being clear whether explosive H-burning sites could be responsible for some s-process-like Xe pattern (Arnould 1985*a*).

In contrast, the association of heavy C with Ne-E(L) might be more readily explained in scenario (III). In particular, a nova origin may be especially well suited in that respect (Arnould & Nørgaard 1981; Wiescher *et al.* 1986).

Pandora's box of isotopic anomalies has been opened more widely recently with the claim that some correlation might exist between C-β, s-Xe, and Ne-E(H), a second identified Ne-E component (Ming *et al.* 1985; Ott *et al.* 1985*b*). If such an association is indeed not a mere artifact of the still incomplete characterization of the ultimate carrier phase(s) of those various anomalies, the science of nucleosynthesis in its present state has probably to confess its inability to provide an explanation. Some correlation between Ne-E(H) and s-Xe might perhaps be accounted for by hydrostatic/explosive C-burning (Arnett & Wefel 1978; Arnould & Nørgaard 1981). However, such a burning phase would produce large amounts of ^{12}C, and not the required ^{13}C enrichment.

The possible origin of the ^{13}C excess would be constrained still further by the characterization of any accompanying nitrogen. If the heavy C is found to be correlated with the ^{14}N-enriched nitrogen isolated from certain meteoritic samples (Lewis & Anders 1983; Lewis *et al.* 1983b), scenarios (I) and (II) would appear to be favoured, the conventional CN(O) cycle yielding both ^{13}C and ^{14}N enrichments. In contrast, the explosive H-burning situations of scenario (III) do not in general produce ^{14}N (an exception can be found in the published calculations of Lazareff *et al.* (1979)). In reality, the latter model would be favoured if ^{13}C excesses are found to be correlated with an anomalous ^{15}N enrichment. Large ^{15}N anomalies have recently been uncovered in a meteorite called Bencubbin (Prombo & Clayton 1985), but these have been shown to be associated with carbon of a more normal isotopic composition (Franchi *et al.* 1986). On the other hand, the isotopically heavy nitrogen found by Lewis *et al.* (1983) could well be associated with the ^{13}C-enriched phase C-α.

Table 1 summarizes the previous discussion schematically.

TABLE 1. SCHEMATIC OVERVIEW OF THE ABILITY OF SCENARIOS I, II AND III (DEFINED IN THE MAIN TEXT) TO ACCOUNT FOR THE ANOMALOUS HEAVY C AND POSSIBLY CORRELATED ANOMALIES

(Symbols + or − are used when a given scenario is promising or unlikely, respectively. The nitrogen which might accompany the reported anomalous components is not well characterized yet.)

anomalous component	scenario		
	I	II	III
C-α+Ne-E(L)	−[1]	−[1]	+[2] or −[3]
C-β+s-Xe	+[2] or −[3]	−	−
C-β+s-Xe+Ne-E(H)	−	−	−

[1] Except if most extreme Ne-E forms (if any) not required; [2], if correlated light N; [3], if correlated heavy N.

4. Ne-E IN METEORITES AND FOSSIL GRAINS

Among the many isotopically distinct neon components identified in meteorites (see, for example, Ott *et al.* 1985a), it is generally considered that the so-called Ne-E is the only one exhibiting an anomalous isotopic composition of nuclear origin (see, for example, Eberhardt *et al.* 1981). It has been observed in three mineral phases of certain carbonaceous chondrites, and is generally interpreted as a pure ^{22}Ne component (Arnould & Nørgaard 1981; Eberhardt *et al.* 1981; Jungck 1982). This isotopic purity is commonly explained by the decay of ^{22}Na in fossil grains (model 2), which could not significantly trap or retain the noble gases. Clearly, the question of the extra-solar ^{22}Na source has to be addressed.

In that respect, several H-burning calculations have been performed. They indicate that such a nuclear processing in nova explosions might be the most promising mechanism of ^{22}Na synthesis (Arnould & Beelen 1974; Arnould & Nørgaard 1981; Wallace & Woosley 1981; Hillebrandt & Thielemann 1982; Wiescher *et al.* 1986). In addition, novae are capable of producing grains that could be the carriers of the newly produced ^{22}Na (see, for example, Clayton & Hoyle 1976). However, the ^{22}Na nova yields are still quite difficult to predict reliably for various reasons (see, for example, Arnould 1985b; Truran 1986).

In contrast to novae, SN explosions do not appear able to produce enough ^{22}Na to account for Ne-E (Arnould & Nørgaard 1981). Explosive SN He-burning is also quite inefficient in that respect (Arnould & Nørgaard 1981). In fact C-burning might be a slightly better candidate (Arnett & Wefel 1978; Arnould & Nørgaard 1981).

As already mentioned in §3, correlations have been observed between Ne-E, heavy C, and possibly s-Xe (along with s-Kr). These correlations put strong constraints on the acceptable nucleosynthetic models for Ne-E, and might even place them in an embarrassing dilemma. The discovery of new correlated isotopic anomalies, or a better identification of the Ne-E carrier(s) (Jungck & Eberhardt 1985; Ming *et al.* 1985; Ott *et al.* 1985*b*) would obviously be of great interest (even if they would make a nucleosynthesis solution still more elusive!).

Before closing this section, let us remark that no ISM γ-ray line from the ^{22}Na decay has now been positively detected (Mahoney *et al.* 1982). In fact, observational limits put on the intensity of that line are consistent with the predicted nova or SN yields. On the other hand, recent observations suggest the existence of an isotopically anomalous ^{22}Ne-enriched GCR component (see, for example, Wiedenbeck 1984; Meyer 1985). A ^{22}Ne (not ^{22}Na!) WR source for such an anomalous component accounts satisfactorily for the observations (Prantzos *et al.* 1986*a, b*).

5. ^{26}Al AND SHORT-LIVED RADIONUCLIDES IN METEORITES

Apart from Ne-E, which possibly arises from the decay of ^{22}Na, there is now strong experimental evidence that ^{26}Al has decayed *in situ* in various meteoritic inclusions, leading to the observation of some anomalous ^{26}Mg excess (see, for example, Wasserburg & Papanastassiou 1982; Papanastassiou 1985). Much excitement in the meteoritic and astrophysical communities has been raised by that discovery. It has been amplified still further by the observation of ^{26}Al in the ISM through the identification in the galactic plane of the 1.809 MeV γ-ray line associated with its β-decay (Mahoney *et al.* 1984; Share *et al.* 1985; Ballmoos *et al.* 1987; see also Prantzos 1986, for a review). That identification of ^{26}Al, and the determination of its (unexpectedly large) interstellar content (*ca.* 3 $M_{\odot}$) are very important in several respects. In particular, that discovery can shed some new light on the possible astrophysical source(s) of ^{26}Al, and on its (their) distribution in the Galaxy. Consequently, new clues might be furnished on the origin of the ^{26}Al-related isotopic anomalies observed in meteorites, and perhaps also on the circumstances and conditions of the solar birth, even if the questions of the origin of the ISM and meteoritic ^{26}Al are not necessarily intimately related.

The meteoritic ^{26}Al data are generally interpreted in terms of the injection of live ^{26}Al (along with other short-lived nuclides?) in the protosolar nebula (model 3). In such views, it is concluded that the abundance ratio $(^{26}\text{Al}/^{27}\text{Al})_0 \approx 5 \times 10^{-5}$ on average at the time T_0 of solidification in the Solar System. However, a large spread seems to exist around that 'canonical' value. In fact, values in the range $0 \lesssim (^{26}\text{Al}/^{27}\text{Al})_0 \lesssim 10^{-3}$ have been reported (see, for example, Wasserburg & Papanastassiou 1982; Papanastassiou 1985). The interpretation of those ^{26}Al data within model 3 provides important clues on the time Δ^* elapsed between the last astrophysical event(s) able to affect the composition of the solar nebula, and the solidification of some of its material. Attempts have also been made to interpret the observed ^{26}Al signature in meteorites in the framework of models 1 (Lee 1978) and 2 (Clayton 1982, 1986), which, contrary to model 3, deny any chronometric virtue to the short-lived radionuclides in general, and to ^{26}Al in particular.

Leaving aside the (important) debate about the relative merits or difficulties of these various models, a quite central question (in both models 2 and 3) concerns the possible thermonuclear process(es) able to produce the ISM and/or meteoritic ^{26}Al. Various mechanisms have been proposed, namely.

(i) Non-explosive H burning in the core of MMLS, followed by stellar-wind ejection. As discussed in detail elsewhere (Arnould & Prantzos 1986; Cassé & Prantzos 1986; Prantzos & Cassé 1986), the ^{26}Al mass ejected in the ISM by such stars between the zero-age main sequence (ZAMS) and the end of the WR stage (identified in practice with the end of central He-burning) is predicted to vary from roughly 10^{-5} to 10^{-4} $M_\odot$ for stars with initial masses M_{ZAMS} of 50–100 $M_\odot$. In such conditions, MMLS might be an important (even the dominant) ^{26}Al contaminant of the Solar System (either in the form of gas or grains, which are known to form around certain WR stars; see §3), following the lines of models 2 or 3. Such stars might also be of key importance for the feeding of the ISM with ^{26}Al at the level required by the 1.809 MeV γ-ray line observations (Prantzos *et al.* 1985; Cassé & Prantzos 1986; Prantzos & Cassé 1986).

(ii) Explosive H-burning in novae. The ^{26}Al production in such a model has been the subject of several investigations (Arnould *et al.* 1980; Wallace & Woosley 1981; Hillebrandt & Thielemann 1982; Arnould 1985 *b*). From those calculations, it has been estimated that novae could constitute a significant source of ^{26}Al in meteorites and the ISM (Leising & Clayton 1985).

(iii) C- or Ne-burning in type-II SNE. The various SN calculations (Arnett & Wefel 1978; Truran & Cameron 1978 Arnould *et al.* 1980; Woosley & Weaver 1986) predict ^{26}Al yields which are generally considered as insufficient to account entirely for the γ-ray line observations (Mahoney *et al.* 1984; Leising & Clayton 1985).

(iv) ^{26}Al production at the bottom of the convective envelope of AGB stars (Nørgaard 1980; Cameron 1984). No detailed calculations for such a mechanism have been performed up to now, so that such a scenario has still to be considered as highly speculative.

A better appraisal of the relative merits of the MMLS, AGB, nova and SN models to account for the meteoritic and ISM ^{26}Al data obviously demands a reduction of the yield uncertainties. This can be achieved through progress in the modelling of the various astrophysical ^{26}Al production sites. In this respect, the MMLS model (just relying on main sequence stars!) appears to be relatively more secure than models for highly complicated evolutionary phases (AGB, nova or SN explosions). In addition, various uncertainties stem from the still more or less poor knowledge of certain key nuclear reaction rates. This aspect of the problem has been reviewed by Arnould (1985 *b*).

To help assess the merits of the various models, the following considerations may be of some interest.

(i) Up to now, no undisputable correlation has been found between the ^{26}Al-related and other isotopic anomalies. In particular, the associated Ca isotopic composition and ^{25}Mg/^{24}Mg ratio are found to be essentially solar (Arnould *et al.* 1980; Wasserburg & Papanastassiou 1982). This observation appears to be accounted for by the MMLS model in a rather wide range of realistic conditions (Arnould & Prantzos 1986). The constraint may be more serious for the nova models, in which case the ^{25}Mg/^{24}Mg production ratio has been predicted recently to deviate from solar values far more than in the massive star case (Wiescher *et al.* 1986). On the other hand, if C or O would accompany ^{26}Al in meteoritic inclusions, their predicted isotopic composition could be completely different in the MMLS and nova models. A WC or WO contamination would indeed lead to a ^{12}C or ^{16}O enrichment, whereas ^{13}C and ^{17}O would be enhanced in the nova case. Finally, the sn model might have difficulties in accounting namely for the normal Ca isotopic compositions (Arnould *et al.* 1980).

(ii) The mmls (and especially WR stars), novae and sne are not expected to have the same spatial distribution and/or frequency in the galactic plane (but large uncertainties still affect those quantities). In principle, observations of the ism γ-ray lines with higher spatial resolution than those achieved up to now should help clarify the nature of the dominant (if any!) ism ^{26}Al source.

In addition to the ^{22}Ne and ^{26}Al excesses interpreted in terms of the *in situ* decay of ^{22}Na and ^{26}Al, respectively, various other anomalies attributable to short-lived species have been detected, like ^{36}Cl, ^{41}Ca, ^{53}Mn, ^{60}Fe, ^{107}Pd, ^{129}I, ^{135}Cs, ^{146}Sm, and ^{205}Pb, some cases being more firmly established than others (see, for example, Arnould 1985 b; Papanastassiou 1985; Arnould & Prantzos 1986, for more details and references). Let us emphasize that the decay products of the mentioned radionuclides have not necessarily been found in the same inclusions, or even in the same meteorites. In particular, no correlation has been discovered up to now between the well-documented anomalies related with the ^{26}Al and ^{107}Pd decays. It has also to be noted that, among the above-mentioned short-lived species, only ^{60}Fe could be of interest for γ-ray line astronomy (Mahoney *et al.* 1982).

As in the ^{22}Na and ^{26}Al cases, the predicted astrophysical yields of those radionuclides are in general still more or less difficult to predict reliably, as a result of combined astrophysical and nuclear physics uncertainties. It is not possible to review this question in detail here. Let us instead limit ourselves to the following brief remarks.

(i) Apart from ^{26}Al, mmls can produce ^{36}Cl, ^{60}Fe, ^{99}Tc, ^{107}Pd, ^{135}Cs and ^{205}Pb to a more or less large extent during the WC phase of their WR evolutionary stage (Arnould & Prantzos 1986). An s-process can indeed develop in such conditions (Prantzos *et al.* 1986 *a*, *b*), part of the synthesized species being ejected into the ism as a result of strong mass losses. Such a model predicts in particular a ^{107}Pd production at a level consistent with the observations. In contrast, ^{60}Fe is expected to be produced at a level which probably lies below the present-day γ-ray line observability threshold (in fact, no positive detection has yet been reported (Mahoney *et al.* 1982)).

(ii) In its present state, the mmls model does not predict the production of other short-lived species of interest, such as ^{53}Mn, ^{129}I, or ^{146}Sm. Note, however, that the production of short-lived species in mmls has been studied only up to central He exhaustion in $M \geqslant 50 \, M_\odot$ stars. The further evolution of the considered stars, which could possibly end (at least for some of them) with a sn explosion, might bring its share of isotopic anomalies and of short-lived radionuclides.

(iii) sne might produce short-lived species other than ^{26}Al (possibly along with other anomalies). However, the modelling of such objects still suffers from various uncertainties, and the relative yields of the short-lived radionuclides cannot be calculated in a given sn model at the level of self-consistency and confidence attained in the mmls model.

(iv) It has not been yet demonstrated that novae can produce significant amounts of short-lived species other than ^{22}Na or ^{26}Al.

(v) It has been suggested (Cameron 1984) that low-mass agb stars could produce at least ^{107}Pd and ^{129}I along with ^{26}Al. However, it has to be emphasized that no detailed calculations exist to substantiate such speculations.

6. Xenology and supernova nucleosynthesis

Xe has perhaps the richest spectrum of isotopic anomalies. One of its components, s-Xe, has already been encountered in §§3 and 4. Another component found in certain carbonaceous chondrites, and referred to as Xe-HL, exhibits a very interesting correlation between excess light (L) and heavy (H) Xe isotopes (see, for example, Lewis & Anders 1983; Fisenko *et al.* 1985, and references therein).

The synthesis of the light isotopes ^{124}Xe and ^{126}Xe is attributed exclusively to the p-process (§2), while the excess of heavy isotopes has been debated for quite a long time. The proposed models include production via the fission of an actinide (like ^{250}Cm; Howard *et al.* 1975), or of a now-extinct superheavy element (see, for example, Anders 1981). Further progress in the characterization of Xe-H and of its complement of isotopic patterns now seems to rule out the fission hypotheses (Lewis *et al.* 1983*a*). As such, models which invoke production during the r-process are now considered more likely (see, for example, Heymann & Dziczkaniec 1981; Lewis *et al.* 1983*a*; Beer *et al.* 1983).

Even if many uncertainties still subsist, the p- and r-processes are possibly related to the pre-sn evolution and sn explosion of massive stars (§2). Indeed, several calculations have shown that an isotopic pattern resembling Xe-HL could be obtained through an adequate selection of pre-sn compositions and sn layers and conditions. However, strong constraints have to be imposed on the sn models, as well as on the post-sn mixing of the relevant layers (see, for example, Heymann & Dziczkaniec 1981).

The various alleged sn Xe components are expected to be accompanied by other isotopic anomalies especially in the neighbouring elements Te, Ba, Nd and Sm. There has been some claim that Te indeed exhibits excesses of light and heavy Te isotopes correlated with Xe-HL, but those data are disputed (for more details and references, see Heymann & Dziczkaniec (1981) and Smith & De Laeter (1986)). On the other hand, no detectable anomalies have been found in the accompanying Ba, Nd and Sm (Lewis *et al.* 1983*a*). Such a dilemma for the nucleosynthesis models can be solved if it is assumed that Ba, Nd and Sm can fractionate from Xe before trapping of the latter in the carriers (Lewis *et al.* 1983*a*).

Another very interesting piece of information is provided by the observation of a possible correlation between Xe-HL, essentially normal C, and isotopically light N (Lewis & Anders 1983; Lewis *et al.* 1983*b*). Such data are far from being obvious to interpret in terms of sn nucleosynthesis. As illustrated by Arnould (1985*a*), a subtle blend of the right amounts of various sn layers might account for those observations. However, such a scenario involves many uncertainties and is completely *ad hoc*, its general astrophysical plausibility being far from obvious.

7. Summary and some additional comments

The discovery of isotopic anomalies in meteorites suggests that the Solar System is made of material from compositionally different and imperfectly mixed reservoirs.

One of those reservoirs, which accounts for the bulk Solar System material, is considered to be made of the well-homogenized ashes of many nucleosynthesis events (Big Bang, red giants, Wolf–Rayet stars, novae, supernovae of various types, galactic cosmic-ray interactions with the ism), the yields of which are the required input to models for the chemical evolution of the Galaxy in the solar neighbourhood. In spite of some success achieved in the interpretation of the bulk Solar System composition, much remains to be done before a reasonably complete and reliable model can be provided.

A more 'granular' aspect of the nucleosynthesis events which have been able to contribute to the Solar System composition might reveal itself through the large array of discovered isotopic anomalies. Several selected examples have helped to emphasize the rich diversity of potential contaminants, which are identified as gas or grains originating from red giants, massive Of or WR stars, or nova and supernova explosions. Many difficulties are still encountered in the attempt to select the most active of those contaminants (if any!). That situation arises from the large uncertainties in the astrophysical modellings, as well as in the incomplete characterization of the isotopic anomalies and of their possible correlations. The uttermost importance of the existence or lack of such correlations for the nucleosynthesis interpretation of the isotopic anomalies clearly shows up in the preceding discussions.

All the remaining problems and unanswered questions certainly make the common adventure of cosmochemistry and nuclear astrophysics very exciting and challenging. Joint efforts may become of growing importance in the future, particularly when comets will reveal their (isotopic) secrets (those privileged witnesses of the ancient solar nebula are discussed in other contributions to this meeting).

Finally, we would like to restate the fact that the existence of isotopic anomalies, in particular those related to the decay of the short-lived species, can tell us something important about the solar birth, at least in the view of live radionuclide injection in the protosolar nebula (model 3 of §1). In that respect, contamination and/or triggering by SN explosions or by low-mass AGB stars have been proposed. In our opinion, those scenarios face some severe difficulties. A model relying on the contamination of a Bok-type globule by one or several massive $(M \gtrsim 40\ M_{\odot})$ Of or WR stars will be presented elsewhere, along with a critical discussion of the other proposed scenarios.

References

Anders, E. 1981 *Proc. R. Soc. Lond.* A **374**, 207.

Anders, E. & Ebihara, M. 1982 *Geochim. cosmochim. Acta* **45**, 2363.

Arnett, W. D. & Wefel, J. P. 1978 *Astrophys. J.* **224**, L139.

Arnett, W. D. & Thielemann, F.-K. 1985 *Astrophys. J.* **295**, 589.

Arnould, M. 1980 *Cahier no. 8*, Physique Nucléaire Théorique (ed. M. Demeur). Université Libre de Bruxelles. Bruxelles.

Arnould, M. 1985a In *Isotopic ratios in the Solar System* (ed. Centre National d'Études Spatiales), p. 283. Toulouse: Cepadues-Éditions.

Arnould, M. 1985b In *Proc. Accelerated Radioactive Beams Workshop* (ed. L. Buchmann & J. D'Auria), p. 29. TRIUMF TRI-85-1.

Arnould, M. 1986a In *Advances in nuclear astrophysics* (ed. E. Vangioni-Flam, J. Audouze, M. Cassé, J.-P. Chièze & J. Tran Thanh Van), p. 113. Gif-sur-Yvette: Editions Frontières.

Arnould, M. 1986b In *Prog. part. nucl. Phys.* **17**, 305.

Arnould, M. & Beelen, W. 1974 *Astron. Astrophys.* **33**, 215.

Arnould, M. & Nørgaard, H. 1978 *Astron. Astrophys.* **64**, 195.

Arnould, M. & Nørgaard, H. 1981 *Comments Astrophys.* **9**, 145.

Arnould, M. & Prantzos, N. 1986 In *Nucleosynthesis and its implications on nuclear and particle physics* (ed. J. Audouze & N. Mathieu), p. 363. Dordrecht: Reidel.

Arnould, M., Nørgaard, H., Thielemann, F.-K. & Hillebrandt, W. 1980 *Astrophys. J.* **237**, 931.

Arrhenius, G., Fitzgerald, R., Markus, S. & Simpson, C. 1978 *Astrophys. Space Sci.* **55**, 285.

Audouze, J. & Reeves, H. 1982 In *Essays in nuclear astrophysics* (ed. C. A. Barnes, D. D. Clayton & D. N. Schramm), p. 355. Cambridge University Press.

Ballmoos, P. V., Diehl, R. & Schönfelder, V. 1987 *Astrophys. J. Lett.* **312**, 134–142.

Beer, H., Käppeler, F., Reffo, G. & Venturini, G. 1983 *Astrophys. Space Sci.* **97**, 95.

Begemann, F. 1980 *Rep. Prog. Phys.* **43**, 1309.

Boesgaard, A. M. & Steigman, G. 1985 *A. Rev. Astr. Astrophys.* **23**, 319.

Cameron, A. G. W. 1982 In *Essays in nuclear astrophysics* (ed. C. A. Barnes, D. D. Clayton & D. N. Schramm), p. 23. Cambridge University Press.

Cameron, A. G. W. 1984 *Icarus* **60**, 416.

Carr, R. H., Wright, I. P., Pillinger, C. T., Lewis, R. S. & Anders, E. 1983 *Meteoritics* **18**, 277.

Cassé, M. & Prantzos, N. 1986 In *Nucleosynthesis and its implications on nuclear and particle physics* (ed. J. Audouze & N. Mathieu), p. 339. Dordrecht: Reidel.

Clayton, D. D. 1979 *Space Sci. Rev.* **24**, 147.

Clayton, D. D. 1982 *Q. Jl R. Astr. Soc.* **23**, 174.

Clayton, D. D. 1986 In *Cosmogonical processes* (ed. W. D. Arnett, C. Hansen, J. Truran & S. Tsuruta). Utrecht: VNU Science Press.

Clayton, D. D. & Hoyle, F. 1976 *Astrophys. J.* **203**, 490.

Cohen, M. 1984 *Phys. Rep.* **116**, 173.

Cohen, J. H. & Vogel, S. H. 1978 *Mon. Not. R. astr. Soc.* **185**, 47.

Consolmagno, G. J. & Cameron, A. G. W. 1980 *Moon Planets* **23**, 3.

Deupree, R. G. 1984a *Astrophys. J.* **282**, 274.

Deupree, R. G. 1984b *Astrophys. J.* **287**, 268.

Dyck, H. M., Simon, T. & Wolstencroft, R. D. 1984 *Astrophys. J.* **257**, 276.

Eberhardt, P., Jungck, M. H. A., Meier, F. O. & Niederer, F. R. 1981 *Geochim. cosmochim. Acta* **45**, 1515.

Fisenko, A. V., Vu Minh, D., Semjonova, L. F., Lavrukhina, A. K. & Shukclyukov, Ju. A. 1985 In *Isotopic ratios in the Solar System* (ed. Centre National d'Études Spatiales), p. 147. Toulouse: Cepadues-Éditions.

Franchi, I. A., Wright, I. P. & Pillinger, C. T. 1986 *Lunar Planet. Sci.* **17**, 233.

Grevesse, N. 1984 *Phys. Scripta* **8**, 49.

Heidenreich III, J. E. & Thiemens, M. H. 1985 *Geochim. cosmochim. Acta* **49**, 1303.

Heymann, D. & Dziczkaniec, M. 1981 *Geochim. cosmochim. Acta* **45**, 1829.

Hillebrandt, W. 1986 In *Cosmogonical processes* (ed. W. D. Arnett, C. Hansen, J. Truran & S. Tsuruta). Utrecht: VNU Science Press.

Hillebrandt, W. & Thielemann, F.-K. 1982 *Astrophys. J.* **255**, 617.

Howard, W. M., Arnould, M. & Truran, J. W. 1975 *Astrophys. Space Sci.* **36**, L1.

Iben, I. Jr. & Renzini, A. 1983 *A. Rev. Astr. Astrophys.* **21**, 271.

Iben, I. Jr., Kalcr, J. B., Truran, J. W. & Renzini, A. 1983 *Astrophys. J.* **264**, 605.

Jungck, M. H. A. 1982 Ph.D. thesis, University of Bern.

Jungck, M. H. A. & Eberhardt, P. 1985 *Meteoritics* **20**, 677.

Lattimer, J. M. 1982 In *Formation of planetary systems* (ed. A. Brahic), p. 189. France: Centre National d'Études Spatiales.

Lazareff, B., Audouze, J., Starrfield, S. & Truran, J. W. 1979 *Astrophys. J.* **228**, 815.

Lee, T. 1978 *Astrophys. J.* **224**, 217.

Leising, M. D. & Clayton, D. D. 1985 *Astrophys. J.* **294**, 591.

Lewis, R. S. & Anders, E. 1983 *Scient. Am.* **249**, 66.

Lewis, R. S., Anders, E., Shimamura, T. & Lugmair, G. W. 1983a *Science, Wash.* **222**, 1013.

Lewis, R. S., Anders, E., Wright, I. P., Norris, S. J. & Pillinger, C. T. 1983b *Nature, Lond.* **305**, 767.

Mahoney, W. A., Ling, J. C., Jacobson, A. S. & Lingenfelter, R. E. 1982 *Astrophys. J.* **262**, 742.

Mahoney, W. A., Ling, J. C., Wheaton, W. A. & Jacobson, A. S. 1984 *Astrophys. J.* **286**, 578.

Manuccia, T. J. & Clark, M. D. 1976 *Appl. Phys. Lett.* **28**, 372.

Mathews, G. J. & Ward, R. A. 1985 *Rep. Prog. Phys.* **48**, 1371.

Meyer, J.-P. 1985 In *19th Int. Cosmic Ray Conf., La Jolla* (ed. F. C. Jones, J. Adams & G. M. Mason), vol. 9, p. 141. NASA Conf. Pub., NASA CP-2376.

Miller, G. E. & Scalo, J. M. 1979 *Astrophys. J. Suppl.* **41**, 513.

Ming, T., Lewis, R. S. & Anders, E. 1985 *Meteoritics* **20**, 712.

Niederer, F. R., Eberhardt, P., Geiss, J. & Lewis, R. 1985 *Meteoritics* **20**, 716.

Nørgaard, H. 1980 *Astrophys. J.* **236**, 895.

Olive, K. A. & Schramm, D. N. 1982 *Astrophys. J.* **257**, 276.

Ott, V., Lohr, H. P. & Begemann, F. 1985a In *Isotopic composition of the Solar System* (ed. Centre National d'Études Spatiales), p. 129. Toulouse: Cepadues-Éditions.

Ott, U., Yang, J. & Epstein, S. 1985b *Meteoritics* **20**, 722.

Papanastassiou, D. A. 1985 In *Isotopic ratios in the Solar System* (ed. Centre National d'Études Spatiales), p. 77. Toulouse: Cepadues-Éditions.

Prantzos, N. 1986 In *Advances in nuclear astrophysics* (ed. E. Vangioni-Flam, J. Audouze, M. Cassé, J.-P. Chièze & J. Tran Thanh Van), p. 321. Gif-sur-Yvette: Editions Frontières.

Prantzos, N., Cassé, M., Gros, M., Doom, C. & Arnould, M. 1985 In *19th Int. Cosmic Ray Conf., La Jolla* (ed. F. C. Jones, J. Adams & G. M. Mason), vol. 1, p. 361. NASA Conf. Pub., NASA CP-2376.

Prantzos, N. & Cassé, M. 1986 *Astrophys. J.* **307**, 324.

Prantzos, N., Arcoragi, J. P. & Arnould, M. 1986*a* In *Nucleosynthesis and its implications on nuclear and particle physics* (ed. J. Audouze & N. Mathieu), p. 293. Dordrecht: Reidel.
Prantzos, N., Arnould, M. & Arcoragi, J.-P. 1986*b* *Astrophys. J.* **315**, 209.
Prantzos, N., de Loore, C., Doom, C. & Arnould, M. 1986*c* In *Nucleosynthesis and its implications on nuclear and particle physics* (ed. J. Audouze & N. Mathieu), p. 197. Dordrecht: Reidel.
Prantzos, N., Doom, C., Arnould, M. & de Loore, C. 1986*d* *Astrophys. J.* **304**, 695.
Prombo, C. A. & Clayton, R. N. 1985 *Science, Wash.* **230**, 935.
Reeves, H. 1978 In *Protostars and planets* (ed. T. Gehrels), p. 399. Tucson: University of Arizona Press.
Reisse, J. & Mullie, F. 1985 In *Isotopic ratios in the Solar System* (ed. Centre National d'Études Spatiales), p. 107. Toulouse: Cepadues-Editions.
Robert, F. 1985 *Meteoritics* **20**, 744.
Share, G. H., Kinzer, R. L., Kurfess, J. D., Forrest, D. J., Chupp, E. L. & Rieger, E. 1985 *Astrophys. J.* **292**, L61.
Shima, M. 1986 *Geochim. cosmochim. Acta* **50**, 577.
Smith, C. L. & De Laeter, J. R. 1986 *Meteoritics* **21**, 133.
Swart, P. K., Grady, M. M., Pillinger, C. T., Lewis, R. S. & Anders, E. 1983 *Science, Wash.* **220**, 406.
Tammann, G. A. 1982 In *Supernovae: a survey of current research* (ed. M. J. Rees & R. J. Stoneham), p. 371. Cambridge University Press.
Thielemann, F.-K. & Arnett, W. D. 1985 *Astrophys. J.* **295**, 604.
Thielemann, F.-K., Nomoto, K. & Yokoi, K. 1985 In *Nucleosynthesis and its implications on nuclear and particle physics* (ed. J. Audouze & N. Mathieu), p. 131. Dordrecht: Reidel.
Thiemens, M. H. & Heidenreich III, J. E. 1983 *Science, Wash.* **219**, 1073.
Thiemens, M. H. & Jackson, T. 1985 *Meteoritics* **20**, 775.
Tinsley, B. M. 1980 *Fundam. cosmic Phys.* **5**, 287.
Truran, J. W. 1984 *A. Rev. nucl. part. Sci.* **34**, 53.
Truran, J. W. 1986 In *Nucleosynthesis and its implications on nuclear and particle physics* (ed. J. Audouze & N. Mathieu), p. 97. Dordrecht: Reidel.
Truran, J. W. & Cameron, A. G. W. 1978 *Astrophys. J.* **219**, 226.
Wallace, R. K. & Woosley, S. E. 1981 *Astrophys. J. Suppl.* **45**, 389.
Wasserburg, G. J. & Papanastassiou, D. A. 1982 In *Essays in nuclear astrophysics* (ed. C. A. Barnes, D. D. Clayton & D. N. Schramm), p. 77. Cambridge University Press.
Wiedenbeck, M. E. 1984 *Adv. Space Res.* **4**, 15.
Wiescher, M., Görres, J., Thielemann, F.-K. & Ritter, H. 1986 In *Nucleosynthesis and its implications on nuclear and particle physics* (ed. J. Audouze & N. Mathieu), p. 105. Dordrecht: Reidel.
Woosley, S. E. 1986 In *Nucleosynthesis and chemical evolution* (ed. B. Hauck, A. Maeder & G. Meynet), p. 1. Observatoire de Genève.
Woosley, S. E. & Weaver, T. A. 1986 In *Nucleosynthesis and its implications on nuclear and particle physics* (ed. J. Audouze & N. Mathieu), p. 145. Dordrecht: Reidel.
Wright, I. P. & Pillinger, C. T. 1985 In *Isotopic ratios in the Solar System* (ed. Centre National d'Étude Spatiales), p. 89. Toulouse: Cepadues-Éditions.
Yang, J. & Epstein, S. 1984 *Nature, Lond.* **311**, 544.

Discussion

Sir William McCrea, F.R.S. (*Astronomy Centre, University of Sussex, U.K.*). In his paper, Professor Arnould did not discuss what he referred to as, I think, 'exotic' abundances. Would he care to summarize their significance in the context of this discussion? Does he consider that they, or any other evidence, support the suggestion that a supernova outburst played some part in the origin or early history of the Solar System?

M. Arnould. The so-called Solar System isotopic anomalies reviewed by Professor Anders and Dr Pillinger provide new clues to many important astrophysical problems, as briefly mentioned in my paper.

It has been quite popular to assign a central role to supernovae in the model I refer to as model 3. For example in the 'compressed-cloud' model (Cameron & Truran 1977), a supernova shock compresses a preexisting cloud causing it to collapse, while injecting the anomalous material. On the other hand, the 'snowplow' model (Herbst & Rajan 1980) calls for the formation of the Solar System within the compressed gas shell swept up by a supernova.

Finally, in the 'Bing-Bang' model (Reeves 1978), the anomalous material is attributed to the accumulated ejecta of many supernovae in an OB association.

From the observational point of view, there is no really strong support for the idea that star formation often occurs in close association with supernova remnants. In that respect, there are also theoretical difficulties, or at least large uncertainties, as well as more purely nucleosynthesis problems (e.g. level of contamination, yield uncertainties, non-observation of certain theoretically expected anomalies). All in all, supernovae may not be the most likely events responsible either for the observed anomalies, nor for the Solar System formation, even if they cannot be firmly ruled out in those respects. In view of that situation, other models have been proposed, such as the trigger by low-mass asymptotic giant-branch stars (Cameron 1984). However, in my opinion, that model faces many difficulties and uncertainties. On the other hand, massive mass-losing stars (of the Of and Wolf–Rayet types) before explosion (if any) might be interesting trigger and/or contaminating agents of Solar System-type structures, as briefly mentioned in my paper. More work is now being done on various aspects of that scenario.

Additional references

Cameron, A. G. W. & Truran, J. W. 1977 *Icarus* **300**, 447.
Herbst, W. & Rajan, R. S. 1980 *Icarus* **42**, 35.

Phil. Trans. R. Soc. Lond. A **323**, 269–286 (1987)

Printed in Great Britain

Interstellar molecules

By D. Smith

Department of Space Research, University of Birmingham, P.O. Box 363,
*Birmingham B*15 2*TT, U.K.*

Some 70 different molecular species have so far been detected variously in diffuse interstellar clouds, dense interstellar clouds and circumstellar shells. Only simple (diatomic and triatomic) species exist in diffuse clouds because of the penetration of destructive ultraviolet radiations, whereas more complex (polyatomic) molecules survive in dense clouds as a result of the shielding against this ultraviolet radiation provided by dust grains. A current·list of interstellar molecules is given together with a few other molecular species that have so far been detected only in circumstellar shells. Also listed are those interstellar species that contain rare isotopes of several elements. The gas phase ion chemistry is outlined via which the observed molecules are synthesized, and the process by which enrichment of the rare isotopes occurs in some interstellar molecules is described. Reference is also made briefly to some very recent work in interstellar ion chemistry. A list of the atomic and molecular species that have been detected in cometary atmospheres is given and attention is drawn to the similarities and differences between interstellar and cometary molecules. The physical and chemical processes by which these observed cometary species may be generated from material that sublimes from the cometary nucleus are discussed.

1. Introduction

Diatomic molecules were first observed in diffuse molecular clouds via their characteristic optical absorption spectra some 50 years ago (Swings & Rosenfeld 1937; Dunham & Adams 1937 *a, b*). During the intervening period, and especially during the past 20 years, following the development of radio astronomical techniques, many complex (polyatomic) molecules have been detected (largely via their characteristic rotational emission spectra) in the cool denser regions of gas and dust in the Milky Way and also in other galaxies (see, for example, Rydbeck & Hjalmarson 1985). Until now about 70 different molecular species have been identified in the interstellar medium; particularly rich sources of molecules are the dense molecular clouds in the Orion, Sagittarius and Taurus regions. Molecular species are also very abundant in some circumstellar shells, notably that of the evolved star $IRC + 10216$. The existence of molecules in all of these regions is of great significance to astrophysics because they are the means of probing the physical state of such regions. Hence a great deal of attention has been devoted to understanding how the molecules are synthesized at the very low temperatures of these regions which are not conducive to normal (neutral–neutral) chemistry. Much progress has been made as a result of close cooperation between research workers in several disciplines including radio-and millimetre-wave astronomers, spectroscopists, laboratory kineticists and theoreticians. It is now accepted that gas-phase ionic reactions are largely responsible for the production of most of the molecules so far detected in interstellar clouds, a conclusion reached on the basis of the agreement between observed abundances and those predicted by gas-phase ion-chemical models of individual molecular clouds (Millar & Freeman 1984). An interesting

[21]

point to note immediately is that the individual molecular clouds do not have the same molecular compositions (Rydbeck & Hjalmarson 1985) and there are interesting implications of this heterogeneity. This is mentioned again in later sections.

Many molecular species have also been detected in cometary atmospheres (Huebner 1985) and some are common to both comets and interstellar clouds (Irvine *et al.* 1986). Other cometary molecules, not observed in interstellar clouds, could be fragments of well-known interstellar molecules (e.g. the cometary NH and NH_2 radicals could well be photofragments of parent NH_3 molecules, see §5). These similarities naturally raise the question as to whether comets are interstellar in origin. Comets could possibly have originated in cold interstellar gas and dust clouds where a wide variety of interstellar molecules existed, or in warmer gas (e.g. in the presolar nebula (Irvine 1983)), which previously had been heated to temperatures sufficiently high to destroy the more fragile polyatomic molecules. In these warmer regions, smaller molecules such as CO, CH_4, N_2, H_2O and NH_3 could survive and become part of the comet, but the most volatile species (He and H_2) would not condense. This latter situation could correspond to the medium out of which the giant planets were formed and in which molecules such as CH_4 and NH_3 were plainly present.

Although it is evident that there are similarities between interstellar molecules and the molecules in the comae and tails of comets, information is obviously required on the composition of cometary nuclei. Most studies of cometary molecules detect only those species that have travelled a considerable distance from the nucleus and that may be photofragments of larger (parent) species or species formed in chemical reactions in the cometary atmosphere. With such data, one can only deduce the composition of the nucleus by attempting to trace back the processes which lead to the observed species. The true chemical composition of cometary nuclei could be an indicator of the physical and chemical composition of the interstellar or presolar gas from which they are formed; perhaps they will even reveal the composition of the nebula from which the Sun was formed.

In anticipation of data on the true composition of cometary nuclei becoming available, it is clearly of value to appreciate the nature of interstellar molecules, the physical conditions and locations of the clouds where they are detected, their modes of synthesis, their stability and their chemistry. These aspects of interstellar molecules are discussed in the next three sections and, in the final section of this paper, the nature of cometary molecules, as they are presently known (before *Giotto*), and their possible relation to interstellar molecules are discussed.

2. MOLECULES DETECTED IN INTERSTELLAR CLOUDS AND CIRCUMSTELLAR SHELLS

Molecules are observed in various types of interstellar gas clouds, including diffuse clouds and dark and dense clouds, and also in the denser atmospheres of evolved stars (circumstellar shells). The diversity of the molecular species observed in each region reflects the different physical conditions (i.e. gas density, temperature, chemical composition and radiation field) in these various regions. A current list of molecules observed in interstellar and circumstellar regions is given in table 1. They range from the diatomic species such as the ubiquitous H_2, which is by far the most abundant (the next most abundant is CO which is typically 10^4 times less abundant than H_2), to the 13-atomic $HC_{11}N$, which so far has only been detected in the envelope surrounding the carbon-rich star IRC+10216 (Bell *et al.* 1982) and in the Taurus molecular cloud (Bell & Matthews 1985).

TABLE 1. MOLECULES DETECTED IN INTERSTELLAR CLOUDS AND CIRCUMSTELLAR SHELLS

(From various sources including Winnewisser $et\ al.$ (1979), Rydbeck & Hjalmarson (1985), Irvine (1987) and Irvine $et\ al.$ (1986).)

hydrogen H_2

molecules containing only C and H

CH	$C{\equiv}C$	$C{\equiv}CCH$	$(C{\equiv}C)_2H$
CH^+	$C{\equiv}CH$	$H_3CC{\equiv}CH$	$H_3C(C{\equiv}C)_2H$
$*CH_4$	$*HC{\equiv}CH$	C	
	$*H_2C{=}CH_2$	$\triangle$	
		$HC{=}CH$	

molecules containing O				molecules containing N			
OH	HCO^+	CH_3CHO	CH_3CO_2H	CN	H_2CN^+	CH_3CH_2CN	$H(C{\equiv}C)_4CN$
H_2O	HOC^+?	CH_3OH	CH_3OCH_3	HCN	NH_2CN	$H_2C{=}CHCN$	$H(C{\equiv}C)_5CN$
CO	H_2CO	CH_3CH_2OH	$HOCO^+$?	HNC	CH_2NH	$HC{\equiv}CCN$	$C{\equiv}CCN$
HCO	CH_2CO	HCO_2H	$C{\equiv}CCO$	N_2H^+	CH_3NH_2	$H(C{\equiv}C)_2CN$	$H_3CC{\equiv}CCN$
				NH_3	CH_3CN	$H(C{\equiv}C)_3CN$	$H_3C(C{\equiv}C)_2CN$?

molecules containing O and N

NO, HNO?, HNCO, HOCN?, NH_2CHO

molecules containing S and Si

SO, SN, CS, H_2S, SO_2, OCS, HCS^+, H_2CS, CH_3SH, HNCS, SiO, SiS, $*SiC_2$, $*SiH_4$

Molecules marked with an asterisk have been detected only in circumstellar shells. The question marks indicate tentative detections. A tentative detection of NaOH has been reported (Hollis & Rhodes 1982). A detection of HCl has also been reported (Blake $et\ al.$ 1985). $HOCO^+$ and HOCN are both possible assignments for the same lines (Thaddeus $et\ al.$ 1981).

Diffuse clouds are so called because they are partly transparent to visible and ultraviolet radiation ($\lambda > 1216$ Å†), a consequence of the low gas-number densities, which are typically $ca.$ 10^2 cm^{-3}. These radiations heat the gas to temperatures that exceed 200 K in some parts of the clouds. It was in diffuse clouds that interstellar molecules (i.e. CH, CH^+, CN) were first detected via their characteristic absorption spectra in the visible region of the spectrum (Swings & Rosenfeld 1937; Dunham & Adams 1937a, b; McKellar 1941). Only simple (diatomic and triatomic) molecules can exist in significant concentrations in diffuse clouds because the ambient ultraviolet radiation would efficiently photodissociate larger, more weakly bound molecules (van Dishoeck 1986). Detailed atomic and molecular compositions of several diffuse cloud have been obtained. Especially thorough studies have been those of the diffuse clouds in Othe direction of ζ-Ophiucus, χ-Ophiucus, ζ-Perseus and o-Perseus (see van Dishoeck & Black (1986) for details). Typical of the molecules observed in these clouds are HD, OH, CO, CH, CH^+, CN and C_2. The observed relative abundances of these species can be reconciled quite well with relative abundances predicted by gas phase ion chemical models (van Dishoeck & Black 1986), apart from the apparent over-abundance of CH^+ in diffuse clouds. This has been attributed to local shock heating of the gas, which promotes the endothermic reaction of C^+ with H_2 producing CH^+ (Elitzur & Watson (1978); Adams $et\ al.$ (1984b) and §3a).

A much greater variety of molecular species is observed in the cooler, denser clouds of gas molecules and dust grains, via radio- and millimetre-wave astronomy (Winnewisser $et\ al.$ 1979;

† Å $= 10^{-1}$ nm $= 10^{-10}$ m.

[23]

Guélin 1985; Rydbeck & Hjalmarson 1985). The greater gas-number densities (which can exceed 10^4–10^6 cm^{-3}) and the lower temperatures (lower than 10 K in some clouds) promote the chemistry that results in the generation of complex molecules (see §3b). The shielding of these molecules from ultraviolet radiation by the dust grains is essential for their existence. Thus, the wide variety of molecules listed in table 1 (and presumably many more, as yet undetected) can exist in dense and dark clouds among which are the much-studied Orion Molecular Cloud (OMC 1), the Taurus Molecular Cloud (TMC 1) and the giant molecular cloud Sagittarius B2 (Sgr B2). Severe gradients of both gas density and temperature exist in these clouds in which gravitational collapse (clumping) and protostar formation are evident (Evans 1978; Andrew 1980). However, it is clear that, on average, OMC 1 ($T \approx 50$ K) is a somewhat warmer cloud than TMC 1 ($T \approx 10$ K). Also, although many of the molecular species are common to all dense clouds, there are interesting indications of chemical heterogeneity. For example, the clouds in the Taurus region (including TMC 1) are significantly enriched in the cyanopolyyne molecules, HC$_n$N, compared with other similar regions in the galaxy (e.g. the dark cloud L134N), whereas more C, H, O bearing molecular species have been detected in OMC 1 than in TMC 1 (Irvine 1983). Indeed, it is clear that within some individual clouds there exist spatial variations in the relative abundances of some molecular species. These observations of spatial heterogeneity may be important in the consideration of the origins of comets. This heterogeneity has also attracted the interest of astrochemists and astrophysicists because it raises the question whether differences in molecular composition are due to differences in the elemental composition of the clouds or to the occurrence of a different chemistry involving different rates of chemical evolution under different physical conditions (Penzias 1980; Crutcher & Watson 1985).

The rare stable isotopes of several elements (i.e. D, ^{13}C, ^{15}N, $^{17, 18}$O, $^{29, 30}$Si and $^{33, 34}$S) have been detected in interstellar molecules. A current list of the rare isotopic variants of interstellar molecules is given in table 2. The abundance ratios of the rare to common isotopic variants of several molecular species have been determined in some interstellar clouds and it is often found that the observed ratios are different from those expected on the basis of terrestrial isotopic ratios. For example, the ^{13}CO/^{12}CO ratio in several dense clouds is about twice the terrestrial ^{13}C/^{12}C ratio (Winnewisser *et al.* 1979). Also the D/H ratio in interstellar molecules invariably exceeds the terrestrial D/H ratio (Winnewisser *et al.* 1979). Isotopic ratios are of great interest because they can provide an indication of the nuclear history of the galaxy. The D/H ratio is particularly important, not least because it is a critical parameter in cosmological models (Clayton 1985). Because isotopic ratios within the various regions of the galaxy (except for the

TABLE 2. RARE ISOTOPIC VARIANTS OF INTERSTELLAR MOLECULES

(From various sources including Winnewisser *et al.* (1979) and Irvine *et al.* (1986).)

isotope	molecules in which isotope detected
D	H_2, H_2O, HCO^+, N_2H^+, HCN, HNC, NH_3, H_2CO, CH_3OH, HC_3N, HC_5N
^{13}C	CO, CS, HCN, HNC, HCO^+, OCS, H_2CO, HC_3N, CH_3CN, CH_3OH
^{15}N	HCN, HNC, NH_3, N_2H^+
^{17}O	CO, HCO^+
^{18}O	CO, OH, H_2O, HCO^+, H_2CO
^{29}Si	SiO, SiS
^{33}S	CS
^{34}S	CS, SO, SO_2, OCS, SiS

small volume of the interstellar medium near to the Solar System) can only be determined by observing molecular emissions, it is important to appreciate to what extent the isotopic ratios in these molecules are representative of the cloud material as a whole. Careful consideration of this problem, together with the wealth of laboratory data recently obtained on the rates of isotope exchange in ion–molecule reactions, has indicated that enrichment of heavy (rare) isotopes can occur in interstellar molecules as a result of 'isotopic fractionation'. This phenomenon is briefly described in §3b.

A rich variety of molecules has also been detected in the atmospheres of envolved stars (notably that of the carbon-rich star IRC+10216, Zuckerman (1980) and Wannier *et al.* (1980)), including many polyatomic species such as the cyanopolyynes. The high pressures and temperatures in the lower atmospheres of such objects ensures that reactions between neutral species are mainly responsible for molecular synthesis, but in the upper atmospheres ionic reactions play an important role (Omont 1987). The first cyclic molecule to be detected in interstellar or circumstellar gas, i.e. SiC_2, was detected in the atmosphere of IRC+10216 (Thaddeus *et al.* 1984), together with the symmetrical molecules CH_4 and SiH_4 (Rydbeck & Hjalmarson 1985), which have not yet been detected in interstellar clouds.

Five molecular positive ions have been identified so far in interstellar clouds (table 1), these being CH^+, HCO^+, N_2H^+, HCS^+ and H_2CN^+. A tentative detection of COH^+ has also been reported, although this identification is unlikely to be confirmed (see §3). A detection of the ion HCO_2^+ has also been reported, but this has not yet been confirmed. However, it is not unreasonable to expect that HCO_2^+ will exist in interstellar clouds, as well as the many other positive ion species that are considered to be involved in the gas-phase ion chemistry leading to many of the observed interstellar molecules.

3. Production of molecules in interstellar clouds

It is now generally accepted that the synthesis of molecules in interstellar clouds occurs largely via gas-phase ionic reactions. Ion–neutral reactions generate complex ions which have to be neutralized to produce the observed wide variety of neutral molecules. The neutralization process is predominantly positive-ion–electron dissociative recombination. Several excellent reviews describe these reaction processes and include many specific examples of reactions that occur in interstellar clouds (Dalgarno & Black 1976; Smith & Adams 1981; Herbst 1985). Production of some molecular species on the surfaces of interstellar dust grains (followed by desorption into the gas phase) cannot be entirely ruled out; indeed, production of the most abundant molecule, H_2, can be quantitatively explained only by invoking grain-surface combination of H atoms followed by desorption. The success of gas-phase ion chemical models in accounting for the relative abundances of most interstellar molecular species is compelling evidence for gas-phase synthesis. Both steady-state (time-independent) and time-dependent models, which involve many hundreds of individual reactions, have been developed to describe the chemical evolution of interstellar molecules (see §3b). The central role of gas-phase reactions is also indicated by the apparent enrichment of the rarer (heavy) isotopes in some interstellar molecules. This so-called 'isotopic fractionation' (see, for example, Crutcher & Watson (1985) and also below) is not likely to occur to the extent observed in interstellar molecules if they are formed catalytically on surfaces. However, that both gas phase and to some degree surface production of interstellar molecules occurs is nicely summed up by the phrase often stated by

A. Dalgarno 'The best evidence for grain surface production is the existence of H_2, the best evidence for gas phase synthesis is the abundance of HD'.

Rapid growth in the understanding of interstellar chemistry has taken place during the past few years. The detection of an increasing number of different interstellar molecular species and the determination of their relative abundances has stimulated the development of laboratory experiments to study gas-phase ionic reactions at appropriately low temperatures. As a result, there has been an explosive increase in the amount of kinetic data that are necessary to satisfy the demands of the increasingly sophisticated ion chemical models. The laboratory experiments have also revealed many important routes for the synthesis of interstellar molecules and have indicated important new classes of reactions specific to interstellar chemistry. An example is radiative association (see below). Notable experimental developments have been the variable-temperature selected-ion flow tube (VT-SIFT) experiments (Smith & Adams 1979) which have provided a very great deal of relevant kinetic data (Adams & Smith 1987a) and the more recent and very promising CRESU (Cinetique de Reaction en Ecoulement Supersonique Uniforme) experiment, which is now producing kinetic data at temperatures as low as 8 K (Rowe $et\ al.$ 1985). Extensive compilations of the rate coefficients and product-ion distributions for ion–neutral reactions, including many reactions relevant to interstellar and cometary chemistry, have been published (Albritton 1978; Anicich & Huntress 1986). Some recent exciting results, especially those obtained by the VT-SIFT and CRESU experiments, are presented in §4. Important new data are also becoming available relating to the process of dissociative recombination following the development of the flowing afterglow–Langmuir probe (FALP) technique (Adams $et\ al.$ (1984a) and see §4).

It is clear that the chemistry occurring in particular interstellar clouds is dependent on the chemical composition and the physical conditions of the clouds. As stated above, it is now known that particular clouds are often very spatially heterogeneous, and much effort is being directed to studying the small-scale variations in composition in particular clouds. More obvious differences are apparent between the types of molecules present in diffuse clouds and in dense clouds, which reflect the obvious differences in the nature of these clouds. Although it is not the intention in this paper to discuss the details of interstellar chemistry, it is instructive to outline the somewhat differing chemistries occurring in diffuse clouds and in dense clouds as these are now understood. Later the elements of cometary chemistry will be discussed.

(a) Diffuse cloud chemistry

Because of the transparency of diffuse clouds to ultraviolet radiation, which can efficiently dissociate molecules, a substantial fraction of all elements in diffuse clouds exist in the atomic form. Thus the major constituents are H atoms, as well as H_2 molecules, with C, N and O atoms as important minor constituents. These atomic species play a major role in the ion chemistry, via which molecules are synthesized, and this is shown schematically in figure 1. The chemistry is generally considered to be initiated by the ionization of H and H_2 by cosmic rays and by the photoionization of C atoms by ultraviolet radiation. The H^+ and H_2^+ ions so produced then initiate the ionic reactions represented in the upper part of figure 1 and the C^+ primary ion initiates that chemistry represented in the lower part of figure 1. Whereas the chemistry initiated by cosmic rays can occur throughout a diffuse cloud (because of the penetrating power of cosmic rays) the chemistry beginning with C^+ may be more important in the outer part of the cloud where the ultraviolet radiation field is most intense. The chemistry is described by

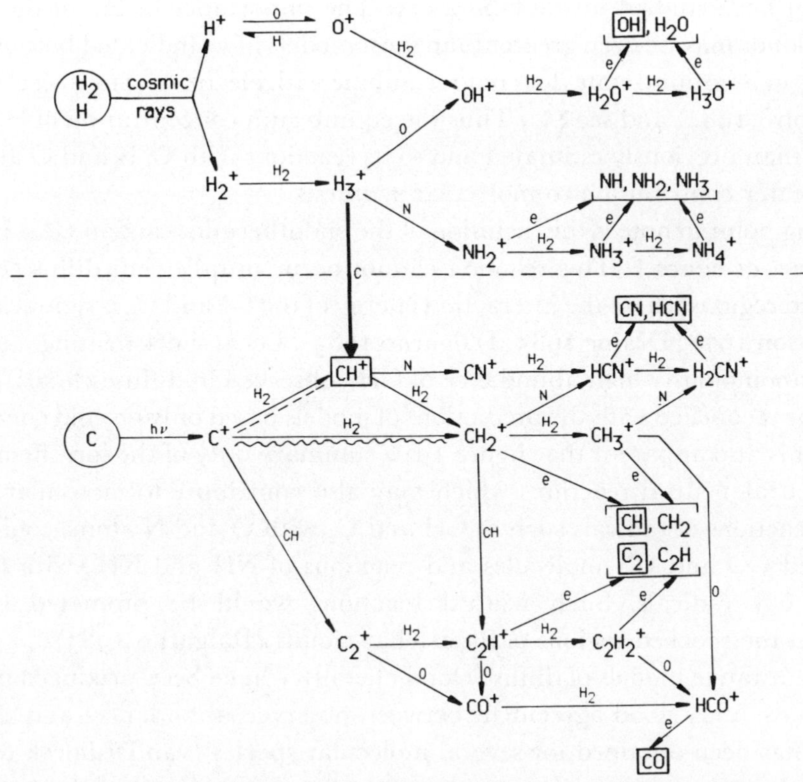

FIGURE 1. A representation of the ion chemistry of diffuse interstellar clouds. The horizontal dashed line divides the chemistry that is initiated by cosmic-ray ionization of H atoms and H_2 molecules from the chemistry that is initiated by ultraviolet photoionization of C atoms. The thick arrow represents the $H_3^+ + C$ reaction which produces CH^+ and which couples these chemistries. The $C^+ + H_2$ reaction, indicated by the dashed line, cannot occur in quiescent clouds and is thought to occur in shocked regions. The $C^+ + H_2$ reaction indicated by the wavy line is a radiative association reaction. The species enclosed by boxes have been positively detected in diffuse clouds. The letter e along an arrow indicates that the ion reacts with electrons undergoing dissociative recombination to generate fragment neutral molecules.

sequences of reactions of ions with atoms (mainly C, N and O) and molecules (mainly H_2 but also small hydrocarbon molecules) generating ions of increasing atomicity which can then react with electrons to produce the fragment neutral molecules shown in figure 1. Again, it is important to note that the concentrations of polyatomic species are limited by photodissociation in the ultraviolet radiation field. Thus, relatively weakly bound species such as NH_3 cannot be present in large concentrations. Note that the ion chemistry leading to nitrogen hydrides shown in figure 1 depends on the reaction of H_3^+ with N atoms which has not yet been observed in laboratory experiments. However, most of the ionic reactions involving H_2, N and O included in figure 1 have been studied experimentally and their rate coefficients measured. A major exception is the very slow, but nevertheless important, radiative association reaction of C^+ with H_2 generating CH_2^+, the rate coefficient for which has had to be obtained by calculation (Black & Dalgarno 1973; Herbst *et al.* 1977). The routes for the production of small hydrocarbon molecules, OH and H_2O and the strongly-bonded CN-bearing molecules and CO are at least qualitatively understood and are as given in figure 1. An important link between the cosmic-ray-initiated chemistry and the ultraviolet-indicated chemistry could be the $H_3^+ + C$ reaction indicated by the thick arrow in figure 1, but this reaction, although expected to be fast, has not yet been studied in the laboratory. The importance of H_3^+ in the ion chemistry of interstellar clouds may be even greater than most models have indicated because of the recent finding that H_3^+ in its ground state does not recombine with electrons (Smith & Adams (1984b), Michels & Hobbs (1984) and see §4). Thus the equilibrium concentration of H_3^+ in the clouds will be greater than previously estimated and so its reactions with C, N and O atoms (figure 1) will make a greater contribution to molecular synthesis.

An interesting point to note is the inclusion of the endothermic reaction $C^+ + H_2 \rightarrow CH^+ + H$ in the lower part of figure 1. This reaction cannot occur in quiescent diffuse clouds but can occur in shocked regions where the interaction energy of the C^+ and H_2 is significantly increased (Elitzur & Watson 1978; Draine 1985; Dalgarno 1985). Local shock heating has been invoked to explain the anomolously high abundance of CH^+ observed in diffuse clouds, an abundance which cannot be reconciled with the predictions of models based only on cold chemistry (Adams *et al.* 1984b). It is also apparent that figure 1 is a summary only of the ion chemistry and does not involve neutral–neutral reactions which may also contribute to molecular synthesis. For example, the reactions of radicals such as CH and C_2 with O and N atoms could generate the strongly bonded CO and CN molecules and reactions of NH and NH_2 with C atoms could also generate CN radicals. Such neutral reactions would be promoted by the higher temperatures in the shocked regions of interstellar clouds (Dalgarno 1985).

Detailed quantitative models of diffuse cloud chemistry have been produced including those for specific clouds. Very good agreement between observed abundances and those predicted by the models has been obtained for several molecular species (van Dishoeck & Black 1986).

(b) Dense cloud chemistry

Dense clouds consist largely of H_2 molecules, He atoms and dust grains, the last shielding the central regions of the clouds from stellar ultraviolet radiation, which would effectively dissociate complex molecules. Grains may also play some role in the synthesis of molecules by surface catalysis (Duley & Williams 1984). The ion chemistry is initiated by the action of cosmic rays on H_2 and the He generating H^+, H_2^+ and He^+ ions. As can be seen by comparing figure 1 with figure 2, which is a schematic representation of the ion chemistry of dense clouds, the initial stages of the ion chemistry of dense and diffuse clouds are similar. One difference is that C^+ ions are generated in dense clouds by the reaction of He^+ with the relatively abundant CO rather than by direct photoionization of C atoms. Again, the synthesis of the key molecules H_2O and CH_4 is initiated by reactions of H_3^+ and C^+. It is also possible that NH_3 is generated in dense clouds via the reaction sequence initiated by the H_3^+ reaction with N atoms (see figure 1) as well as by the route shown in figure 2. It is not yet clear which of these two routes predominates. What is more clear (as indicated in figure 2), is the important role of CH_3^+ ions in the chemistry of dense clouds (compared to diffuse clouds). The lower temperatures, higher gas-number densities and the weaker radiation fields in dense clouds promote the production of large molecular ions and hence, via recombination, the production of the observed complex neutral molecules.

To facilitate discussion, the overall ion chemistry represented by figure 2 is roughly divided into left-hand, central and right-hand columns which respectively describe the production of carboxy molecules, hydrocarbon molecules and cyano and amino molecules. The ion species common to each of these chemistries is CH_3^+. This is because, although the reaction of CH_3^+ with H_2 is a vital link in the chain of reactions which generate hydrocarbon molecules, this

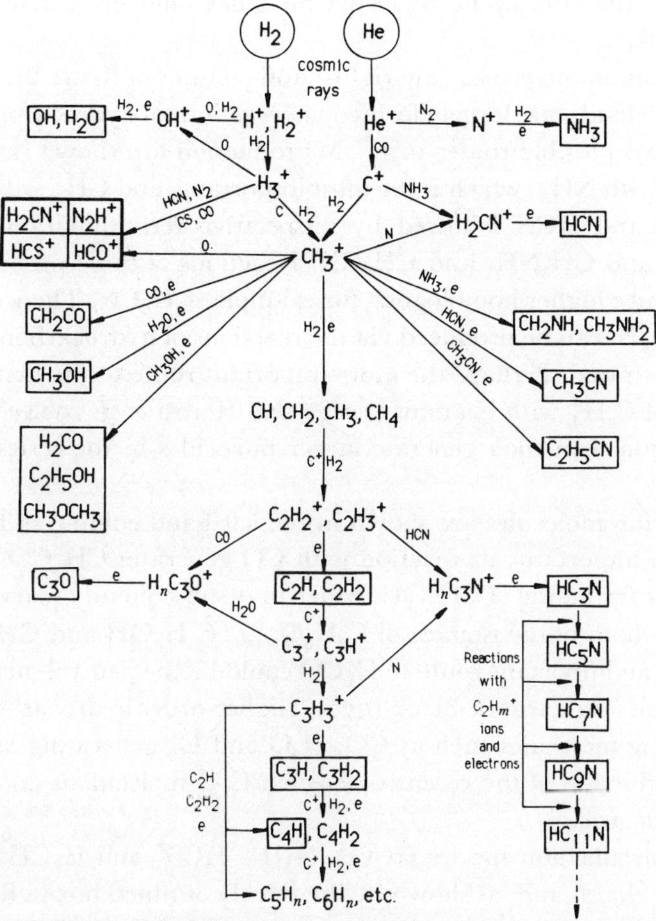

Figure 2. A representation of the ion chemistry of dense interstellar clouds. The ion chemistry is initiated by cosmic-ray ionization of H_2 and He. The production of hydrocarbon molecules is indicated by the central column of the diagram. The production of O-containing and N-containing molecules are indicated in the left-hand and right-hand columns respectively. Where a molecule and an electron lie along an arrow connecting two species, this implies that first an ion–molecule reaction occurs generating a product molecular ion which then undergoes dissociative recombination with electrons to form the neutral molecules indicated. The molecules enclosed by boxes have been detected in dense interstellar clouds. These include the positive-ion species in the thickly lined box.

reaction is relatively slow and therefore some of the CH_3^+ ions can react with minority molecules such as H_2O, NH_3, HCN, etc.

The hydrocarbon reaction chain develops first via the radiative association reaction of CH_3^+ with H_2 (discussed below) which produces CH_5^+ ions. Dissociative recombination of CH_3^+ and CH_5^+ with electrons can then generate small neutral hydrocarbon molecules including CH_4. Reactions of these species with C^+ (and CH_3^+) ions generate ions containing two carbon atoms, reactions known as carbon insertion reactions. Dissociative recombination of the product ions generates C_2H and C_2H_2 molecules. Further carbon insertion reactions generate higher order hydrocarbons as illustrated in figure 2. The $C_3H_3^+$ ions formed in the reaction of C_3H^+ with H_2 are especially interesting because it is likely that a fraction of them are the cyclic isomer and therefore probably the precursor ions of c-C_3H_2, the first cyclic neutral molecule detected in interstellar clouds (Matthews & Irvine (1985) and Thaddeus *et al.* 1985). This is discussed further in §4. Note that the cyclic molecule SiC_2 has only been detected in IRC+10216 (Thaddeus *et al.* 1984).

The cyano and amino molecules (the right-hand column of figure 2), which are so evident in dense interstellar clouds, are largely formed via reactions of hydrocarbon ions with N atoms, NH_3 and HCN. Two possible routes to HCN production are shown in figure 2, these being the reactions of C^+ with NH_3 (which is the most important) and CH_3^+ with N atoms. Reactions of CH_3^+ with NH_3 and HCN followed by dissociative recombination generate the amino molecules CH_2NH and CH_3NH_2 and CH_3CN. Reactions of these product neutral molecules with CH_3^+ can produce higher homologues, for example C_2H_5CN. The well-known interstellar cyanopolyynes, HC_nN, can be produced via the reactions of hydrocarbon ions with both HCN and N atoms as illustrated. Perhaps the more important route to the first of the series, HC_3N, is via the reaction of $C_3H_3^+$ with N atoms (see §4 and Herbst *et al.* 1984a). Reactions of HC_3N with hydrocarbon ions can then generate larger molecules in the series, HC_5N, HC_7N, etc. (Knight *et al.* 1986).

The oxygen-bearing molecules are shown in the left-hand column of figure 2. Reactions of CH_3^+ ions are again important; its reaction with CO generates CH_2CO and its reaction with H_2O leads first to the formation of CH_3OH, which may subsequently react with CH_3^+ to produce H_2CO and possibly both of the isomers of C_2H_6O, i.e. C_2H_5OH and CH_3OCH_3. However, it must be stated that an important route to H_2CO could be the neutral–neutral reaction of CH_3 radicals with O atoms (Dalgarno & Black 1976). Higher-order hydrocarbon ions can react with small oxygen-bearing molecules such as CO, H_2O and O_2 generating larger oxygen-bearing molecules. The production of the recently-detected C_3O molecule is an example as shown in figure 2 (Herbst *et al.* 1984b).

Four positive molecular ion species HCO^+, N_2H^+, HCS^+ and H_2CH^+ have been detected in dense interstellar clouds and, as shown by the thickly outlined box in figure 2, they are most probably produced in proton transfer reactions of H_3^+ with CO, N_2, CS and HCN. A detection of COH^+ (an isomer of HCO^+) has been tentatively reported (Woods *et al.* 1983), although it is certain that this species cannot be present in appreciable concentrations in dense clouds because it is rapidly converted to HCO^+ in reaction with H_2 (M. J. McEwan 1986 personal communication). Recent work relating to the detection and abundances of HCS^+ and H_2CN^+ is briefly discussed in §4, together with some comments on other sulphur-bearing interstellar molecules that are not included in figure 2 because the routes to their synthesis are uncertain.

Detailed quantitative models of dense clouds (including steady-state and time-dependent models) have been produced involving large numbers of parallel and sequential ionic reactions (see, for example, Millar & Freeman 1984; Leung *et al.* 1984; Millar & Nejad 1985; Millar 1985). Useful review articles are available that describe in detail the gas phase ion chemistry and refer to specific reaction types (Dalgarno & Black 1976; Smith & Adams 1981; Herbst 1985). Some neutral–neutral reactions are included in the models and arguments have been given for grain surface production of some molecules including NH_3 and sulphur-bearing molecules, the production of which cannot yet be convincingly explained via gas phase models of quiescent dense clouds (Millar 1982; Duley & Williams 1984).

It was stated in §2 that the heavy isotopes of some elements were apparently enriched in some interstellar molecules. Deuterium enhancement is particularly obvious. Detailed laboratory studies of isotope exchange in ion–neutral reactions have shown that fractionation of heavy isotopes can occur in these reactions, especially at low temperatures. The elementary reaction

$$D^+ + H_2 \rightleftharpoons H^+ + HD \tag{1}$$

is a good illustration of this phenomenon. As a result of the difference in the zero-point energies of H_2 and HD, (1) is significantly exothermic to the right. Consequently, the endothermic reaction $H^+ + HD$ cannot occur at a significant rate at the low temperatures of interstellar clouds and thus it is expected that much of the deuterium in dense interstellar clouds is contained in HD molecules. Similarly, the reaction

$$H_3^+ + HD \rightleftharpoons H_2D^+ + H_2 \tag{2}$$

fractionates deuterium into H_2D^+ and subsequent reactions of H_2D^+ with other neutral interstellar molecules, for example, the reaction of H_2D^+ with HCN generating both H_2CN^+ and $HDCN^+$ results in an enhanced abundance of deuterium (relative to hydrogen) in the product molecules. This explains why deuterium is significantly fractionated into DCN relative to HCN. Fractionation of ^{13}C into CO can occur via the reaction

$$^{13}C^+ + {}^{12}CO \rightleftharpoons {}^{12}C^+ + {}^{13}CO \tag{3}$$

and this is probably the reaction responsible for the observed enhanced abundance of ^{13}C in the CO in several dense clouds (Smith & Adams 1980). Detailed laboratory studies of isotope fractionation in ion–neutral reactions have been carried out (Smith & Adams 1984a) and the interstellar significance of this phenomenon has also received considerable attention (see, for example, Smith *et al.* (1982) and Crutcher & Watson (1985)). Very recently, following laboratory studies of the reactions

$$C_2H_2^+ + HD \rightleftharpoons C_2HD^+ + H_2, \tag{4}$$

the enhanced abundance of C_2D and the very different C_2D/C_2H abundance ratios observed in OMC 1 and TMC 1 have been explained in terms of the different rates of deuterium fractionation into C_2D (which is formed by dissociative recombination of C_2HD^+ ions) and the temperature dependences of the radiative association rates of $C_2H_2^+$ and C_2HD^+ with H_2 (Herbst *et al.* 1987). It is now clear that isotope fractionation must be taken into account when observed molecular abundances are being exploited, for example, to estimate galactic gradients of isotopic ratios of the elements.

4. SOME RECENT ADVANCES IN INTERSTELLAR CHEMISTRY

Much of the ion chemistry of diffuse and dense clouds, summarized by figures 1 and 2 respectively, is not contentious, because it has been substantiated both by laboratory studies of many of the reactions and by comprehensive ion–chemical modelling with these laboratory data. However, it is still necessary to make certain assumptions in tracing the synthesis of some observed interstellar molecules which, if shown to be invalid, would require a radical rethink of much of the chemistry. Thus research is proceeding to clarify several aspects of the chemistry, hand-in-hand with more thorough astronomical observations and searches for other interstellar species. Questions being asked include the following. Are the rate coefficients for ionic reactions, which have largely been determined in laboratory experiments at temperatures at 80 K or greater, also appropriate to the lower-temperature conditions of interstellar clouds? How widespread are hydrodynamic and magnetohydrodynamic (MHD) shocks in interstellar clouds and what influence do such shocks have on the chemistry? What is the role of kinetically excited ions? How important is the process of radiative association that is commonly invoked for the synthesis of interstellar molecules? How important are ion–atom reactions and neutral–neutral reactions in the chemistry? What are the products of the dissociative recombination reactions that are so central to current ion–chemical schemes of molecular synthesis? These questions and others are being considered in relation to specific problems such as: the apparent overabundance of HCS^+ and the underabundance of HCl in dense clouds; the unknown routes for the production of cyclic C_3H_2 and the cyanopolyynes; and the recombination rate of the very important H_3^+ ion under interstellar conditions. Some recent progress has been made on each of these fronts and this will now be briefly referred to here.

An important advance has been made recently in the appreciation of interstellar reaction kinetics. Following theoretical predictions (Clary 1985), laboratory experiments have been carried out to measure the rate coefficients for the reactions of ions with polar molecules at low temperatures. These laboratory studies have shown, in accordance with these theoretical predictions, that rapid increases occur in the rate coefficients at low temperatures (Clary *et al.* 1985 and Marquette *et al.* 1985). For reactions involving very polar molecules (such as HCN, HC_3N and CS (see below)), the rate coefficients are more than ten times greater at 10 K than they are at 300 K (Adams *et al.* 1985). This discovery has major implications to model predictions for the rates of production and loss of some interstellar species. An immediate consequence of this work has been to explain the apparently anomalous overabundance of HCS^+ (relative to CS) in dense interstellar clouds (Millar *et al.* 1985). The abundance ratio, HCS^+/CS, predicted by ion–chemical models was a factor of 10–100 lower than the observed abundances, which, unlike the analogous HCO^+/CO ratios, were very variable among interstellar clouds. This worrying problem was resolved by the realization that reactions of the very polar molecule CS proceed much more rapidly than had previously been appreciated and consequently the reassessed production rate of HCS^+ (formed mainly by the reaction of H_3^+ with CS, see figure 2) and the loss rate of CS were both much greater. Following this, the predicted HCS^+/CS abundance ratios now agree with the observed ratios which vary from cloud-to-cloud as a result of the different rate coefficients for the reactions of CS at the different cloud temperatures. The abundance of H_2CN^+ that has recently been detected in Sgr B2 (Ziurys & Turner 1986) is quite consistent with the predicted rapid formation of this ion in the reaction of H_3^+ with HCN (Millar *et al.* 1985).

A great deal of effort is currently being made towards gaining an understanding of molecular synthesis in the shocked regions of interstellar gas. As was previously mentioned (§3*a*), the apparent overabundance of CH^+ in diffuse interstellar clouds has been attributed to shock chemistry (Adams *et al.* 1984*b* and Dalgarno 1985). Efforts have also been made to explain the production of H_2S in an analogous way, i.e. via the sequential reactions

$$S^+ \xrightarrow{H_2} SH^+ \xrightarrow{H_2} SH_2^+ \xrightarrow{H_2} SH_3^+ \xrightarrow{e} H_2S. \tag{5}$$

The three ion–molecule reactions in this sequence are all endothermic and hence cannot occur in cold gas but can occur in shocked gas (Adams *et al.* 1984*b* and Millar *et al.* 1986). Indeed, it has been predicted that the column density of SH^+ ions could be as large as that of CH^+ ions in some MHD shocks although not in purely hydrodynamic shocks. Such an observation could therefore be an important indicator of MHD shocked regions. The influence of kinetic excitation on the rates of some important interstellar reactions, including those leading to the production of NH_3, have been discussed by Adams *et al.* (1984*b*).

As previously stated (§3*b*), and as is clear from figure 2, radiative association is considered to be an important process for the synthesis of many molecules particularly in dense, cold clouds. Yet, until recently, this reaction process had not been observed directly in a laboratory experiment, although there were very strong theoretical and experimental indicators of its importance. However, the radiative association reaction

$$CH_3^+ + H_2 \rightarrow CH_5^+ + h\nu \tag{6}$$

has now been directly observed in the laboratory (Barlow *et al.* 1984). Furthermore, the measured rate coefficient for the binary reaction (6) is in reasonable agreement with that estimated on the basis of the measured rate coefficient for the analogous ternary association reaction

$$CH_3^+ + H_2 + He \rightarrow CH_5^+ + He. \tag{7}$$

This is an important step forward in that the rate coefficients for many of the radiative association reactions included in interstellar ion–chemical models have been derived from their ternary association reaction analogues (Smith & Adams 1978; Herbst *et al.* 1983, 1984*a*). Following this work, it has been reasoned that the cyclic interstellar species c-C_3H_2 is formed in the reaction

$$C_3H^+ + H_2 \rightarrow c\text{-}C_3H_3^+ + h\nu \tag{8}$$

followed by the recombination reaction

$$c\text{-}C_3H_3^+ + e \rightarrow c\text{-}C_3H_2 + H. \tag{9}$$

Although further work is required before this can be confirmed, it has been shown (Adams & Smith 1987*b*) that the ternary analogue of (8) does result in the production of c-$C_3H_3^+$. However, whether (9) results in the production of c-C_3H_2 remains to be determined.

Radiative association has also been invoked to explain the production of the cyanopolyyne series of molecules (Herbst *et al.* 1984*a* and Knight *et al.* 1986). As stated above, the production of the first of the series, HC_3N, probably proceeds via the recently studied reaction (Federer *et al.* 1986)

$$C_3H_3^+ + N \rightarrow H_2C_3N^+ + H \tag{10}$$

followed by the recombination reaction $H_2C_3N^+ + e \rightarrow HC_3N$. (Atom reactions are technically more difficult to study than reactions with stable molecules but important developments in technique are facilitating such studies (Adams & Smith 1985; Federer *et al.* 1986.)) An extensive study of the reactions of HC_3N has been carried out recently and reactions of the type

$$C_2H_2^+ + HC_3N + He \rightarrow H_3C_5N^+ + He \tag{11}$$

have been observed (Knight *et al.* 1985). It has also been postulated that other members at the cyanopolyyne series may be synthesized in interstellar clouds by radiative association analogues of reactions such as (11) (Schiff & Bohme 1979; Knight *et al.* 1985*b*).

Concerning dissociative recombination reactions of interstellar ions, a recent result worthy of note is the finding that the major interstellar ion H_3^+ (when in its ground state) does not recombine with electrons at a significant rate (Adams *et al.* 1984a; Michels & Hobbs 1984). The implications of this to interstellar physics and chemistry have been discussed recently (Smith & Adams 1984b). Work has also begun to determine the neutral products of dissociative recombination reactions of interstellar ions (Quéffelec *et al.* 1985; Vallée *et al.* 1985).

5. Cometary molecules and cometary chemistry

Molecules are observed in both the comae and the tails of comets and include several ionic species. A list of the atomic and molecular species so far observed in cometary atmospheres is given in table 3. Some of the species in the list are rather recent, somewhat tentative identifications. Compared to interstellar molecules, there is a preponderance of diatomic and triatomic species including several radical species which are probably fragments of more stable parent molecules that have sublimed from the cometary nuclei. Primary ions are formed by the action of solar photons and the solar wind on these molecules. Other ions can be formed by ion–molecule reactions (see below). Some cometary molecules have not been observed in interstellar clouds, including NH and NH_2, which are presumably photofragments of NH_3 parent molecules. The important question to be answered is, given the relative concentrations of the observed species in the comae and/or the tails, can the nature of the material in the nuclei be deduced? To answer this question requires a great deal of information including the rate coefficients for many chemical reactions and the photodissociation and photoionization cross sections for many species over the broad wavelength range of the solar spectrum (Oppenheimer 1975; Huntress *et al.* 1980; Lüst 1981; Mitchell *et al.* 1981; Biermann *et al.* 1982; Huebner 1985). The lack of data on the latter process is particularly frustrating to progress (van Dishoeck 1987). Concerning the chemistry, a good deal of ionic reaction rate data are available but kinetic data on relevant radical–radical reactions are much more scarce. Thus, the quantitative aspects of cometary chemistry are not very advanced.

TABLE 3. ATOMIC AND MOLECULAR SPECIES OBSERVED IN COMETARY ATMOSPHERES

(List compiled from various sources including Lüst (1981) and Irvine *et al.* (1986).)

C and H species
$H, C, C^+, CH, CH^+, C_2, C_3$

species containing O	species containing N
$O, OH, OH^+, H_2O?, H_2O^+,$	$NH, NH_2, NH_3?, N_2^+, CN,$
CO, CO^+, HCO, CO_2^+	$CN^+, HCN?, CH_3CN?, HC_3N?$

S and Si species
$S, SH^+, H_2S^+, CS, CS^+, S_2, Si$

metals
$Na, K, Ca, Ca^+, Y, Cr, Mn, Fe, Co, Ni, Cu$

The question marks indicate tentative detections.

Figure 3 is a schematic representation of some of the likely ionic reactions occurring in cometary atmospheres. It is constructed on the assumption that the parent molecules originating from the nucleus are H_2O, CO_2, NH_3, CH_4 and H_2S, that is the kind of species that are expected to be present in the presolar nebula. The primary processes are largely photodissociation and photoionization which result in fragment radicals and ions. As can be seen in figure 3,

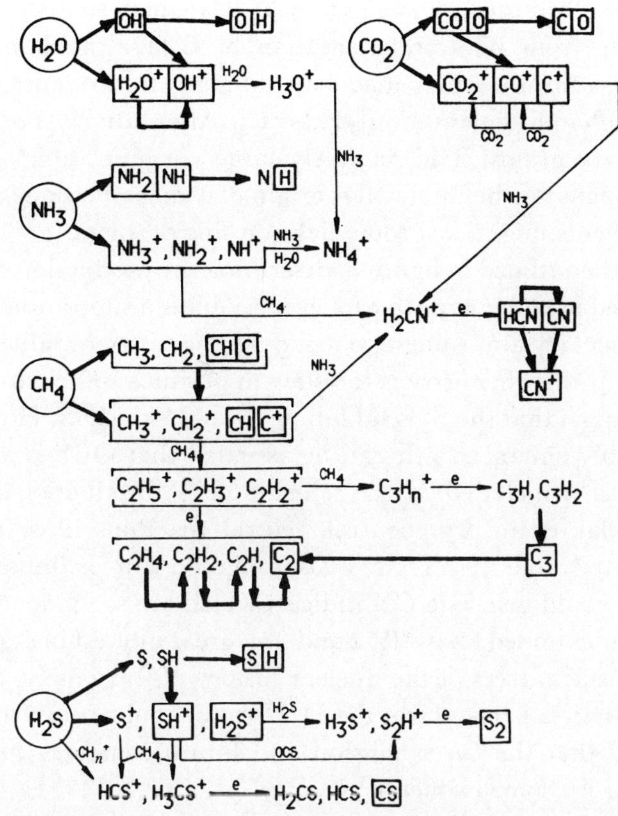

FIGURE 3. A representation of some of the routes to the production of some observed cometary species (which are enclosed in boxes). It is assumed that the species enclosed by circles are subliming from the cometary nucleus. The thick arrows indicate photodissociation and photoionozation. Thus, for example, H_2O can be photo-dissociated to OH or photoionized to H_2O^+ and OH^+. The thin arrows indicate ion–neutral reactions and e above the arrows signifies dissociative recombination leading to the indicated neutral molecules. Note the coupling of the NH_3, CH_4 and CO_2 chemistries which produce HCN. Other couplings must occur but are not represented here.

many of the species actually detected in cometary atmospheres can be produced by photo-fragmentation of these assumed parent molecules. Other observed species can be produced via gas phase ionic reactions similar to those discussed previously in the context of diffuse and dense interstellar cloud chemistry. A detailed discussion of the ion chemistry is unwarranted here, but a few comments are in order.

The ion chemistries of the species derived from H_2O, CO_2 and NH_3 are straightforward (see figure 3). The chemistry of methane-derived species is similar to that occurring in dense clouds but, because of photofragmentation, more radical species are probably involved. Significant in this chemistry is the production of C_2 and C_3 molecules which are both observed in comets. (C_3 has not yet been detected in interstellar clouds.) If $C_3H_n^+$ ions are formed in reactions of $C_2H_n^+$ ions with hydrocarbons and if C_2 and C_3 molecules are formed by dissociative recombination of $C_2H_n^+$ and $C_3H_n^+$ ions, then it is reasonable to expect that C_2 would be present in greater concentration than C_3 because of the additional reaction step required to produce C_3. That the reverse is so casts some doubt on the production scheme outlined in figure 3 for C_2 and C_3. However, without detailed quantitative modelling which must include reliable photodissociation crosssections, this is not yet a convincing argument against this reaction scheme. It has, however, prompted the suggestion (A. Dalgarno 1986 personal communication) that C_3 may originate from photofragmentation of C_3H_2 (the known cyclic interstellar molecule) or other organic molecules may be subliming from cometary nuclei. C_2 is then generated by further photofragmentation of the C_3. Alternatively both C_2 and C_3 may be sputtered from cometary grains. The relatively large concentration of C_3 relative to C_2 is perhaps a strong argument for the interstellar origin of comets. Perhaps the *in situ* observations from the *Giotto* instruments might cast some light on this problem.

The sulphur chemistry outlined in figure 3, describing the production of S_2 and CS molecules which are both observed in cometary atmospheres, requires a substantial concentration of H_2S to be present in the cometary atmosphere to drive the chemistry rapidly because S_2 is detected only near the nucleus. If this chemistry is too slow to produce sufficient S_2 then the inevitable conclusion has to be drawn that the S_2 is subliming from the nucleus. However, the production rate of S_2 is considerably enhanced if it can be assumed that OCS is available in significant concentrations and this is included in the scheme for S_2 production in figure 3. As in the discussions of interstellar chemistry, neutral–neutral reactions have not been included in figure 3, but such are possible in cometary atmospheres. For example, the reaction of NH radicals with C atoms could generate CN radicals which are so evident in comets.

Finally, as has been mentioned above (§§2 and 3*b*), great interest has centred around isotopic ratios in interstellar gas as tracers of the nuclear history of the galaxy. Similarly, the isotopic ratios in cometary material, particularly the D/H ratio, could be an indicator of the origins of comets. It is hoped that the *Giotto* mission and future cometary missions might provide accurate isotopic ratios for some elements.

I am indebted to my colleague Dr N. G. Adams for his invaluable assistance in the preparation of this paper.

References

Adams, N. G. & Smith, D. 1985 *Astrophys. J.* **294**, L63–L65.
Adams, N. G. & Smith, D. 1987*a* In *Astrochemistry* (ed. M. S. Vardya & S. P. Tarafdar), pp. 1–18. Dordrecht: D. Reidel.
Adams, N. G. & Smith, D. 1987*b* **317**, L25–L27.
Adams, N. G., Smith, D. & Alge, E. 1984*a* *J. Chem. Phys.* **81**, 1778–1784.
Adams, N. G., Smith, D. & Clary, D. C. 1985 *Astrophys. J. Lett.* **296**, L31–L34.
Adams, N. G., Smith, D. & Millar, T. J. 1984*b* *Mon. Not. R. astr. Soc.* **211**, 857–865.
Albritton, D. L. 1978 *Atom. Data nucl. Data Tabl.* **22**, 1–101.
Andrew, B. H. (ed.) 1980 *Interstellar molecules*. Dordrecht: D. Reidel.
Anicich, V. G. & Huntress, W. T. Jr 1986 *Astrophys. J. Suppl.* **62**, 553–672.
Barlow, S. E., Dunn, G. H. & Schauer, M. 1984 *Phys. Rev. Lett.* **52**, 902–905.
Bell, M. B., Feldman, P. A., Kwok, S. & Matthews, H. E. 1982 *Nature, Lond.* **295**, 389–391.
Bell, M. B. & Matthews, H. E. 1985 *Astrophys. J. Lett.* **291**, L63–L65.
Biermann, L., Giguere, P. T. & Huebner, W. F. 1982 *Astron. Astrophys.* **108**, 221–226.
Black, J. H. & Dalgarno, A. 1973 *Astrophys. Lett.* **15**, 79–82.
Blake, G. A., Keene, J. & Phillips, T. G. 1985 *Astrophys. J.* **295**, 501–506.
Clary, D. C. 1985 *Mol. Phys.* **54**, 605–618.
Clary, D. C., Smith, D. & Adams, N. G. 1985 *Chem. Phys. Lett.* **119**, 320–326.
Clayton, D. D. 1985 *Astrophys. J.* **290**, 428–432.
Crutcher, R. M. & Watson, W. D. 1985 In *Molecular astrophysics: state of the art and future directions* (ed. G. H. F. Diercksen, W. F. Huebner & P. W. Langhoff), pp. 255–280. Dordrecht: D. Reidel.
Dalgarno, A. 1985 In *Molecular astrophysics: state of the art and future directions* (ed. G. H. F. Diercksen, W. F. Huebner & P. W. Langhoff), pp. 281–293. Dordrecht: D. Reidel.
Dalgarno, A. & Black, J. H. 1976 *Rep. Prog. Phys.* **39**, 573–612.

van Dishoeck, E. F. 1987 In *Astrochemistry* (ed. M. S. Vardya & S. P. Tarafdar), pp. 51–65. Dordrecht: D. Reidel.
  (In the press.)
van Dishoeck, E. F. & Black, J. H. 1986 *Astrophys. J. Suppl.* **62**, 109–145.
Draine, B. T. 1985 In *Molecular astrophysics: state of the art and future directions* (ed. G. H. F. Diercksen, W. F. Huebner
  & P. W. Langhoff), pp. 295–310. Dordrecht: D. Reidel.
Duley, W. W. & Williams, D. A. 1984 *Interstellar chemistry*, pp. 86–114. London: Academic Press.
Dunham, T. Jr & Adams, W. S. 1937*a* *Am. astr. Soc. Publs* **9**, 5.
Dunham, T. Jr & Adams, W. S. 1937*b* *Publ. astr. Soc. Pacif.* **49**, 26.
Elitzur, M. & Watson, W. D. 1978 *Astrophys. J. Lett.* **222**, L141–L144.
Evans, N. J. 1978 In *Protostars and planets* (ed. T. Gehrels), pp. 153–164. Tucson: University of Arizona Press.
Federer, W., Villinger, H., Lindinger, W. & Ferguson, E. E. 1986 *Chem. Phys. Lett.* **123**, 12–16.
Guélin, M. 1985 In *Molecular astrophysics: state of the art and future directions* (ed. G. H. F. Diercksen, W. F. Huebner
  & P. W. Langhoff), pp. 23–44. Dordrecht: D. Reidel.
Herbst, E. 1985 In *Molecular astrophysics: state of the art and future directions* (ed. G. H. F. Diercksen, W. F. Huebner
  & P. W. Langhoff), pp. 237–254. Dordrecht: D. Reidel.
Herbst, E., Adams, N. G. & Smith, D. 1983 *Astrophys. J.* **269**, 329–333.
Herbst, E., Adams, N. G. & Smith, D. 1984*a* *Astrophys. J.* **285**, 618–621.
Herbst, E., Adams, N. G., Smith, D. & Defrees, D. J. 1986 *Astrophys. J.* **312**, 351–357.
Herbst, E., Schubert, J. G. & Certain, P. R. 1977 *Astrophys. J.* **213**, 696–704.
Herbst, E., Smith, D. & Adams, N. G. 1984*b* *Astron. Astrophys.* **138**, L13–L14.
Hollis, J. M. & Rhodes, P. J. 1982 *Astrophys. J. Lett.* **262**, L1–L5.
Huebner, W. F. 1985 In *Molecular astrophysics: state of the art and future directions* (ed. G. H. F. Diercksen,
  W. F. Huebner & P. W. Langhoff), pp. 311–330. Dordrecht: D. Reidel.
Huntress, W. T. Jr, McEwan, M. J., Karpas, Z. & Anicich, V. G. 1980 *Astrophys. J. Suppl.* **44**, 481–488.
Irvine, W. M. 1983 In *Proceedings of the international conference on cometary exploration* (ed. K. Szego). Budapest: Central
  Research Institute for Physics.
Irvine, W. M. 1987 In *Astrochemistry* (ed. M. S. Vardya & S. P. Tarafdar). Dordrecht: D. Reidel, pp. 245–252.
Irvine, W. M., Schloeb, F. P., Hjalmarson, A. & Herbst, E. 1986 Five College Radio Astronomy Observatory
  Report No. 238, University of Massachusetts.
Knight, J. S., Freeman, C. G., McEwan, M. J., Adams, N. G. & Smith, D. 1985 *Int. J. Mass. Spectrum. Ion. Process.*
  67, 317–330.
Knight, J. S., Freeman, C. G., McEwan, M. J., Smith, S. C., Adams, N. G. & Smith, D. 1986 *Mon. Not. R. astr.
  Soc.* **219**, 89–94.
Leung, C. M., Herbst, E. & Huebner, W. F. 1984 *Astrophys. J. Suppl.* **56**, 231–256.
Lüst, R. 1981 In *Topics in current chemistry*, vol. 99, pp. 73–98. Berlin: Springer-Verlag.
Marquette, J. B., Rowe, B. R., Dupeyrat, G., Poissant, G. & Rebrion, C. 1985 *Chem. Phys. Lett.* **122**, 431–435.
Matthews, H. E. & Irvine, W. M. 1985 *Astrophys. J. Lett.* **298**, L61–L65.
McKellar, A. 1941 *Publ. Dom. astrophys. Obs.* **7**, 251.
Michels, H. H. & Hobbs, R. H. 1984 *Astrophys. J. Lett.* **286**, L27–L29.
Millar, T. J. 1982 *Mon. Not. R. astr. Soc.* **199**, 309–319.
Millar, T. J. 1985 In *Molecular astrophysics: state of the art and future directions* (ed. G. H. F. Dierckson, W. F. Huebner
  & P. W. Langhoff), pp. 613–620. Dordrecht: D. Reidel.
Millar, T. J., Adams, N. G., Smith, D. & Clary, D. C. 1985 *Mon. Not. R. astr. Soc.* **216**, 1025–1031.
Millar, T. J., Adams, N. G., Smith, D., Lindinger, W. & Villinger, H. 1986 *Mon. Not. R. astr. Soc.* **221**, 673–678.
Millar, T. J. & Freeman, A. 1984 *Mon. Not. R. astr. Soc.* **207**, 405–432.
Millar, T. J. & Nejad, L. A. M. 1985 *Mon. Not. R. astr. Soc.* **217**, 507–522.
Mitchell, G. F., Prasad, S. S. & Huntress, W. T. Jr 1981 *Astrophys. J.* **244**, 1087–1093.
Omont, A. 1987 In *Astrochemistry* (ed. M. S. Vardya & S. P. Tarafdar), pp. 357–367. Dordrecht: D. Reidel.
Oppenheimer, M. 1975 *Astrophys. J.* **196**, 251–259.
Penzias, A. A. 1980 In *Interstellar molecules* (ed. B. H. Andrew), pp. 397–404. Dordrecht: D. Reidel.
Quéffelec, J. L., Rowe, B. R., Morlais, M., Gomet, J. C. & Vallee, F. 1985 *Planet. Space Sci.* **33**, 263–270.
Rowe, B. R., Marquette, J. B., Dupeyrat, G. & Ferguson, E. E. 1985 *Chem. Phys. Lett.* **113**, 403–406.
Rydbeck, O. E. H. & Hjalmarson, A. 1985 In *Molecular astrophysics: state of the art and future directions*
  (ed. G. H. F. Dierecksen, W. F. Huebner & P. W. Langhoff), pp. 45–175. Dordrecht: D. Reidel.
Schiff, H. I. & Bohme, D. K. 1979 *Astrophys. J.* **232**, 740–746.
Smith, D. & Adams, N. G. 1978 *Astrophys. J. Lett.* **220**, L87–L92.
Smith, D. & Adams, N. G. 1979 In *Gas phase ion chemistry*, vol. 1 (ed. M. T. Bowers), pp. 1–44. New York: Academic
  Press.
Smith, D. & Adams, N. G. 1980 *Astrophys. J.* **242**, 424–431.
Smith, D. & Adams, N. G. 1981 *Int. Rev. Phys. Chem.* **1**, 271–307.
Smith, D. & Adams, N. G. 1984*a* In *Ionic processes in the gas phase* (ed. M. A. Almoster-Ferreira), pp. 41–66.
  Dordrecht: D. Reidel.
Smith, D. & Adams, N. G. 1984*b* *Astrophys. J. Lett.* **284**, L13–L16.

Smith, D., Adams, N. G. & Alge, E. 1982 *Astrophys. J.* **263**, 123–129.
Swings, P. & Rosenfeld, L. 1937 *Astrophys. J.* **86**, 483.
Thaddeus, P., Guélin, M. & Linke, R. A. 1981 *Astrophys. J. Lett.* **246**, L41–L45.
Thaddeus, P., Cummins, S. E. & Linke, R. A. 1984 *Astrophys. J. Lett.* **283**, L45–L48.
Thaddeus, P., Vrtilek, J. M. & Gottlieb, C. A. 1985 *Astrophys. J. Lett.* **299**, L63–L66.
Vallee, F., Rowe, B. R., Gomet, J. C., Quéffelec, J. L. & Morlais, M. 1985 *Chem. Phys. Lett.* **124**, 317–320.
Wannier, P. G., Redman, R. O., Phillips, T. G., Leighton, R. B., Knapp, G. R. & Huggins, P. J. 1980 In *Interstellar molecules* (ed. B. H. Andrew), pp. 487–493. Dordrecht: D. Reidel.
Winnewisser, G., Churchwell, E. & Walmsley, C. M. 1979 In *Modern aspects of microwave spectroscopy* (ed. G. W. Chantry), pp. 313–501. New York: Academic Press.
Woods, R. C., Gudeman, C. S., Dickman, R. L., Goldsmith, P. F., Huguenin, G. R. & Irvine, W. M. 1983 *Astrophys. J.* **270**, 583–588.
Ziurys, L. M. & Turner, B. E. 1986 *Astrophys. J. Lett.* **302**, L31–L36.
Zuckerman, B. 1980 In *Interstellar molecules* (ed. B. H. Andrew), pp. 479–486. Dordrecht: D. Reidel.

Discussion

M. K. WALLIS (*Department of Applied Mathematics and Astronomy, University College Cardiff, U.K.*). Extending Professor Anders's point, detailed modelling by Mitchell *et al.* and Huebner *et al.* (1982) has shown that sufficient complex molecules cannot be built up in cometary atmospheres, for time is too short in the expanding coma for equilibrium chemical thermodynamics to apply. The icy-clathrate hypothesis (Delsemme & Swings 1958) that supposes the nucleus is composed of ices of simple gases, has failed; in particular the fraction of NH_3 must be low to permit the observed H_2O ions. The long-observed C_3 and recent S_2 compounds must derive by breakdown of complex parents. My question is, might interstellar molecules give the sulphur dimer S_2, or must it be of mineral origin?

Additional references

Delsemme, A. H. & Swings, P. 1958 *Astron. Astrophys.* **15**, 1.
Huebner, W. F. *et al.* 1982 In *Comets* (ed. L. L. Wilkening), p. 496. Tucson: University of Arizona Press.

D. SMITH. S_2 molecules can be synthesized in intersteller clouds via ion–molecule reactions (producing S_2H^+) followed by proton transfer (e.g. $S_2H^+ + NH_3 \rightarrow S_2 + NH_4^+$) or dissociative recombination ($S_2H^+ + e \rightarrow S_2 + H$). Therefore S_2 could be present in cometary nuclei perhaps in the form of $(S_2)_n$ oligomers which can then sublime into the comae to be dissociated to the observed S_2 molecules.

F. L. WHIPPLE (*Smithsonian Astrophysical Observatory, Cambridge, Massachusetts, U.S.A.*). I wonder how stable Dr Kroto's C_{60} molecules would be in the interstellar medium or the solar nebulae under the conditions in which comets may have formed.

D. SMITH. In Kroto's laboratory experiments, the C_{60} molecules are ejected from a solid surface by a pulse of laser light. I would guess, therefore that C_{60} would be stable in cold interstellar clouds or in the cooler regions of the presolar nebula.

Phil. Trans. R. Soc. Lond. A **323**, 287–304 (1987)
Printed in Great Britain

Local and exotic components of primitive meteorites, and their origin

By E. Anders

Enrico Fermi Institute and Department of Chemistry, University of Chicago, Chicago, Illinois 60637,
U.S.A.

Most of the material of chondrites was heavily reprocessed in the early Solar System, and hence retains only a dim memory of its interstellar origin even in the least altered meteorites. Opaque matrix, the most primitive material, seems to have taken up Fe^{2+} and changed its mineralogy and texture. Chondrules and Ca, Al-rich inclusions have been further altered by melting, oxidation or reduction, loss of volatiles, etc. None the less, small amounts of exotic components have survived, as indicated by isotopic anomalies. A dust component enriched in ^{16}O is the most abundant and widespread. It survives in Ca, Al-rich inclusions as anomalous spinel grains, but is recognizable even in highly evolved meteorites and planets from variations in bulk oxygen isotopic composition. A few inclusions show small nucleosynthetic anomalies for many elements (Si, Ca, Ti, Cr, Sr, Ba, Nd, Sm), always coupled with mass fractionation of several of these elements, as well as O and Mg. Seven extinct radionuclides (^{26}Al, ^{41}Ca, ^{53}Mn, ^{107}Pd, ^{129}I, ^{146}Sm, and ^{244}Pu) have been recognized from their decay products, and provide clues to the chronology and nucleosynthetic sources of the early Solar System.

Highly volatile elements, such as C, N and the noble gases, show especially large and numerous isotopic anomalies. The noble-gas components include Ne-E (monoisotopic ^{22}Ne from the β^+ decay of 2.6 a ^{22}Na), Xe-HL (enriched 2-fold in the light and heavy isotopes), and Xe-S (enriched in the even-numbered, middle isotopes 128, 130 and 132). They are located in carriers that themselves are anomalous, e.g. carbon enriched up to 2.4-fold in ^{13}C or depleted by more than 30 % in ^{15}N, or spinel enriched in ^{16}O and ^{13}C. Other anomalies include nitrogen enriched 2-fold in ^{15}N and hydrogen enriched 5-fold in D; probably relict interstellar molecules. A variety of astronomical sources seem to be required: novae, red giants, supernovae and molecular clouds.

1. Introduction

All meteoritic matter originally came from outside the Solar System and thus is exotic in the broadest sense of the term. However, much of this material was reprocessed in the early Solar System (by vaporization, melting, mixing, isotopic exchange, chemical reactions, etc.), and became isotopically homogeneous. Such commonplace material is properly called 'local', leaving the term 'exotic' or 'presolar' for material that still retains an anomalous isotopic signature. Exotic material sometimes occurs as discrete, potentially separable grains, but often has been assimilated by large amounts of local matter, and is recognizable only by its anomalous isotopic composition.

Qualitatively, a similar picture may be expected for comets. But as comets formed in more distant, cooler parts of the Solar System, reprocessing may have been less intensive and pervasive. Thus the local material may be less processed and less dominant, with the exotic material more abundant and diverse. None the less, primitive meteorites are a good frame of reference for interpretation of cometary data. I shall therefore first review the properties and classification of chondrites (§2) and then their local and exotic components (§§3 and 4).

2. Properties and classification of chondrites

Chondrites are stony meteorites containing chondrules, millimetre-sized silicate spherules that appear to be frozen droplets of a melt. They consist largely of *olivine* [$(Mg,Fe)_2SiO_4$], *orthopyroxene* [$(Mg,Fe)SiO_3$], and *plagioclase feldspar* [solid solution of $CaAl_2Si_2O_8$ and $NaAlSi_3O_8$]. In the more primitive chondrites, glass occurs in place of crystalline feldspar.

Chondrules are embedded in a ground-mass or *matrix*. In the less primitive chondrites, the matrix is somewhat finer-grained than the chondrules, but otherwise has the same mineralogy and composition. Millimetre-sized grains of *metal* (nickel-iron with 5–60 % Ni) and *troilite* (FeS) are also present. In the more primitive chondrites, at least part of the matrix is very fine-grained (to *ca.* 10^{-6} cm) and richer in Fe^{2+} than are the chondrules.

Five chondrite classes are recognized, differing in the proportions of oxidized to reduced iron. *Enstatite* (E) chondrites are highly reduced, containing iron only as metal and FeS. *Carbonaceous* (C) chondrites are highly oxidized, containing mainly Fe^{2+}, Fe^{3+}, and little or no Fe^0. The middle ground is occupied by three classes of intermediate oxidation state and variable iron content: H, L, and LL chondrites. (The letters refer to total iron content: high, low and low-low.) Collectively, they are often called *ordinary* (O) chondrites.

Van Schmus & Wood (1967) have further subdivided each of these classes into '*petrologic types*', numbered from 1 to 6. These types were originally designed to reflect increasing chondrule-to-matrix intergrowth (probably due to thermal metamorphism in the meteorite parent bodies, at T up to 1100 °C), but have turned out to correlate well with compositional trends, e.g. volatile content. Probably these types represent an accretion time sequence, the higher types having accreted earlier, and therefore being located in deeper and warmer locations in the parent body.

Only the most primitive chondrites, up to petrologic type 3, are of interest to us, as they are least affected by metamorphism and hence have best preserved their exotic components as well as a record of processes in the solar nebula. Regrettably, even these meteorites are not pristine. Types 1 and 2 – found only among C-chondrites – have undergone hydrothermal alteration in their parent bodies (DuFresne & Anders 1962; R. N. Clayton & Mayeda 1984). Type 3, in turn, has been slightly metamorphosed, at T up to *ca.* 300–400 °C. Fortunately the least metamorphosed type 3s can be recognized from their low thermoluminescence sensitivity and other criteria, and on this basis, type 3s have been divided into 10 subclasses, 3.0 to 3.9 (Sears *et al.* 1980; Anders & Zadnik, 1985).

C3 chondrites contain a minor but very important textural component: millimetre-sized, often irregular *inclusions* of refractory, Ca, Al-rich but Si-poor minerals such as melilite, $Ca_2(Al_2,MgSi)SiO_7$ or *spinel*, $MgAl_2O_4$. They are known by the acronym CAI.

3. 'Local' components of chondrites and their origins

To a first approximation, chondrites are condensates from the solar nebula, as first recognized by Wood (1958). Thus a convenient if oversimplified framework for interpreting chondrites is the equilibrium condensation sequence of a solar gas (Larimer 1967; Grossman 1972; Lewis 1972; Grossman & Larimer 1974). Figure 1 shows this sequence. A more detailed version of the 300–1800 K portion is given in Figure 2. As chemical equilibrium is independent of the path, these diagrams apply equally well to isobaric heating or cooling and to isothermal compression.

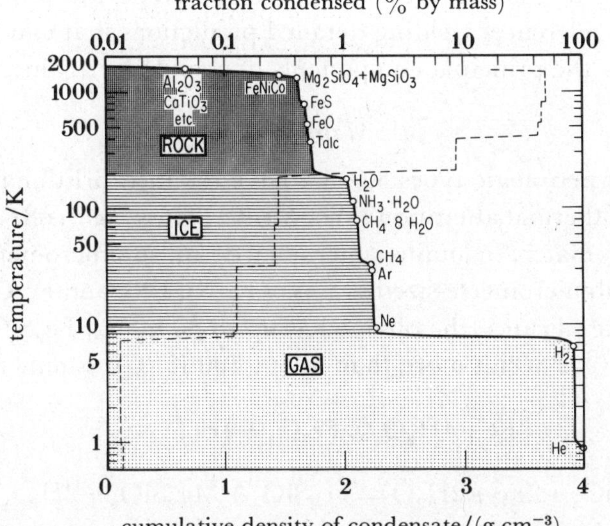

FIGURE 1. Condensation of solar gas at 10^{-4} atm† (Lewis 1972). Equilibrium condensation of a solar gas explains the composition of planets, at least to first order. Inner planets consist of rock, comets and satellites of outer planets consist of rock and ice, and the outer planets consist of all three.

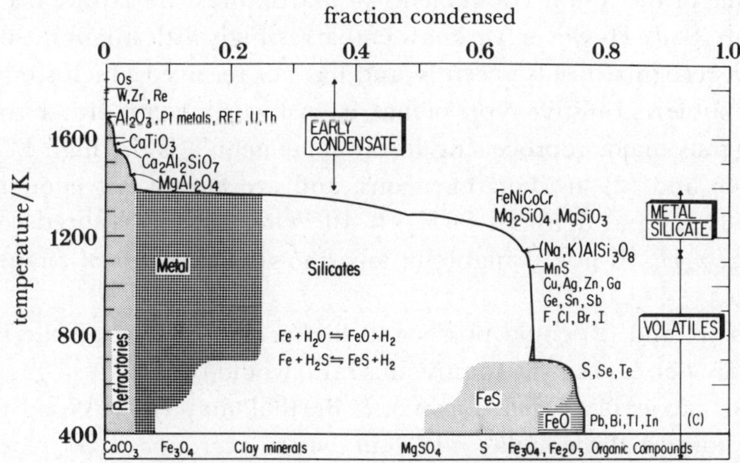

FIGURE 2. 'Rock' portion of the diagram in figure 1, on linear scales Below *ca.* 700 K, metal begins to react with H_2O and H_2S in the nebula, yielding FeO (which enters silicates) and FeS. The mineral assemblage at 400–500 K resembles that of ordinary chondrites. Below 400 K, many major changes take place, yielding a mineral assemblage resembling that of carbonaceous chondrites.

Figure 1 helps rationalize the composition of planets and other small bodies in the solar system. The inner planets and the asteroids consist only of *rock*; comets and the satellites of the outer planets consist of *rock and ice*, whereas the outer planets consist of *rock, ice, and gas*.

This appealing picture is oversimplified in several respects. (1) The various chondrite classes differ somewhat in bulk composition (Fe/Si, Mg/Si, Ca/Si, etc.), and so at best only one class can be a *total* condensate. All others must have been fractionated, presumably by gain or loss of major condensates (Larimer & Anders 1970). (2) The nebula was not wholly gaseous; some presolar solids survived (R. N. Clayton *et al.* 1973, see also §4). (3) Chemical equilibrium may not have been reached, especially at lower temperatures. (4) The composition may have

† 1 atm = 101 325 Pa.

deviated from solar, owing to gas-dust fractionations, etc. None the less, the condensation diagram is a good frame of reference, yielding detailed predictions that can be compared with observations. I shall review the principal components of chondrites against this framework.

(a) Matrix

Only a few meteorites of petrologic types 3.0–3.4 have retained pristine matrix, unaffected by metamorphism or hydrothermal alteration (Wood 1962; Huss *et al.* 1981; Nagahara 1984). This opaque (or 'Huss') matrix contains micrometre- to submicrometre-silicate grains, amorphous material and submicrometre-sized grains of Fe-Ni, FeS, and Fe_3O_4. The Fe_3O_4 and the high Fe^{2+}-content of the silicates (the olivine has up to 50 mol% Fe_2SiO_4), are consistent with equilibrium in a solar gas at not more than about 400 K (Grossman 1972):

$$3Fe + 4H_2O \rightleftharpoons Fe_3O_4 + 4H_2, \tag{1}$$

$$2MgSiO_3 + 2Fe + 2H_2O \rightleftharpoons Fe_2SiO_4 + Mg_2SiO_4 + 2H_2. \tag{2}$$

In a solar gas, reaction (1) goes to the right below 400 K, whereas reaction (2) reaches 50 mol% Fe_2SiO_4 at 470 ± 40 K.

Was the matrix totally reprocessed in the solar nebula or has it retained some of its exotic features? At least some of the minor components of matrix must be exotic, because they are isotopically anomalous (§4d). However, we know embarrassingly little about the bulk of opaque matrix, as it has been recognized only recently and has not received much study; in contrast to the more common but less primitive types of matrix (Ashworth 1977; Hutchison *et al.* 1979). One feature that suggests major reprocessing in the solar nebula is the high Fe^{2+} and Fe_3O_4 content. Reactions (1) and (2) are heterogeneous, and are barely fast enough at the high temperatures (400–500 K) and densities [$n(H_2) \approx 10^{14}$ cm^{-3}]. The required combination of moderate T and large $n(H_2)$ is not available in any known astrophysical environment other than the solar nebula.†

Actually, reactions (1) and (2) could proceed at higher temperatures if the H_2O/H_2 ratio were greater than solar. The easiest way to raise this ratio would be by a dust:gas fractionation, because dust is a major carrier of oxygen (Larimer & Bartholomay 1983; Wood 1984). As dust would tend to settle toward the nebular midplane where accretion takes place, some such fractionation is expected. We shall return to this problem in §3b.

Granted that Fe^{2+} and Fe_3O_4 contents of matrix were established in the solar nebula, it does not follow that all other features were, too. Very little work has been done on matrix; in particular, no systematic search has been made for surviving interstellar features. Several isotopically anomalous, hence presolar components have been found in chondrites [C, H, N and noble gases; §4d] and though none of them have been seen *in situ* – or otherwise conclusively linked to matrix – their presence shows that at least some heat-sensitive presolar components got into meteorites without serious damage. Thus it is quite conceivable that matrix of primitive meteorites is only lightly reprocessed interstellar dust (Wood 1973, 1981; Anders 1965; Cameron 1973; D. D. Clayton 1982; Huss 1987). This is a testable proposition that should be checked experimentally.

---

† There is a minor loophole in this argument. If the iron in interstellar grains had been very finely dispersed (monolayer or individual atoms) at some stage of its history, then it may have become oxidized before its arrival in the Solar System. However, such ultrafine material would react with the nebular gases during its reorganization to larger crystals, and so its final oxidation state would still be established in the nebula.

(b) Chondrules

Chondrules formed by flash melting of pre-existing solids, not by direct condensation from gas (Nagahara 1981; Wood 1984). The melting time was short (seconds to minutes), judging from the failure of three reactions to go to completion: loss of Na and other volatiles (Schmitt *et al.* 1965; Tsuchiyama *et al.* 1981), reduction of Fe^{2+} to Fe by nebular hydrogen (Wood 1984), and isotopic exchange with oxygen of the nebula (R. N. Clayton 1981). The cooling time likewise was short (seconds to hours, depending on textural type), as shown by laboratory experiments (Tsuchiyama *et al.* 1980; Planner & Keil 1982; Nagahara 1983).

The short cooling times imply that the chondrule-forming region was *small*, allowing rapid heat loss by radiation. The existence of five major chondrite classes requires at least five separate domains, differing in oxidation state, bulk chemistry and oxygen isotope composition.

Chondrules are complementary to other major components of the meteorite (metal, *total* matrix), being depleted in siderophiles and volatiles but enriched in refractories (Grossman & Wasson 1983, 1985; Wood 1984). Interestingly, the bulk chondrite always is close to solar composition, which suggests that these components are cogenetic and accreted *comprehensively*. The parent material may have been the opaque matrix, perhaps after a preliminary stage of reprocessing (Rambaldi & Wasson 1984). It is important to distinguish this primitive, opaque matrix from more fractionated and recrystallized types, which probably represent 'failed chondrules': material that overshot or undershot the melting range (Wood 1984).

Chondrule formation seems to have involved some recycling, as demonstrated by relict crystals (Nagahara 1981; Rambaldi *et al.* 1983) and compound chondrules (Gooding & Keil 1980). It was followed by accretion of rims of matrix-like material (Allen *et al.* 1980).

It seems that chondrule formation was a pervasive process that occurred throughout the inner solar system and reprocessed some 50–90% of the material. The inner planets are depleted in some of the same elements as the chondrites (e.g. alkalis, S, Se, Ga, Ge, etc.) and by Occam's Razor, it seems reasonable to invoke the same process (Morgan & Anders 1980).

The mechanism of chondrule formation has not yet been conclusively identified. The complementarity of chondrules and matrix speaks against models that make chondrules and matrix in different places and mix them afterwards. The most promising model involves heating of interstellar matter as it falls into the solar nebula. Direct 'infall heating' by aerodynamic drag (Wood 1984, 1985) turns out to be quantitatively inadequate, but heating by viscous interaction with the nebula seems more promising (Wood 1986). The chondrule-forming zone in this model is a thin, hot layer of the nebula, as implied by the short cooling times of chondrules. However, this model is very recent, and still has to pass many quantitative tests.

The redox state (i.e. Fe^{2+} content of olivine) of primitive chondrules is quite variable even within a single chondrule, ranging from less than 1–40 at.%. At 1500 K in a solar gas, the equilibrium value would be only 0.001%, and so either the gas was non-solar or the parent material was highly oxidized and did not fully equilibrate with the ambient gas upon melting. Wood (1984), who favours the former alternative, has shown that Fe^{2+} contents of 0.5–30 at.% could be attained if dust were enriched 10^2 to 10^4-fold relative to gas, thus raising the H_2O/H_2 ratio in the system. However, it seems unlikely that chondrules fully equilibrated with the nebula, as they did not reach uniform Fe^{2+} contents, did not lose all their volatiles, and did not isotopically equilibrate their oxygen. It is more likely that chondrules formed from an Fe^{2+}-rich precursor (opaque matrix or a reprocessed derivative thereof), which was *partially* reduced and devolatilized on melting.

[43]

Even for the highly reduced E-chondrites, it appears that the precursor material was relatively oxidized (Rambaldi *et al.* 1984). Conditions must have become more reducing with time (Grossman *et al.* 1985), presumably because of a dust-gas fractionation in the nebula that raised the C/O ratio above the critical value of 0.8 (Larimer 1975; Larimer & Bartholomay 1979, 1983).

(c) Ca, Al-*rich inclusions* ('CAI')†

Chemically and mineralogically, CAI look like an *early condensate* from a solar gas (Larimer & Anders 1970; Marvin *et al.* 1970; Grossman 1972). They contain the suite of refractory minerals and trace elements expected to condense in the first *ca.* 6 % of the 'rock' fraction, between 1800 and 1400 K (figure 2). However, figure 2 applies equally well to evaporation and to condensation, and so some authors have argued that CAI are *evaporation residues* (of presolar matter) rather than condensates (Kurat 1970; D. D. Clayton 1975; Wood 1981).

It turns out that the truth lies in between, perhaps closer to the latter view. In the most conclusive test thus far, Niederer & Papanastassiou (1984) have shown that the isotopic fractionation patterns of Ca and Mg require a complex sequence of both processes. Of course, both are natural extensions of the chondrule-forming process. In some cases, this process overshot the mark, vaporizing part of the chondrule and leaving a residue enriched in refractories such as Ca, Al, and Sc (Osborn *et al.* 1974). As the vapour would have recondensed eventually in the (generally cool) nebula, the chondrule-forming process thus would yield condensates, in addition to evaporation residues and fully or partially melted chondrules.

Two main classes of inclusions are recognized on the basis of grain size. *Coarse-grained* inclusions are uniformly enriched in refractory elements to about 17 times that of Cl-chondrite levels (Grossman *et al.* 1977), and hence may be interpreted as a *total condensate* of the first 6 % of 'rock' (figure 2) or as an *equilibrium evaporation residue* of the last 6 %. The latter view seems to be closer to the truth, judging from the presence of isotopically anomalous spinel (§4*a*), which could not have condensed from or equilibrated with the nebula.

Fine-grained inclusions have highly fractionated abundance patterns, being enriched in refractories (but by variable factors) and in volatiles such as alkalis and halogens (Grossman & Ganapathy 1976). They contain not only the usual refractory minerals, but also feldspathoids such as nepheline ($NaAlSiO_4$) and sodalite ($3NaAlSiO_4.NaCl$). In terms of the condensation diagram (figure 2), they can be described as discontinuous condensates, which lost material condensing between 1800 and *ca.* 1600 K (heavy lanthanides and other highly refractory elements) and 1400–1200 K (metal and silicate), but collected elements condensing in other intervals, sometimes disproportionately so.

Some important if ill-understood clues to the formation conditions of coarse-grained CAI come from 'Fremdlinge'‡: 10–100 µm aggregates containing metal; alloys of Pt metals, Ga, Ge, Mo, Sn, and Re; Ca-phosphates; and sulphide or oxide minerals of other rare elementss such as Sc, V, Zr, Nb, Mo and W (El Goresy *et al.* 1978; Armstrong *et al.* 1985). Some alloys can be rationalized in terms of condensation temperature, structure, or chemical reactivity; others, such as Os, Ru, Rh, cannot (T_{cond} = 1910, 1650 and 1470 K at 10^{-3} atm; structure HCP for Os, Ru but FCC for Rh).

---

† The literature on this topic is vast. For a more thorough treatment, see Grossman (1980), Wood (1981), Wark & Lovering (1982), Kornacki & Wood (1984) and Huss (1987).
‡ German for 'strangers', to connote a separate, possibly extrasolar, origin.

It seems that these grains condensed under conditions highly conducive to nucleation and hence low supersaturation. Thus atoms could choose their preferred condensation sites and even a rare element such as Re was able to form its own phase rather than alloying with the previously condensed, structurally similar and more abundant Os. Yet owing to the rarity of these metals (solar $Os/H = 2 \times 10^{-11}$), it would take some 600 years for a 1 μm Os grain to form at $P = 10^{-4}$ atm, correspondingly less at higher P (Palme & Wlotzka 1976). It is hard to see how the required high temperatures were sustained for so long a time, and how they were regulated to maintain low supersaturation. But the growth time is proportional to $1/P$, and so shorter times would suffice at higher pressures. Perhaps this is an argument for models postulating higher pressures (protoplanets, Cameron (1978); denser gas regions surrounding planetesimals, Hayashi *et al.* (1985)).

To complicate matters, the oxide, sulphide, and metal phases in Fremdlinge require a very wide range of oxygen fugacities and temperatures (Armstrong *et al.* 1985). Most of the trace elements in question are about as rare as Os, and so the same dilemma about growth time applies, but with the added complication of a wide spectrum of 'microenvironments', differing in oxygen fugacity and gas composition. The problem cannot be solved by relegating it to some presolar stage, as no known astronomical object offers the required combination of temperature, density, time and composition. Supernovae, in particular, are too tenuous and short-lived for significant grain growth. Though *abundant* elements such as Mg or Al can yield micrometre-sized condensates in the few months available for condensation (Lattimer *et al.* 1978), *rare* elements such as Os cannot; an average Os atom will have only 10^{-12} collisions with another Os atom during the entire cooling interval from 1600 to 0 K.

By default, it seems necessary to invoke the solar nebula or the meteorite parent bodies as the site of the 'microenvironments'. If the alloys formed by condensation, then the required chemical variety (especially the high oxygen fugacities) could be achieved either in the nebula itself or in the protoatmosphere of growing asteroids or protoplanets (Cameron 1978; Hayashi *et al.* 1985). In the nebula, high oxygen fugacities could be produced by gas-dust fractionation; in a protoatmosphere, by evaporation of accreting ice. In both settings, further variety would result from settling of dust through gas regions depleted in certain elements by prior condensation.

On the other hand, if the alloys formed by exsolution during metamorphism in the meteorite parent body (Armstrong *et al.* 1985), then the required changes in chemical environment could be caused by migration and escape of major volatiles, such as CO, H_2O, H_2S, etc. In any event, it seems that the 'Fremdlinge' are not so 'fremd', after all.

(d) *Carbon and organic matter*

Primitive meteorites contain up to 4 % C, with organic carbon dominant in C1s and C2s, but elemental carbon dominant in C3s (table 1). Abundances of total C, H, and N also decline in this order, with a sharp drop beyond C3. (As C-chondrites of higher petrologic types are very rare, H4 to H6 chondrites have been substituted.) The condensation temperatures from the solar nebula have been estimated from various 'cosmothermometers', such as isotopic fractionations or trace element abundances (Larimer 1978).

The organic carbon consists mainly of an intractable, aromatic polymer similar to coal in its structural units (Hayatsu *et al.* 1977), and a variety of extractable organic compounds, including alkanes, alkenes, arenes, alcohols, carboxylic acids, amino acids, nitrogen

TABLE 1. CARBON IN CHONDRITES

type	temperature/K condensation	metamorphism	carbon total (ppm)	organic (%)	$C\dagger$ (%)	CO_3^{2-} (%)	H (ppm)	N (ppm)
C1	360	—	36000	95	~ 1	~ 4	7900	2800
C2	400	—	23000	95	~ 1	~ 2	8900	1500
C3V	420	600	6700	$\leqslant 20$	> 80	—	500	61
H4	450	900	1300	< 10	> 90	—	5	47
H5	470	1000	1100	< 10	> 90	—	4	43
H6	490	1200	1060	< 10	> 90	—	4	50

$\dagger$ Elemental C or solid solution of C in γ-Fe.

heterocyclics (including the purine and pyrimidine bases of DNA and RNA), etc. (Hayatsu & Anders 1981).

The alkanes consist mainly of the normal, straight-chain isomers, with lesser amounts of four or five slightly branched isomers (mono- and di-methyl). As there exist many thousands of possible isomers (e.g. 10359 at C_{16}), the dominance of only five or six implies a highly selective process: presumably the Fischer–Tropsch synthesis, which involves catalytic hydrogenation of CO. Indeed, laboratory experiments show that all organic compounds reliably identified in meteorites can be produced from CO, H_2 and NH_3 in the presence of catalytically active meteoritic minerals, such as Fe_3O_4 and clays (Studier *et al.* 1968; Hayatsu & Anders 1981).

Figure 3 illustrates the thermodynamics of this process in a solar gas. Let us consider isobaric cooling at 10^{-5} atm, although isothermal compression at *ca.* 400 K would give similar results.

At high temperatures carbon exists mainly in the form of CO. Near 600 K, hydrogenation to CH_4 should commence, and be 50 % complete at 590 K. However, this reaction is very slow in the absence of catalysts, and as the dominant minerals at this temperature (Fe, Mg-silicates, FeNi coated with FeS) are very poor catalysts (Anders *et al.* 1973), most of the CO survives metastably.

At 520 K, disproportionation to C and CO_2 should set in. Again, good catalysts are lacking, and so much of the CO survives. Finally at 400 K, the anhydrous mineral assemblage dominant up to that point (figure 2) transforms to clays and Fe_3O_4, both of which are good catalysts for the hydrogenation of CO. However, at this low temperature, CH_4 – though still the most stable product – is no longer the only possible one. Many other organic compounds can form metastably, and actually do, being favoured by the reaction mechanism.

The formation of organic compounds thus is triggered by the appearance of catalysts at 400 K. Indeed, meteorites rich in clay minerals (C1, C2) are rich in organic compounds, whereas meteorites poor in clay minerals (C3) have little or no organic matter, containing most of their carbon as elemental C (table 1).

The aromatic polymer forms as a secondary product in this scheme, by aromatization of the primary aliphatic products. Such aromatization is known to occur on Fe_2O_3 (Galwey 1972), and has also been seen in a Fischer–Tropsch synthesis extended over six months (Hayatsu *et al.* 1977).

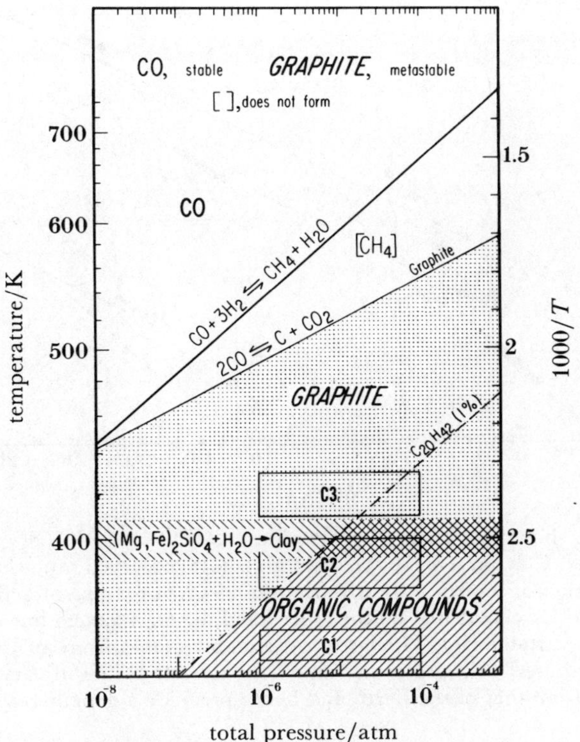

FIGURE 3. Chemical state of carbon in the solar nebula (after Hayatsu & Anders 1981). Each solid line gives the temperatures at which 50 % of the carbon should have reacted according to the equilibrium shown. In each field, the principal stable products are shown in Roman type and metastable products, in italics; those that do not form for kinetic reasons are enclosed in brackets. This diagram applies to isobaric cooling as well as to isothermal compression.

CO survives down to 400 K, owing to the lack of good catalysts. At this temperature, clays form from anhydrous silicates, catalysing hydrogenation of CO to complex organic compounds (Fischer–Tropsch reaction).

The chemical state of carbon in C-chondrites agrees with that predicted from their formation conditions (indicated by boxes), as inferred from various cosmothermometers and -barometers (Larimer 1978). C3 chondrites contain mainly elemental carbon, whereas C1 and C2 chondrites contain mainly organic carbon (table 1).

4. EXOTIC COMPONENTS OF CHONDRITES

(a) Oxygen-16

The most abundant and ubiquitous exotic component is ^{16}O (R. N. Clayton *et al.* 1973; D. D. Clayton 1981). Its presence can be demonstrated by a 3-isotope diagram, a variant of which is shown in fig. 4. On this diagram, all samples related by *mass fractionation* lie along a line of slope $\frac{1}{2}$ (reflecting the mass differences $^{17}O - ^{16}O$ and $^{18}O - ^{16}O$), whereas *mixtures of independent components* lie along 'mixing lines' of variable slope joining the two components.

All terrestrial and some meteoritic samples lie on the fractionation line, but many meteoritic samples do not, suggesting the presence of an independent component. This component is most clearly characterized by the anhydrous minerals of C2 and C3 chondrites (especially CAI), which lie on a line of slope *ca.* 1 and hence must be mixtures of two components of nearly identical $^{17}O/^{18}O$ ratio but different ^{16}O content. The ^{16}O-rich endmember is represented by spinel and pyroxene of $\delta^{18}O = -40‰$ and $\delta^{17}O = -42‰$. The same endmember seems to be required for ordinary chondrites and for certain anomalous inclusions [Clayton 1981, §4*b*], and by

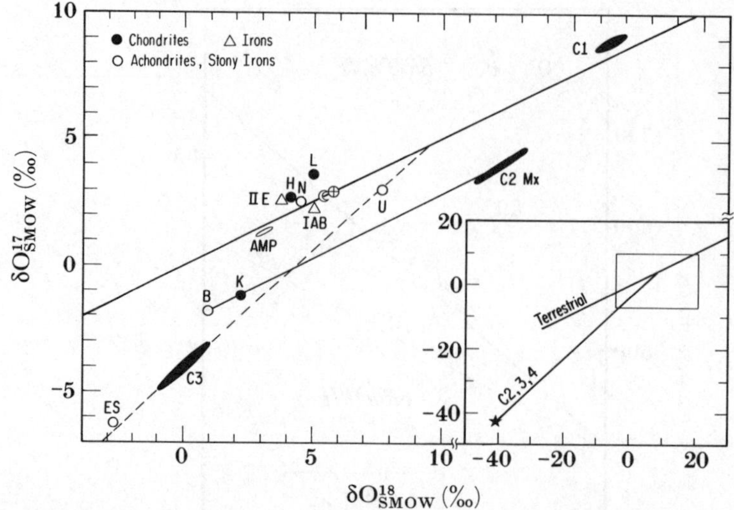

Figure 4. Three-isotope plot shows variation of meteoritic $^{18}O/^{16}O$ and $^{17}O/^{16}O$ relative to standard mean ocean water (smow) (R. N. Clayton *et al.* 1973). Inset gives total observed range for individual minerals; main graph shows bulk meteorites or matrix. All terrestrial samples lie on a mass-fractionation line of slope $\frac{1}{2}$. The anhydrous minerals of C2–C4 chondrites, on the other hand, lie on a mixing line of slope *ca.* 1, corresponding to constant $^{17}O/^{18}O$ but variable ^{16}O. The latter presumably comes from an ^{16}O-rich component, typified by Allende spinel (star in inset). All meteorites fall off the terrestrial line to varying degrees, indicating that their ^{16}O contents differ from that of the Earth and hence preclude a genetic relation.

Occam's Razor, other meteorite classes. Meteorites lying below the terrestrial line contain more of the ^{16}O-rich component than does the Earth, whereas those above the line contain less. The lateral dispersion may reflect mass fractionation between solid and nebular gas. For hydrothermally altered meteorites, i.e. C1s and C2s, there seem to have been further fractionations involving liquid water (Clayton & Mayeda 1984).

This diagram may be used to trace genetic relations among meteorites and planets. Objects lying on the same fractionation line (e.g. Earth, Moon) may be linked by simple mass fractionation and hence *may* have a common origin, either in the same body or in a homogeneous region of the nebula. Objects that do not lie on the same fractionation line contain different amounts of ^{16}O and hence presumably come from different bodies. There is no simple relation between ^{16}O content and primitiveness or (presumed) heliocentric distance. C1 and C2 chondrites, both more primitive than the Earth, lie on opposite sides of the terrestrial line.

(*b*) Mg, Si, Ca, Ti, Cr, Sr, Cd, Ba, Nd, Sm

The literature on the subjects of §4*b*, *c* is vast, and cannot be fully covered in this paper. The reader may wish to consult the reviews by Arnould (this symposium), Shima (1986), Wood (1981), Begemann (1980) and Wasserburg *et al.* (1980).

Mg and Si in cai often show evidence of slight mass fractionation, apparently reflecting condensation, evaporation, or both (Lee & Papanastassiou 1974; Yeh & Epstein 1978; Molini-Velsko *et al.* 1983). So does Cd, a volatile trace element (Rosman *et al.* 1980).

All cai from Allende & Leoville show small (*ca.* 0.8‰) excesses of ^{50}Ti (Niederer *et al.* 1981; Niemeyer & Lugmair 1981, Fahey *et al.* 1985). Presumably a separate dust component was present, enriched in ^{50}Ti (D. D. Clayton 1981). Similar enrichments have been found for Ca^{48} (Jungck *et al.* 1984) and ^{54}Cr (Birck & Allègre 1984).

The remaining elements are isotopically normal in most CAI, but show fractionation and unidentified nuclear effects in a few special inclusions, designated by the acronym 'FUN' (Wasserburg *et al.* 1977). The FUN inclusions also have unusual oxygen isotope compositions, lying on mixing lines between a common (gas?) composition (H in figure 5) and several points (B, C, D, E) along a mass fractionation line AG. This line emanates from the ^{16}O-rich composition A, representing the anomalous dust component in the nebula. Two models have been proposed for the FUN inclusions.

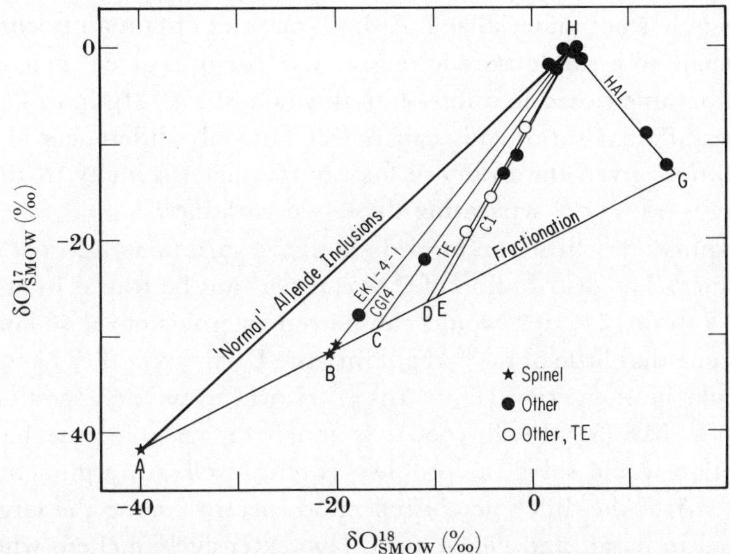

FIGURE 5. Oxygen in FUN inclusions (fractionation and unidentified nuclear effects) from Allende (R. N. Clayton *et al.* 1984). A few exceptional inclusions fall off the normal mixing line, and show FUN effects. Bulk samples and separated minerals of each inclusion form linear arrays converging on a common point H. Presumably these inclusions first formed by mass fractionation of dust (composition A) along the line AG, and later partially exchanged with nebular gas H. Spinel, with the slowest exchange rate, remained on the original fractionation line, whereas more reactive minerals (melilite, pyroxene) moved toward H.

(1) Dust component A underwent mass fractionation, yielding solids B, C, D, etc. Each of these polymineralic solids then reacted with the nebular gas H, under conditions where only the most reactive minerals achieved exchange equilibrium (Blander & Fuchs 1975). This left the minerals strung out along mixing lines BH, CH, etc. (R. N. Clayton & Mayeda 1977; Wasserburg *et al.* 1977; Lee *et al.* 1980).

(2) Solids A, B, C, etc. are mixtures of two mass fractionated components in a presolar locale, gas to the left of A and dust to the right of G. Upon arrival in the Solar System, these solids would partially equilibrate with nebular gas H, as above (D. D. Clayton 1981).

On either model, some *ad hoc* reason must be found why only the highly mass fractionated solids B, C, retained nuclear anomalies; in other words, why is F coupled with UN? One possibility is that a presolar melting process greatly coarsened B, C, D, etc. compared with A, thus minimizing uptake of late, fine-grained ingredients of the Solar System mix of heavy elements.

(c) *Extinct radionuclides*

Seven extinct radionuclides, with the half-lives given in parentheses, have left detectable decay products in meteorites: ^{41}Ca (1.0×10^5 a), ^{26}Al (7.2×10^5 a), ^{53}Mn (3.7×10^6 a), ^{107}Pd

$(6.5 \times 10^6$ a), ^{129}I $(1.6 \times 10^7$ a), ^{244}Pu $(8.1 \times 10^7$ a), and ^{146}Sm $(1.03 \times 10^8$ a).† Some $(^{41}$Ca, ^{26}Al, ^{53}Mn) have been found only in CAI and others only in irons $(^{107}$Pd), but this reflects mainly detectability, not necessarily limited distribution. These nuclides typically occur at only 10^{-2}–10^{-5} the abundance of neighbouring nuclides, and so their decay products will be detectable only if parent and daughter element have been chemically fractionated to an appreciable degree. For this reason, extinct nuclides are rarely found in matrix, which – though quite ancient – is the least fractionated part of meteorites.

The original hope that extinct radionuclides would provide a detailed, relative chronology of the early Solar System has not materialized. At best, one can obtain an isochron giving the ratio of an extinct nuclide to a stable isotope (e.g. ^{26}Al/^{27}Al) in a given meteorite, referring to the time the system became closed to diffusion of the daughter (^{26}Mg in this example). But if two meteorites have different ratios, this can reflect not only differences in age but also differences in initial ratios, given the evidence for isotopic heterogeneity in the early Solar System (§4a, b). There is no way of separating these two variables.

In view of this ambiguity, very little can be said about the variation of extinct radionuclides with heliocentric distance. The distribution of ^{26}Al can perhaps be traced by its heat effects, as the usual ^{26}Al/^{27}Al ratio of 5×10^{-5} would cause even asteroids of $ca.$ 10 km to melt. On this basis, one could argue that little or no ^{26}Al got into the Earth and other terrestrial planets, as the heat effects would be staggering (Urey 1955). However, as the accretion time of the terrestrial planets ($ca.$ 10 Ma (Wetherill 1980)) is much longer than the half-life of ^{26}Al (0.72 Ma), slow accretion would solve the problem equally well. An argument in favour of patchy distribution of ^{26}Al is the difference in reflection spectra among the largest asteroids. 4 Vesta is differentiated to basalt and hence must have extensively melted, whereas 1 Ceres and 2 Pallas have relatively primitive, olivine–magnetite surfaces, implying little or no differentiation (Gaffey & McCord 1978).

(d) *Noble gases*, C, N, *and* H

Several isotopically anomalous components of these elements have been found in meteorites. The noble-gas components, though present at concentrations of only 10^8–10^{10} atoms g^{-1}, were the first to be identified, and then served as tracers in the search for their (likewise exotic) carrier phases. Most of these carriers are carbonaceous, but have not yet been fully characterized or isolated in pure form. They differ from each other in noble-gas release temperature, grain size, density, combustibility and isotopic composition. Pending further characterization, they are non-committally designated by Greek letters.

Table 2 lists the exotic noble-gas components and their carriers, as well as two exotic components of H and N that are not associated with noble gases.

Two types of Ne-E are distinguished, differing in release temperature and carrier. Ne-E(L) is virtually pure (not less than 99%) Ne^{22}, and though the experimental limit for Ne-E(H) is less stringent (not less than 83%), it, too, probably is pure ^{22}Ne. The only plausible source of monoisotopic ^{22}Ne is β$^+$-decay of ^{22}Na ($t_{\frac{1}{2}} = 2.6$ a). The short half-life of ^{22}Na implies that it was synthesized and trapped on an equally short timescale, i.e. under catastrophic conditions, presumably in a nova (Clayton & Hoyle 1976; Arnould & Beelen 1974). The high δ^{13}C values

---

† For references, see Begemann (1980), Wasserburg *et al.* (1980), Birck & Allègre (1985), Crabb & Anders (1982), Lugmair *et al.* (1983) and Hutcheon *et al.* (1984).

TABLE 2. EXOTIC COMPONENTS OF NOBLE GASES, C, H, AND N

component	carrier	enriched in	release T °C	δC^{13} (‰)	δN^{15} (‰)	abundance (ppm)	size μm	probable source	reference
Ne-E(H)	spinel	Ne^{22}	1200	> 2000		1700	1–10	nova?	a,e,i
Ne-E(L)	Cα	Ne^{22}	600	340	> 0	5	1–10	nova	a,c,e,o
Xe-S	Cβ	Xe^{130}	1400	1400	> 0	2	0.1–3	red giant	a,j,n,o,p
Xe-HL	Cδ (diamond)	$Xe^{124,136}$	1000	−38	−330	400	0.005	supernova	g,h,l,o,q
D	polymer†	D	400	?	?	300	?	molecular cloud	d,f,m,p
Heavy N	?				973	< 60		nova?	k

† Minor (acetylenic?) component of meteoritic polymer.
(a) Alaerts et al. (1980); (b) Black & Pepin (1969); (c) Carr et al. (1983); (d) Hayatsu et al. (1983); (e) Jungck & Eberhardt (1979); (f) Kolodny et al. (1980); (g) Lewis et al. (1975); (h) Lewis et al. (1983); (i) Niederer et al. (1985); (j) Ott et al. (1985); (k) Prombo & Clayton (1985); (l) Reynolds & Turner (1964); (m) Robert et al. (1979); (n) Srinivasan & Anders (1978); (o) Swart et al. (1983); (p) Yang & Epstein (1984); (q) Lewis et al. (1987).

certainly are consistent with a nova origin, but the expected high $\delta^{15}N$ is yet to be confirmed, as the samples analysed were not pure enough (Lewis *et al.* 1983).

The association of Ne-E(H) with spinel is less certain now than it was two years ago. Niederer *et al.* (1985) have found heavy C of $\delta^{13}C > 2000\%_0$ in Ne-E-bearing spinel fractions. It is not clear whether this heavy carbon is present as discrete grains clinging to or enclosed in spinel or whether it is dissolved in spinel on an atomic scale.

Xenon-S shows the distinct signature of the s-process: strong predominance of even isotopes over odd ones (128, 130, 132 compared with 129, 131), rise from 128 to 132, and absence of Xe isotopes that do not lie on the s-process path (124, 126, 134, 136). Its carrier, $C\beta$, also shows the expected enrichment in ^{13}C. As the s-process is known to take place in red giants, Xe-S and $C\beta$ must have formed in a red giant. However, it is not known whether they were expelled from the star during the red-giant stage (red giants lose mass at a rate of 10^{-7}–10^{-8} $M_\odot$ a^{-1}, forming circumstellar shells of carbon grains), or only later. Nor is it known whether $C\beta$ represents general galactic background from many red giants or material from a single star in the OB association from which the Solar System formed (Reeves 1979).

The nucleosynthetic origin of Xe-HL is less well understood, though a supernova, with high rates of neutron capture and photonuclear reactions, may provide the right environments for simultaneous enrichment of heavy and light isotopes (Heymann & Dziczkaniec 1979). One interesting clue is the lack of enrichment at mass 129 relative to 128 or 130, although mass 129 – in contrast to the others – originally forms not as xenon but as a long-lived iodine isotope ($t_\frac{1}{2} = 16$ Ma). Apparently the atoms were trapped by a chemically non-selective process, such as ion implantation (Lewis & Anders 1981). This could happen if the supernova ejecta, travelling at *ca.* 10^4 km s^{-1}, overtook a shell of carbon grains ejected at the presupernova stage (D. D. Clayton 1982). The carrier, $C\delta$ has been identified as diamond (Lewis *et al.* 1987), which presumably condensed metastably in the presupernova dust shell. This supports an earlier suggestion by Saslaw & Gaustad (1969) that diamond is a major constituent of interstellar dust.

Deuterium enrichments of up to $3000\%_0$ have been found in stepped combustion or pyrolysis of the organic polymer from primitive chondrites (Kolodny *et al.* 1981). Although such enrichments could, in principle, be achieved by equilibrium fractionations, rates are far too low at the required temperatures of *ca.* 100 K (Geiss & Reeves 1981). More probably, these enrichments were caused by ion-molecule reactions (Kolodny *et al.* 1980; Geiss & Reeves 1981), which are believed to be responsible for the even greater D-enrichments in interstellar molecules (Smith, this symposium). The observed enrichment factor of *ca.* 50 relative to galactic hydrogen (figure 6) might represent either formation of the entire polymer by ion–molecule reactions at a rather high temperature (*ca.* 60 K) and consequently small fractionation, or admixture of a minor amount of highly enriched material, i.e. polymerized interstellar molecules.

The latter possibility seems more likely. On acid treatment or oxidation, the D-rich material reacts preferentially, which suggests that it contributes a minor fraction of the total (McNaughton *et al.* 1982; Yang & Epstein 1984). Detailed structural analyses of the polymer show that while most of it is aromatic, as discussed in §3*d*, a minor part (*ca.* 1 %) is acetylenic (Hayatsu *et al.* 1983). The latter component yields fragments such as C_mH ($m = 3$–10)and C_mCN ($m = 3$–6), which are closely related to polycyanoacetylenes $[H(C{\equiv}C)_nCN]$, a prominent class of interstellar molecules ranging up to $n = 6$.

A strikingly large enrichment of ^{15}N (up to $973\%_0$) has been found in the brecciated stony

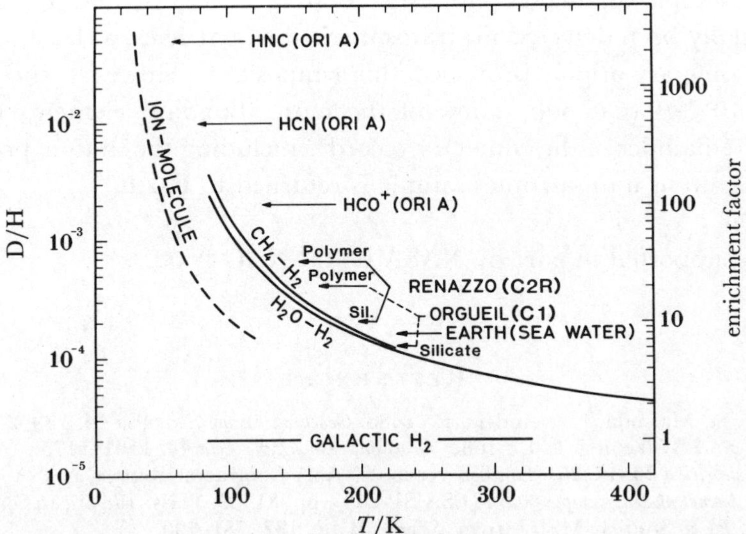

FIGURE 6. (Adapted from Geiss & Reeves (1981).) Organic polymer fractions from C-chondrites are strongly enriched in deuterium, up to 32 times relative to galactic hydrogen. Equilibrium isotope fractionations can produce such enrichments, but only at temperatures of *ca.* 100 K, where rates are prohibitively slow (e.g. the half-time of the Fischer–Tropsch reaction at 110 K is *ca.* 10^{30} years!) More probably, the fractionation is due to ion–molecule reactions, which are thought to be responsible for the large (up to 10^{5}-fold) D-enrichments in interstellar molecules. Only a minor, reactive part of the meteoritic polymer seems to be enriched in D, and probably represents relict interstellar molecules. The major part appears to be of local origin, as outlined in figure 3.

irons Bencubbin and Weatherford (Prombo & Clayton 1985). The nature of the carrier is not known. By their oxygen isotope composition the silicates may be related to C2R and C2M chondrites, and there are suggestions that the metal clasts may be direct nebular condensates (Newsom & Drake 1979). But otherwise these meteorites are far from primitive, having undergone severe brecciation and perhaps impact melting (Kallemeyn *et al.* 1978). At present, it is not possible to decide whether the nitrogen anomaly is nuclear or chemical, i.e. produced by ion–molecule reactions in a molecular cloud (Prombo & Clayton 1985).

5. COMETS

We already know that the 'rock' fraction of comets is approximately chondritic, and that the 'ice' fraction differs appreciably from the equilibrium composition predicted by figure 1, because it contains mainly CO_2 rather than CH_4 as well as formaldehyde and other 'interstellar' molecules (Delsemme 1982; Huebner *et al.* 1982). For further characterization of comets, we would like to know to what extent they shared the 'reprocessing' of chondrites (Anders 1986). Do comets contain chondrules, CAI, Fe^{2+}-rich silicates, hydrated minerals, magnetite, etc.? Were they ever exposed to liquid water? Have they been metamorphosed? What kinds of organic compounds do they contain? Some clues have already been obtained from meteor spectra (Wilkening 1975) or stratospheric dust particles (Brownlee, this symposium; Fraundorf & Shirck 1979), and others will undoubtedly come from the Vega and Giotto missions. But most of the answers will require a sample return mission.

Such a mission is even more essential if we want to study the isotopic record of comets, particularly its presolar part, which ought to be clearer and more diverse than that in meteorites. Two isotopic anomalies – for Mg (Esat *et al.* 1979), and H (McKeegan *et al.* 1985) – have actually been detected in stratospheric dust particles, at least some of which are thought to be of cometary origin (Brownlee, this symposium). However, these particles are so small (less than 10^{-9} g) that only a few of the most abundant eiements can be analysed isotopically. The remainder of the cometary record – including the unique presolar part – will remain inaccessible to us until a comet sample is returned to Earth.

This work was supported in part by NASA Grant NAG 9–52.

References

Alaerts, L., Lewis, R. S., Matsuda, J. & Anders, E. 1980 *Geochim. cosmochim. Acta* **44**, 189–209.
Allen, J. S., Nozette, S. & Wilkening, L. L. 1980 *Geochim. cosmochim. Acta* **44**, 1161–1175.
Anders, E. 1965 *Meteoritika* **26**, 17–25. (English version: NASA Contractor Report, NASA CR-299.)
Anders, E. 1986 In *Comet nucleus sample return* (ESA SP-249), pp. 31–39. Paris: European Space Agency.
Anders, E., Hayatsu, R. & Studier, M. H. 1973 *Science, Wash.* **182**, 781–790.
Anders, E. & Zadnik, M. G. 1985 *Geochim. cosmochim. Acta* **49**, 1281–1291.
Armstrong, J. T., El Goresy, A. & Wasserburg, G. J. 1985 *Geochim. cosmochim. Acta* **49**, 1001–1022.
Arnould, M. & Beelen, W. 1974 *Astron. Astrophys.* **33**, 215–230.
Ashworth, J. R. 1977 *Earth planet. Sci. Lett.* **35**, 25–34.
Begemann, F. 1980 *Rep. Prog. Phys.* **43**, 1309–1356.
Birck, J. L. & Allègre, C. J. 1984 *Geophys. Res. Lett.* **11**, 943–946.
Birck, J. L. & Allègre, C. J. 1985 *Meteoritics* **20**, 609.
Black, D. C. & Pepin, R. O. 1969 *Earth planet. Sci. Lett.* **6**, 395–405.
Blander, M. & Fuchs, L. H. 1975 *Geochim. cosmochim. Acta* **39**, 1605–1619.
Cameron, A. G. W. 1973 In *Interstellar dust and related topics* (ed. J. M. Greenberg & H. C. van de Hulst), pp. 545–547. Dordrecht: Reidel.
Cameron, A. G. W. 1978 In *Protostars and planets* (ed. T. Gehrels), pp. 453–487. Tucson: University of Arizona Press.
Carr, R. H., Wright, I. P., Pillinger, C. T., Lewis, R. S. & Anders, E. 1983 *Meteoritics* **18**, 277.
Clayton, D. D. 1975 *Astrophys. J.* **199**, 765–769.
Clayton, D. D. 1981 *Astrophys. J.* **251**, 374–386.
Clayton, D. D. 1982 *Q. Jl R. astr. Soc.* **23**, 174–212.
Clayton, D. D. & Hoyle, F. 1976 *Astrophys. J.* **203**, 490–496.
Clayton, R. N. 1981 *Phil. Trans. R. Soc. Lond.* A **303**, 339–349.
Clayton, R. N., Grossman, L. & Mayeda, T. K. 1973 *Science, Wash.* **182**, 485–488.
Clayton, R. N., MacPherson, G. J., Hutcheon, I. D., Davis, A. M., Grossman, L., Mayeda, T. K., Molini-Velsko, C., Allen, J. M. & El Goresy, A. 1984 *Geochim. cosmochim. Acta* **48**, 535–548.
Clayton, R. N. & Mayeda, T. K. 1977 *Geophys. Res. Lett.* **4**, 295–298.
Clayton, R. N. & Mayeda, T. K. 1984 *Earth planet. Sci. Lett.* **67**, 151–161.
Crabb, J., Lewis, R. S. & Anders E. 1982 *Geochim. cosmochim. Acta* **46**, 2511–2526.
Delsemme, A. H. 1982 In *Comets* (ed. L. L. Wilkening), pp. 85–130. Tucson: University of Arizona Press.
DuFresne, E. R. & Anders, E. 1962 *Geochim. cosmochim. Acta* **26**, 1085–1114.
El Goresy, A., Nagel, K. & Ramdohr, P. 1978 *Proc. Lunar Sci. Conf.* **9**, 1279–1303.
Esat, M., Brownlee, D. E., Papanastassiou, D. A. & Wasserburg, G. J. 1979 *Science, Wash.* **206**, 190–197.
Fahey, A., Goswami, J. N., McKeegan, K. D. & Zinner, E. 1985 *Astrophys. J.* **296**, L17–L20.
Fraundorf, P. & Shirck, J. 1979 *Proc. Lunar Planet. Sci. Conf.* **10**, 951–976.
Gaffey, M. J. & McCord, T. B. 1978 *Space Sci. Rev.* **21**, 555–628.
Galwey, A. 1972 *Geochim. cosmochim. Acta* **36**, 1115–1130.
Geiss, J. & Reeves, H. 1981 *Astron. Astrophys.* **93**, 189–199.
Gooding, J. L. & Keil, K. 1980 *Meteoritics* **16**, 17–43.
Grossman, J. N., Rubin, A. E., Rambaldi, E. R., Rajan, R. S. & Wasson, J. T. 1985 *Geochim. cosmochim. Acta* **49**, 1781–1795.
Grossman, J. N. & Wasson, J. T. 1983 *Geochim. cosmochim. Acta* **47**, 759–771.
Grossman, J. N. & Wasson, J. T. 1985 *Geochim. cosmochim. Acta* **49**, 925–939.
Grossman, L. 1972 *Geochim. cosmochim. Acta* **36**, 597–619.
Grossman, L. 1980 *A. Rev. Earth Planet. Sci.* **8**, 559–608.
Grossman, L. & Ganapathy, R. 1976 *Geochim. cosmochim. Acta* **40**, 967–977.
Grossman, L., Ganapathy, R. & Davis, A. M. 1977 *Geochim. cosmochim. Acta* **41**, 1647–1664.

Grossman, L. & Larimer, J. W. 1974 *Rev. Geophys. space Phys.* **12**, 71–101.

Hayashi, C., Nakazawa, K. & Nakagawa, Y. 1985 In *Protostars and planets II* (ed. D. C. Black & M. S. Matthews), pp. 1100–1153.

Hayatsu, R. & Anders, E. 1981 *Top. curr. chem.* **99**, 1–34.

Hayatsu, R., Matsuoka, S., Scott, R. G., Studier, M. H. & Anders, E. 1977 *Geochim. cosmochim. Acta* **41**, 1325–1339.

Hayatsu, R., Scott, R. G. & Winans, R. E. 1983 *Meteoritics* **18**, 310.

Heymann, D. & Dziczkaniec, M. 1979 *Proc. Lunar Planet Sci. Conf.* **10**, 1943–1959.

Huebner, W. F., Giguere, P. T. & Slattery, W. L. 1982 In *Comets* (ed. L. L. Wilkening), pp. 496–515. Tucson: University of Arizona Press.

Huss, G. R. 1987 *Icarus.* (In the press.)

Huss, G. R., Keil, K. & Taylor, G. J. 1981 *Geochim. cosmochim. Acta* **45**, 33–51.

Hutcheon, I. D., Armstrong, J. T. & Wasserburg, G. J. 1984 *Lunar planet. Sci.* **15**, 387–388.

Hutchison, R., Bevan, A. W. R., Agrell, S. O. & Ashworth, J. R. 1979 *Nature, Lond.* **280**, 116–119.

Jungck, M. H. A. & Eberhardt, P. 1979 *Meteoritics* **14**, 439–440.

Jungck, M. H. A., Shimamura, T. & Lugmair, G. W. 1984 *Geochim. cosmochim. Acta* **48**, 2651–2658.

Kallemeyn, G. W., Boynton, W. V., Willis, J. & Wasson, J. T. 1978 *Geochim. cosmochim. Acta* **42**, 507–515.

Kolodny, Y., Kerridge, J. F. & Kaplan, I. R. 1980 *Earth planet. Sci. Lett.* **46**, 149–158.

Kornacki, A. S. & Wood, J. A. 1984 *J. Geophys. Res.* **89**, B573–B587.

Kurat, G. 1970 *Earth planet. Sci. Lett.* **9**, 225–231.

Larimer, J. W. 1967 *Geochim. cosmochim. Acta* **31**, 1215–1238.

Larimer, J. W. 1975 *Geochim. cosmochim. Acta* **39**, 389–392.

Larimer, J. W. 1978 In *The Origin of the Solar System* (ed. S. F. Dermott), pp. 347–393. New York: Wiley.

Larimer, J. W. & Anders, E. 1967 *Geochim. cosmochim. Acta* **31**, 1239–1270.

Larimer, J. W. & Anders, E. 1970 *Geochim. cosmochim. Acta* **34**, 367–388.

Larimer, J. W. & Bartholomay, H. A. 1983 *Lunar planet. Sci* **14**, 423.

Larimer, J. W. & Bartholomay, M. 1979 *Geochim. cosmochim. Acta* **43**, 1455–1466.

Lattimer, J. M., Schramm, D. N. & Grossman, L. 1978 *Astrophys. J.* **219**, 230–249.

Lee, T., Mayeda, T. K. & Clayton, R. N. 1980 *Geophys. Res. Lett.* **7**, 493–496.

Lee, T. & Papanastassiou, D. A. 1974 *Geophys. Res. Lett.* **1**, 225–228.

Lewis, J. S. 1972 *Icarus* **16**, 241–252.

Lewis, R. S. & Anders, E. 1981 *Astrophys. J.* **247**, 1122–1124.

Lewis, R. S., Anders, E., Wright, I. P., Norris, S. J. & Pillinger, C. T. 1983 *Nature, Lond.* **305**, 767–771.

Lewis, R. S., Srinivasan, B. & Anders, E. 1975 *Science, Wash.* **190**, 1251–1262.

Lewis, R. S., Tang, M., Wacker, J. F., Anders, E. & Steel, E. 1987 *Nature, Lond.* **326**, 160–162.

Lugmair, G. W., Shimamura, T., Lewis, R. S. & Anders, E. 1983 *Science, Wash.* **222**, 1015–1018.

Marvin, U. B., Wood, J. A. & Dickey, J. S., Jr. 1970 *Earth Planet. Sci. Lett.* **7**, 346–350.

McKeegan, K. D., Walker, R. & Zinner, E. 1985 *Geochim. cosmochim. Acta* **49**, 1971–1987.

McNaughton, N. J., Fallick, A. E. & Pillinger, C. T. 1982 *J. Geophys. Res.* **87**, A297–A302.

Molini-Velsko, C., Mayeda, T. K. & Clayton, R. N. 1983 *Lunar planet. Sci* **14**, 509–510.

Morgan, J. W. & Anders, E. 1980 *Proc. Natn. Acad. Sci. U.S.A.* **77**, 6973–6977.

Nagahara, H. 1981 *Nature, Lond.* **292**, 135–136.

Nagahara, H. 1983 In *Proc. 8th Symp. Antarctic Meteorites*, 61–84. Tokyo: National Institute of Polar Research.

Nagahara, H. 1984 *Geochim. cosmochim. Acta* **48**, 2581–2595.

Newsom, H. E. & Drake, M. 1979 *Geochim. cosmochim. Acta* **43**, 689–707.

Niederer, F. R., Eberhardt, P., Geiss, J. & Lewis, R. S. 1985 *Meteoritics* **20**, 716–717.

Niederer, F. R. & Papanastassiou, D. A. 1984 *Geochim. cosmochim. Acta* **48**, 1279–1293.

Niederer, F. R., Papanastassiou, D. A. & Wasserburg, G. J. 1981 *Geochim. cosmochim. Acta* **45**, 1017–1031.

Niemeyer, S. & Lugmair, G. W. 1981 *Earth planet. Sci. Lett.* **53**, 211–225.

Osborn, T. W., Warren, R. G., Smith, R. H., Wakita, H., Zellmer, D. L. & Schmitt, R. A. 1974 *Geochim. cosmochim. Acta* **38**, 1359–1378.

Ott, U., Yang, J. & Epstein, S. 1985 *Meteoritics* **20**, 722–723.

Palme, H. & Wlotzka, F. 1976 *Earth planet. Sci. Lett.* **33**, 45–60.

Planner, H. N. & Keil, K. 1982 *Geochim. cosmochim. Acta* **46**, 317–330.

Prombo, C. A. & Clayton, R. N. 1985 *Science, Wash.* **230**, 935–937.

Rambaldi, E. R., Housley, R. M. & Rajan, R. S. 1984 *Nature* **311**, 138–140.

Rambaldi, E. R., Rajan, R. S., Wang, D. & Housley, R. M. 1983 *Earth planet. Sci. Lett.* **66**, 11–24.

Rambaldi, E. R. & Wasson, J. T. 1984 *Geochim. cosmochim. Acta* **48**, 1885–1897.

Reeves, H. 1979 *Astrophys. J.* **231**, 229–235.

Reynolds, J. H. & Turner, G. 1964 *J. Geophys. Res.* **69**, 3263–3281.

Robert, F., Merlivat, L. & Javoy, M. 1979 *Nature, Lond.* **282**, 785–789.

Rosman, K. J. R., deLaeter, J. R. & Gorton, M. P. 1980 *Earth planet. Sci. Lett.* **48**, 166–170.

Saslaw, W. C. & Gaustad, J. E. 1969 *Nature, Lond.* **221**, 160–162.

Schmitt, R. A., Smith, R. H. & Goles, G. G. 1965 *J. Geophys. Res.* **70**, 2419–2444.

Sears, D. W., Grossman, J. N., Melcher, C. L., Ross, L. M. & Mills, A. A. 1980 *Nature, Lond.* **287**, 791–795.
Shima, M. 1986 *Geochim. cosmochim. Acta* **50**, 577–584.
Srinivasan, B. & Anders, E. 1978 *Science, Wash.* **201**, 51–56.
Studier, M. H., Hayatsu, R. & Anders, E. 1968 *Geochim. cosmochim. Acta* **32**, 151–174.
Swart, P. K., Grady, M. M., Pillinger, C. T., Lewis, R. S. & Anders, E. 1983 *Science, Wash.* **220**, 406–410.
Tsuchiyama, A., Nagahara, H. & Kushiro, I. 1980 *Earth planet. Sci. Lett.* **48**, 155–165.
Tsuchiyama, A., Nagahara, H. & Kushiro, I. 1981 *Geochim. cosmochim. Acta* **45**, 1357–1367.
Urey, H. C. 1955 *Proc. Natn. Acad. Sci. U.S.A.* **41**, 127–144.
Van Schmus, W. R. & Wood, J. A. 1967 *Geochim. cosmochim. Acta* **31**, 747–765.
Wark, D. A. & Lovering, J. F. 1982 *Geochim. cosmochim. Acta* **46**, 2581–2594.
Wasserburg, G. J., Lee, T., & Papanastassiou, D. A. 1977 *Geophys. Res. Lett.* **4**, 299–302.
Wasserburg, G. J., Papanastassiou, D. A. & Lee, T. 1980 In *Early Solar System processes and the present Solar System*, pp. 144–191. Soc. Italiana di Fisica.
Wetherill, G. W. 1980 *A. Rev. Astron. Astrophys.* **18**, 77–113.
Wilkening, L. L. 1975 *Nature, Lond.* **258**, 689–690.
Wood, J. A. 1958 Smithsonian Astrophysics Observatory Technical Report no. 10.
Wood, J. A. 1962 *Geochim. cosmochim. Acta* **26**, 739–749.
Wood, J. A. 1973 *Icarus* **2**, 152–180.
Wood, J. A. 1981 *Earth planet. Sci. Lett.* **56**, 32–44.
Wood, J. A. 1984 *Earth planet. Sci. Lett.* **70**, 11–26.
Wood, J. A. 1985 *Meteoritics* **20**, 787–788.
Wood, J. A. 1986 *Lunar planet. Sci.* **17**, 956–957.
Yang, J. & Epstein, S. 1984 *Nature, Lond.* **311**, 544–547.
Yeh, H. W. & Epstein, S. 1978 *Lunar planet. Sci.* **9**, 1289–1291.

Phil. Trans. R. Soc. Lond. A **323**, 305–311 (1987)

Printed in Great Britain

Morphological, chemical and mineralogical studies of cosmic dust

By D. E. Brownlee

Department of Astronomy, University of Washington, Seattle, Washington 98195, *U.S.A.*

Significant numbers of 5 μm–1 mm particles of interplanetary dust have been collected and subjected to laboratory analysis. The extraterrestrial origin of selected samples has been established by detection of space-exposure effects such as implanted solar wind and tracks of solar cosmic rays. The collected samples should contain both cometary and asteroidal particles. The asteroidal component is probably a representative sample of typical main-belt asteroids. The elemental composition of the majority of particles is similar to CI and CM chondrites. Most particles can be grouped into one of two general classes, those that contain hydrated minerals and those that are anhydrous. Some of the hydrated particles may be samples of CI/CM matrix whereas most of the anhydrous particles are clearly unrelated to any known chondrite class. Some of the anhydrous particles are very porous aggregates that at least superficially resemble models of cometary meteors. Future studies of cosmic dust will probably lead to criteria for distinguishing between asteroidal and cometary particles.

Introduction

The submillimetre particles that permeate the interplanetary medium are commonly referred to as cosmic dust. The particles crater lunar rocks and spacecraft and they enter the atmosphere at an annual rate of 10^4 t. The flux of 10 μm dust particles near the Earth is relatively high, about 1 m^2 d^{-1} but the spatial density is less than 1 km^{-3}. Dust-impact craters found on buried lunar samples suggest that the current dust density has remained roughly constant over much of the Solar System's history. This is remarkable in light of processes that destroy interplanetary dust at a rate of 10 t s^{-1}. Dust particles are lost from the interplanetary medium on timescales of 1000–100000 years because of collisions and orbital decay caused by the drag component of sunlight pressure (Grün *et al.* 1985). Steady-state survival of the Solar System dust cloud requires continual injection of fresh particles. For most of the Solar System's history the major suppliers of new dust have been comets and asteroids. Comets are widely believed to be the major dust source but the infrared emission from asteroid dust measured by the IRAS satellite indicates that dust from asteroid collisions may also be a significant component (Zook & McKay 1986).

Recovered samples of cosmic dust particles are of major scientific interest because they are samples of bodies believed to be relic planetismals preserved since the earliest history of the Solar System. Analysis of collected dust samples is an interesting compliment to studies of conventional meteorites because there are classes of extraterrestrial material that are only collectable in the form of dust. Unlike meteorites, collectable dust samples may be rather representative samples of the asteroids and comets in the planetary system. Conventional meteorites are wonderful samples for investigating early Solar System history but they are limited because only certain types of extraterrestrial materials can become meteorites. To become a conventional meteorite an object must be relatively large (so that it can be found on the ground), it must have undergone a rare set of gravitational perturbations to reach an earthcrossing orbit and lastly it must be sufficiently strong to survive the mechanical and thermal stresses experienced by rock-sized objects entering the atmosphere at hypervelocity. Dust particles, on the other hand, have fewer orbital restrictions because they diffuse through the Solar System because of light pressure drag (the Poynting–Robertson (P–R) effect) and they do not have to be strong to survive atmospheric entry. Dust particles decelerate at altitudes where the air density is low and the maximum dynamic ram pressure is orders of magnitude lower than that experienced deeper in the atmosphere by larger meteorites travelling at hypervelocity. Thus atmospheric entry is a filter that allows weak materials to survive as dust but not as objects larger than a centimetre. This filter may be critical for cometary materials because observation of the entry of visual cometary meteors indicates that typical cometary samples are more fragile than the weakest recovered conventional meteorite specimens.

COLLECTION

Collection of 2–50 μm particles is done routinely in the stratosphere by using impactors mounted on U2 aircraft and particles up to millimetre size have been recovered in abundance from the ocean floor and from blue ice lakes in Greenland. Some particles have been collected from orbiting spacecraft but most have been appreciably altered because the high velocity of collection. The most pristine particles are those collected in the stratosphere. They are small and hence minimally heated and they have not undergone chemical reactions with terrestrial materials. Typical stratospheric particles are heated (uniformly) to about 500 °C for a few seconds during entry. Particles entering at low velocity and at near grazing incidence angle can survive without being heated above 300 °C. Most particles larger than 100 μm are heated to their melting points and they form spheres often called 'cosmic spheres'. Because of magnetite formation most of the spheres are magnetic and they can be collected from deep sea sediments and Greenland ice lakes with simple magnets. These larger particles are too rare to collect in the atmosphere. The first deep-sea cosmic spheres were collected during the *Challenger* expeditions of the past century.

The majority of all collected particles have chondritic elemental compositions and their distinction from terrestrial contamination is usually quite straightforward (Brownlee 1985). Individual particles have been proven to be extraterrestrial on the basis of implanted solar wind, solar-flare tracks, trace-element composition and cosmogenic isotopes produced by cosmic-ray irradiation. Both the spheres and the smaller unmelted particles (micrometeorites) are important sources of information on interplanetary dust. The small particles are the best preserved but the spheres are individually more massive and they can be collected in unlimited quantities. The largest deep-sea spheres contain over 10^6 times the mass of the typical particle collected in the stratosphere.

CHEMICAL COMPOSITION OF THE DUST PROGENITORS

For the refractory elements, the average dust elemental compositions are best determined from the massive cosmic spheres larger than 300 μm in size. Analysis of hundreds of unweathered particles has shown that 85% of the particles have Mg, Al, Si, Ca, Ti and Mn compositions compatible with those of the CI and CM meteorites (Bates 1986). Volatile elements and many of the siderophile elements are lost from cosmic spheres during atmospheric entry. The analyses indicate that only a minor fraction of the samples could have been derived from ordinary chondrites or achondrites and it is clear that the relative abundance of meteoroid composition types in the dust differs radically from the meteorites. Only 2.5% of meteorite falls are of the CI or CM type (Wasson 1974).

If dust particles truly are a representative sampling of the Solar System's minor bodies, then the particle analyses imply that most of these bodies have unfractionated major and minor refractory element abundances similar to these very rare meteorite types. If, as expected, most of the spheres are cometary then the major dust producing comets must also have this composition. Except for special effects attributable to single mineral grain precursors there are no major subdivisions among the chondritic composition spheres.

The small unmelted particles collected in the stratosphere do not appear to have suffered loss of volatiles during entry. It is likely that the small particles are genetically related to the larger dust fragments and their volatile abundances are probably also representative of the sphere producing particles before they entered the atmosphere. Unfortunately the analysis of minor elements in the stratospheric particles is not straightforward and because of smaller sample size there is more dispersion in the data. The existing volatile measurements for C, S, Zn and Br all indicate high volatile abundances similar to CI meteorites (van der Stap *et al.* 1986) but the data are limited to a small set of particles.

Some of the collected particles are single mineral grains or are clusters of a limited number of grains and these samples do not have chondritic elemental abundances. The striking result, however, from years of work is that the bulk of all particles in the 5–1000 µm size have compositions similar to CI or CM meteorites. This rather monotonous result indicates that the particles must be fine grained and not substantially fractionated from what is believed to be solar composition. It is significant that no meteorite type including the CIs and CMs have this homogenous composition on a size scale as small as the interplanetary particles.

Mineralogical and morphological dust types

Some of the deep-sea and Greenland samples contain unmelted relict mineral grains. The grains are normally contained in a matrix that is usually either recrystallized melt or unmelted fine grained material. The general abundance of large mafic grains is higher than observed in CI meteorites. Commonly the relict grains are forsterite, enstatite and pyrrhotite. Rarer phases include FeNi metal, spinel, diopside and perovskite. The trace element content of the forsterite grains indicates a similarity to the CM meteorites (Steele *et al.* 1985).

Unfortunately the deep-sea samples that were not strongly altered by heating during atmospheric entry are altered in the ocean sediment environment and so little can really be said about their original morphology of the nature of their submicrometre mineral grains. Sea-floor weathering does not harm many of the large grains but its effect on the small ones makes direct comparison with the stratospheric particles difficult. The true nature of the larger particles cannot be studied directly.

The stratospheric particles are well preserved and mineral grains can be studied well into the submicrometre-size region. Most of the particles that do not have chondritic elemental compositions appear to be single mineral grains or grain clumps that were previously imbedded in fine grained material of chondritic composition. Most of these particles are composed of olivine, pyroxene or sulphide and they are encrusted with black chondritic material. They appear to be mere fragments from chondritic host materials. The most important materials to study are the chondritic particles where mineral chemistry and structure can be studied for coexisting grains in a given particle.

Although there may be a variety of subgroups the major subdivisions of chondritic dust particles are those that contain hydrous minerals and those that do not. These are truly distinct particle types that differ in morphology and mineralogy. The hydrated particles are largely composed of hydrated silicates with lesser amounts of sulphide, magnetite and anhydrous mafic minerals (Tomeoka & Buseck 1986).

In some cases the dominant hydrous phases are serpentines with basal spacings of 7.2 Å† and in others they are smectites with spacings of 11 Å. Although there are some exceptions, most of the hydrated particles are compact masses of platey phyllosilicate sheets and fibres. Their porosity is low and externally they often are smooth at the micrometre level. The anhydrous particles are aggregates of fairly equidimensional grains ranging in size from the submicrometres to micrometres. As viewed in the SEM some of these particles resemble porous grape clusters where the individual grapes are typically 0.3 μm in size. The most porous of these particles have densities of less than unity and they are the most porous meteoritic materials (Bradley & Brownlee 1986).

The individual grains in the anhydrous samples are of three types; single mineral grains, carbonaceous matter and 'tar balls'. The tar balls are collections of 100–1000 Å mineral grains imbedded in a yet uncharacterized carbonaceous material. The tar balls are unique to the anhydrous particle type. Individual phases in the anhydrous particles include enstatite, olivine, NiFe carbide, pyrrhotite, pentlandite, magnetite, albite and trace amounts of metal and other phases.

The hydrated particles are similar in many ways to CI and CM meteorites while the anhydrous particles are distinct from all established meteorite groups. Many of the hydrated types could come from the same parent materials that produce CI and CM meteorites although the H isotopic data (see elsewhere in this symposium) and the existence of smectites argues against a simple direct genetic relation. The porosity of the anhydrous particles is quite distinct from the chondrite classes which are all solid rocks with minimal pore space. The porosity of these samples is consistent with models for cometary grains. Unfortunately it is not known if such porous fine grained material could survive in the internal environment of asteroids that contain no ice to fill pore spaces and have extensive histories of collisional modification.

ORIGINS

A major thrust of future laboratory studies of interplanetary dust will be to determine the origins of some of the different classes of particles. This will involve synthesis of laboratory analytical data, astronomical observations and data from cometary missions. Even the general distinction of cometary and asteroidal particles might provide valuable insights into early Solar System processes. The asteroidal dust particles that have been collected are probably representative samples of the typical main-belt asteroids and are therefore presumably typical

† 1 Å = 10^{-1} nm = 10^{-10} m.

of the materials that accreted in the Mars–Jupiter region of the solar nebula. The cometary dust samples may or may not be particles from typical comets but at least they must be samples of materials that accreted well beyond the orbit of Saturn. The relative properties of samples from these two regions should relate to differences in the temperatures, pressures, gas compositions and processes that occurred in the inner and outer regions of the solar nebula. For example, if the porous anhydrous particles were shown to be cometary and the compact hydrated particles were shown to be asteroidal, then this would give direct insight into the nature of the processes that produced water bearing silicates in the nebula. In addition to the distinction of asteroidal and cometary particles, a major goal of future activity will be to identify possible presolar components that might exist in either particle type. Because comets formed in very cold regions of the nebula it is likely that presolar grains in cometary solids might have survived Solar System processes with minimal alteration.

A key to the distinction of cometary particles from asteroidal ones is the analysis of dust on cometary missions. The recent data on the elemental composition of individual Halley particles (Kissel *et al.* 1986) combined with future elemental, isotopic and morphological analysis on proposed comet rendezvous missions may provide definitive criteria for identifying cometary particulates. The ultimate information source for this work, of course, will be direct cometary sample return, an ESA cornerstone mission. Even when there have been several comet missions, including a sample return mission, the interplanetary particles will still be of value because they are samples of a larger set of parent bodies than could be visited by spacecraft. In addition to the mission results there are other possible methods of distinguishing asteroidal and cometary particles. Sandford (1986) has discussed a technique involving the densities of solar-flare tracks in crystalline grains. Particles spiralling in from the asteroid belt by P–R drag all have exposure times and hence track densities above a minimum value while cometary particles have a broader range of exposure including some that can be very short. Another cosmic-ray exposure technique involves the concentration of ^{10}Be a cosmogenic isotope produced primarily by exposure to galactic cosmic rays. Raisbeck & Yiou (1985) found a ^{10}Be concentration in a single deep-sea sphere that was above the saturation value that occurs at 1 AU and they suggested that the particle was irradiated in the outer Solar System where the galactic cosmic ray flux was less modulated by the solar magnetic field. Other observable properties such as spectral reflectivity might also be useful for matching particle types with parent bodies.

The search for presolar components is an active area in meteorite research and in dust studies. Possible ways of connecting meteoritic materials with interstellar grains are to search for isotopic or spectral features that are unique to interstellar grains. The strongest isotopic link for the laboratory dust samples are the high D/H values for several micrometeorites measured with the ion microprobe by McKeegan *et al.* (1985). The high D/H ratios are suggestive of a link to the ISM because such strong fractionation is unknown in the Solar System but it is common in molecular clouds, where it is believed to be produced by ion–molecule reactions. Spectral links to particular particle types might be made by the analysis of well-known interstellar features in the IR, UV and visual. These include the shape of the 9.7 μm 'silicate feature', the 2175 Å bump, the diffuse bands and an assortment of IR features. High quality IR spectra for individual 10 μm particles have been reported by Sandford & Walker (1985) and Sandford (1986). These authors made an interesting comparison of the IR features in a hydrated micrometeorite with the highly reddened protosellar object W33 A. In addition to a good fit

310
D. E. BROWNLEE

for the silicate feature at 9.7 μm there is also a good correspondence for the features at 6 and
6.8 μm. As an interesting case of scientific synergism, the 6.8 μm feature in the dust particle
was shown to be due to carbonates because electron microscopy showed that carbonates were
abundant in the particle, and that after acid treatment the carbonates and the 6.8 μm feature
were gone. Although this does not prove that the dust in W33 A contains carbonates it suggests
this and it does demonstrate that there are fascinating possibilities for linking purely
astronomical observations with studies of meteoritical materials.

References

Bates, B. 1986 Ph.D. thesis, University of Washington.
Bradley, J. B. & Brownlee, D. E. 1986 *Science, Wash.* **231**, 1542–1544.
Brownlee, D. E. 1985 *A. Rev. Earth planet. Sci.* **13**, 147–173.
Grün, E., Zook, H. A., Fechtig, H. & Giese, R. H. 1985 *Icarus* **62**, 244–272.
Kissel, J. *et al.* 1986 *Nature, Lond.* **321**, 336–337.
McKeegan, K. D., Walker, R. M. & Zinner, E. 1985 *Geochim. cosmochim. Acta.* **49**, 1971–1987.
Raisbeck, G. M. & Yiou, F. 1985 *Meteoritics* **20**, 734–735.
Sandford, S. A. 1986 *Science, Wash.* **231**, 1540–1541.
Sandford, S. A. & Walker, R. M. 1985 *Astrophys. J.* **291**, 838–851.
Steele, I. M., Smith, J. V. & Brownlee, D. E. 1985 *Nature, Lond.* **313**, 297–299.
van der Stap, C. C. A. H., Vis, R. D. & Verheul, H. 1986 *Lunar planet. Sci.* **17**, 1013–1014.
Tomeoka, K. & Buseck, P. R. 1985 *Nature, Lond.* **314**, 338–340.
Wasson, J. T. 1974 *Meteorites.* (316 pages.) New York: Springer Verlag.
Zook, H. A. & MacKay, D. S. 1986 *Lunar planet. Sci.* **17** 977–978.

Discussion

J. DARIUS (*University College, London and Science Museum, U.K.*). Because some of the submicrometre particles Professor Brownlee described have high carbon content, can he distinguish
decisively between anhydrous grains of low exposure age and particles of terrestrial origin such
as volcanic ash?

D. E. BROWNLEE. All of the high-carbon grains mentioned here are constituents of larger
particles that have chondritic elemental abundances. To my knowledge no terrestrial particle
in this size range matches chondritic composition for the elements that are routinely measured
with the electron microscope (Na, Mg, Al, Si, S, Ca, Ti, Cr, Mn, Fe and Ni). Volcanic ash is
collected in the stratosphere following some eruptions and it is usually straightforward to identify
on the basis of composition and morphology.

M. K. WALLIS (*Department of Applied Mathematics and Astronomy, University College, Cardiff, U.K.*).
May I dispute the long-standing supposition that P–R drag removes the interplanetary dust?
  Firstly, lifetimes of 100 μm grains to P–R drag are so long that destructive collisions with
the smaller grains measured by spacecraft are dominant (Leinert *et al.* 1983; Grün *et al.* 1985).
  Secondly, the mean radiation pressure due to geometric asymmetries gives tangential forces
on elongated grains (Voshchinnikov & Il'in 1983) that are much stronger than the P–R drag
calculated for spherical grains, which is small through the 10^{-4} relativistic factor.
  Thirdly, electromagnetic forces on the charged grains and fragments under 1 μm in size cause
mean diffusion out of the Solar System, overwhelming the inwards P–R drag (Wallis 1986).
  So please reorient your thinking away from the P–R transport and loss concept, which neither
fits the interplanetary data nor accords with theory.

Additional references

Leinert, C., Röser, S. & Buitrago, J. 1983 *Astron. Astrophys.* **118**, 345.
Voshchinnikov, N. V. & Il'in, V. B. 1983 *Soviet Astr. Lett.* **9**, 101.
Wallis, M. K. 1986 *Nature, Lond.* **320**, 146.

D. E. BROWNLEE. Lifetimes of micrometre-sized dust in the interplanetary medium are limited by the effects of collisions, radiation pressure drag (the P–R effect) and radiation pressure 'blow out'. Sputtering and Lorentz effects are probably only important in the sub-micrometre region. Collisions have generally been considered to be the dominant destructive effect for particles larger than 100 μm and P–R drag the major effect for the few micrometre to 100 μm range. This is clearly shown in figures 6 and 8 in Grün *et al.* (1985). For 10 μm particles, the size most commonly collected in the stratosphere, the P–R lifetime is a factor of 100 shorter than the collisional lifetime. The P–R lifetimes for this size are consistent with particle exposure ages calculated from the measured densities of solar-flare tracks in collected micrometeorites. If 10 μm particles were only removed by collisions, their million-year exposures would result in track densities high enough to obliterate internal crystal structure. If the anisotropy effect described by Voshchinnikoav & Il'in is important for real interplanetary particles, as well and the cylinders described in their paper, then this effect could considerably effect the orbital evolution of interplanetary dust.

G. W. WETHERILL (*Department of Terrestrial Magnetism, Carnegie Institute of Washington, U.S.A.*). Assuming these particles are derived from comets, what would one expect their dynamical history to be; for example the relative importance of Poynting–Robertson effect, collisional destruction and radiation pressure?

D. E. BROWNLEE. I do not think that we really understand the evolutionary histories of interplanetary dust. If the model of Grün *et al.* (1985) is correct then we would expect that the typical 10 μm cometary-dust particle at 1 AU would be a fragment of a millimetre-sized meteoroid. The model predicts that the meteoroids of less than 100 μm are produced by collisional fragmentation of millimetre-sized particles. The collisional destruction of the larger particles is predicted to occur on a timescale of only 10 000 years. The 10 μm particles liberated by this process would have additional lifetimes of the order of 10 000 years before the P–R drag causes sufficient orbital decay that they are destroyed near the Sun. During orbital decay, their orbits would become partially circularized. The major discrepancy between this model and observations, is the short lifetime of the millimetre-sized particles. The measured cosmic ray exposure times of the millimetre-sized spheres collected from the ocean floor and Greenland are a 100 times the predicted collisional lifetimes. The fragmentation laws used in the model assume that the particles fragment in a manner similar to the large pieces of basalt that have been experimentally well studied with hypervelocity impacts in the laboratory. Real meteoroids may respond quite differently. So far no shock effects have been recognized in collected samples of interplanetary dust.

Phil. Trans. R. Soc. Lond. A **323**, 313–322 (1987)

Printed in Great Britain

313

Stable isotope measurements of meteorites and cosmic dust grains

By C. T. Pillinger

*Planetary Sciences Unit, Department of Earth Sciences, The Open University,
Milton Keynes, MK7 6AA, U.K.*

Although it was well known that a high ^{13}C abundance was a common feature of the spectra of evolved stars, it took over 50 years to find evidence of carbonaceous instellar dust, which might have been ejected from such objects, in the Solar System. However, it is now established that dust probably produced in novae and red giants can be located in primitive meteorites and the latest state of knowledge in respect of such components is reviewed herein. Nitrogen isotopic measurements have been helpful in distinguishing another form of dust that is carbonaceous but does not have a distinctive ^{13}C abundance. Likewise they suggest a non-carbonaceous material (possibly a sulphide) present in the meteorite Bencubbin could be a relict of supernovae outbursts. None of the components seen in meteorites can be detected in deep-sea spheres or stratospheric grains to provide a link between interstellar matter and comets. Until now interstellar dust has been the realm of observing astronomers and theoretians; stable isotope measurements are responsible for recognizing a material which it should be possible to isolate and study in the laboratory.

Introduction

Within weeks of the discovery of ^{13}C by optical spectrographic means (King & Birge 1929) astronomical investigations of the relative abundance of carbon isotopes had begun. The spectra of carbon-rich stars were soon found to contain lines corresponding to both ^{12}C–^{13}C (Sandford 1929) and ^{13}C–^{13}C bonding (Menzel 1930); the latter immediately demonstrating the very high ^{13}C abundance in environments beyond the Solar System. In fact, the major carbon components in terrestrial, lunar samples and bulk meteorites all have ^{12}C/^{13}C ratios close to 89 (table 1) whereas, because of nuclear processing via the CNO cycle, values for late-type stars can be as low as 4 although high values (greater than 100) are also possible. Thus it is obvious that interstellar gas and dust ejected from evolved stars ought to contain components of characteristically high ^{13}C abundance. Indeed, in the interstellar medium as a whole, the relative proportions of ^{12}C and ^{13}C are between 60 and 70 to 1 (Wannier 1980; Penzias 1980).

Among the first astronomical objects considered for the presence of ^{13}C was Halley's Comet, the ^{13}C–^{12}C line being found in spectra recorded during the 1910 apparition (Bobrovinikoff 1930). Although no isotopic composition was estimated, the author did suggest the need for similar studies on meteorites to determine any connection between these and comets. Appropriate meteorite data are now available but attempts to obtain well-defined isotopic compositions from comets are few and very imprecise (Vanysek & Rahe 1978); the studies performed seem to point towards an abundance of ^{13}C that is less than the Solar System as a whole (table 1). Measurements made for cosmic dust grains thought to be the terrestrial remains of cometary material (D. Brownlee, this symposium) will be discussed later in this paper.

[65]

TABLE 1

(Source: Vanysek (1977) and Vanysek & Rahe (1978).)

object	$^{12}C/^{13}C$
Earth	89^{+10}_{-1}
Moon	89^{+3}_{-2}
bulk meteorites	$89^{+3}_{-0.5}$
comets	
Ikeya 1963 I	70 ± 15
Tayo–Sato–Kosaka 1969 IX	100 ± 20
Kohoutek 1973 XII	115^{+30}_{-20}
	135^{+60}_{-45}
Kobayshi–Berger–Milon 1975	110^{+20}_{-30}

Trofimov (1950) seems to have been the first investigator to consider the possibility of finding variations in the isotopic composition of meteoritic carbon to parallel the known wide-scale distribution in stars. The spread encountered, 89.6–91.8, was even narrower than known for terrestrial samples at the time (89.0 marine carbonate, 92.4 petroleum) thus Trofimov was forced to conclude that terrestrial matter and the 35 meteorites he studied 'belonged to a single system of creation and a single nuclear genesis'. However, once isotopic anomalies[†] had been discovered for other elements, e.g. oxygen (Clayton *et al.* 1972) and noble gases (Reynolds & Turner 1964; Black & Pepin 1969; Lewis *et al.* 1975) it was inevitable that they would eventually be found for carbon. Nevertheless, it was still another decade before material with $^{13}C/^{12}C$ ratios high enough to be beyond the reach of fractionation mechanisms at realistic temperatures was discovered (Swart *et al.* 1983 a). These values when considered together with isotopic measurements made for noble gases and nitrogen, thought to be trapped within the host carbon, most strongly imply that dust grains from various stellar sources can be isolated from primitive meteorites to provide an opportunity for laboratory studies as a compliment to astronomical observations.

EXPERIMENTAL METHODS

Of paramount importance in the discovery of interstellar dust grains in meteorites has been the application of stepped combustion extraction procedures. In this technique, samples are subjected to a sequence of increasing temperatures in the presence of molecular oxygen. There are several early literature references to burning extraterrestrial specimens in two or three stages (Briggs & Ketto 1962; Otting & Zahringer 1967; DesMarais 1978) but Ott *et al.* (1981) were the first authors to increase combustion temperatures systematically as a means of obtaining carbon dioxide for carbon isotopic measurement. Unfortunately, the progressive rise of $\delta^{13}C$[‡] from $-34.1\permil$ at 200 °C to $-4.1\permil$ at 1100 °C, during the analysis of an acid residue from the Allende meteorite, was dismissed as kinetic isotope fractionation. Swart *et al.* (1982),

---

[†] An isotopic anomaly is normally defined as a composition that cannot be accredited to mixing or fractionation and thus is attributable to a specific process of nucleosynthesis. The fact that non-mass-dependent isotopic effects occur during photochemical dissociation of oxygen (Thiemens & Heidenreich 1983) is ignored herein, because so far it has not been demonstrated that such reactions have a role in terms of carbon and nitrogen isotopic systematics in meteorites. This may, however, change.

[‡] $\delta(\permil) = 1000 \dfrac{[(H/L)_{sample} - (H/L)_{standard}]}{(H/L)_{standard}}$, where H represents heavy isotope; L represents light isotope.

[66]

however, showed that different carbon phases burned at specific temperatures without an isotopic fractionation of greater than $2\%_0$. A compilation of stepped combustion data for various standard materials is given as figure 1. The temperature at which elemental carbon combusts is dependent upon its crystallinity with amorphous carbon being least stable and diamond most (Swart *et al.* 1983 *b*; Grady *et al.* 1985). No systematic investigation into particle

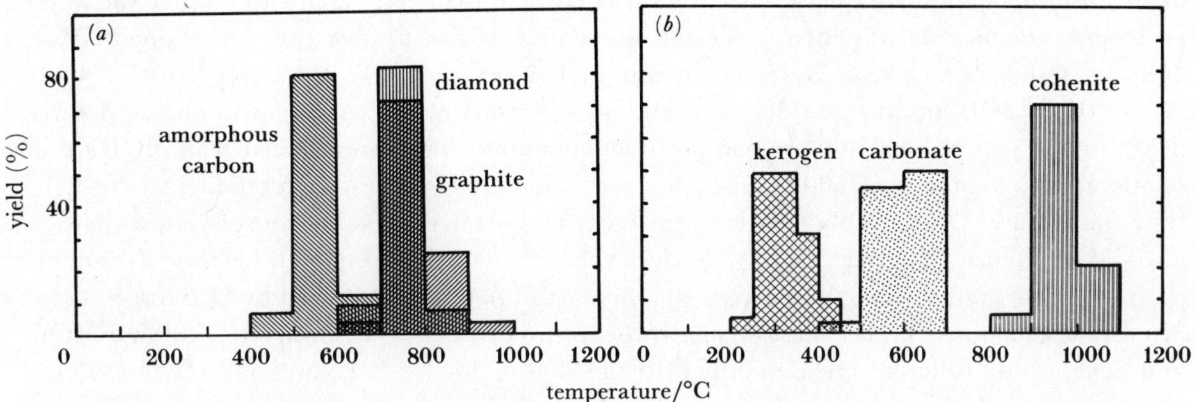

FIGURE 1. Carbon release during combustion. (*a*) Elemental forms of carbon; (*b*) carbon compounds.

size has yet been attempted to establish the effect of surface area, although intuitively this must be important. Organic compounds are particularly unstable in the presence of oxygen, oil source rock kerogens being predominently destroyed at 450 °C (Gilmour & Pillinger 1985) although a continuum exists for samples of increasing maturity, which may be considered as being progressively more graphitized (I. Gilmour, C. T. O'Donnell & C. T. Pillinger, unpublished results). The most stable compounds appear to be carbides: the meteoritic carbide cohenite surviving until at least 1000 °C (Grady *et al.* 1986). More refractory carbides have not been tested but are likely to be even more stable, bearing in mind that SiC is commonly used as a heating element in furnaces designed for use up to 1600 °C.

It must be stressed that the data in figure 1 are for pure substances. Inevitably 'protection' in a mineral matrix could prevent ignition until dissociation of the mineral or diffusion to the surface took place. Equally catalytic effects involving alkali metals (Bergman 1972) or promoted combusion of one phase coexisting with another by exothermic reaction (G. Eglinton, personal communication) cannot be ruled out when dealing with meteorites or other geological materials. Nevertheless high-resolution stepped combustion (less than 100 °C temperature increments) is vital in two respects in distinguishing interstellar dust components. At low temperatures it is capable of removing (i) terrestrial contaminants introduced into meteorites during museum storage and in sample handling procedures and (ii) indigenous macromolecular (kerogen-like) material, which constitutes the vast majority of the carbon budget of primitive meteorites. The improvement of mass-spectrometric measurement techniques (Carr *et al.* 1986*a*) has meant that stepped combustion can be conducted with much smaller steps (frequently 25 °C) to provide better information as to the nature of the carbon involved. The limit of the experiment presently is as much time as sample-availability.

Interstellar dust components in meteorites

Interstellar carbon of high ^{13}C abundance was discovered (Swart *et al.* 1983*a*) as a result of stepped combustion experiments on highly processed residues obtained from the Murchison meteorite for the purpose of identifying exotic noble-gas components (Alaerts *et al.* 1980). Firstly the meteorite was demineralized with HCl and HF to leave only carbonaceous, and probably oxide and minor sulphide, phases and then treated with oxidizing agents to remove the latter and some organics. It was then separated according to size to give three fractions 2C1Oc, 2C1Om and 2C1Of (coarse, 3–10 μm; medium, 1–3 μm; and fine, 1 μm respectively). Only two of these, 2C1Om and 2C1Of, were studied by Swart *et al.* (1983*a*) who observed three peaks in the sequence of isotopic compositions measured for gas extracted from 2C1Om at temperatures greater than 500 °C. Because two anomalous noble-gas components, Ne-E(L) (essentially pure ^{22}Ne, possibly derived from decay of short lived ^{22}Na, residing in a low-density, low-stability phase (Eberhardt 1974; Anders 1981)) and s-Xe (so-named because its isotopic pattern (Srinivasan *et al.* 1978) matches the theoretical pattern calculated by Clayton & Ward (1978) for xenon produced by a slow-neutron-capture process) were known to exist in 2C1Om and have grossly different release temperatures – 600 °C for Ne-E(L) and 1400 °C for s-XE – it was considered plausible that the low- (600 °C) and high- (1200 °C) temperature ^{13}C peaks might correspond to decomposition of the respective host phases. A peak δ^{13}C value of $+1100\text{‰}$ (^{12}C/^{13}C = 42) measured at 1000 °C in the 2C1Of, known to contain relatively more s-Xe than 2C1Om, supported this interpretation. The view is further endorsed by the circumstantial evidence that the s-process is a feature of red-giant stars characterized by a ^{12}C/^{13}C ratio as low as 4 and that novae thought to the direct provenance of ^{22}Ne or its precursor ^{22}Na are ^{13}C-enriched. Thus the carbon phases were dubbed Cα and Cβ according to nomenclature previously introduced by Anders and his co-workers (see Anders 1981).

Swart *et al.* (1983*a*) also obtained isotopic measurements for another carbon phase found in 2C1Of. This is the carbon termed Cδ, which burns from 400–500 °C and is thought to be the host of a noble gas component, originally named carbonaceous chondrite fission (ccf) -Xe by Reynolds & Turner (1964) and suggested to be from an extinct superheavy element (Anders & Heymann 1969). The same component is also referred to as DME-Xe, but its currently accepted nomenclature is Xe(HL) because of the distinctive enrichment in both heavy and light isotopes. The δ^{13}C values measured for Cδ from Murchison (-38‰) and another meteorite, Allende (-31‰) do nothing to betray that Cδ might be an interstellar dust component. However, nitrogen isotopic compositions obtained on the same samples show ^{14}N/^{15}N ratios up to 406 compared to the value of 272 measured for air and suggest that stellar synthesis, possibly in a supernova, is involved. The host carbon and Xe(HL) are likely to be from the same supernova source.

The existence of heavy carbon in carbonaceous chondrites has been confirmed by several other groups using stepped combustion (Kerridge 1983; Yang & Epstein 1984; Halbout *et al.* 1986). Only the latter authors, however, have succeeded in obtaining data to suggest that more than one component exists. Ion-probe studies (Niederer *et al.* 1985; Zinner & Epstein 1986; E. Zinner personal communication) have raised the maximum δ^{13}C value obtained to *ca.* 7000‰. Work with ion probes is only just beginning and it is too early to know whether this technique will be able to distinguish between different forms of interstellar carbon (see below). So far there is no independent confirmation concerning Cδ probably because no other conjoint carbon, nitrogen, noble gas studies have been made.

Carbon alpha

Work by my group in collaboration with the University of Chicago has continued and we are now able to refine the interpretation made by Swart *et al.* (1983*a*). An important facet in confirming that low temperature, heavy carbon is indeed Cα the host of NeE(L) has been the stepped combustion of 2C1Oc a residue not studied by Swart *et al.* (1983*a*). The major heavy carbon component in this sample is combustible over the well defined temperature interval 600–700 °C with a maximum δ^{13}C of $+338\%_0$ and no δ^{13}C values greater than $+78\%_0$ at higher temperatures (Carr *et al.* 1983). Residue 2C1Oc is known to contain neon with a high ^{22}Ne/^{20}Ne and it is considered that Ne-E is present in rather pure form (Alaerts *et al.* 1980) albeit as both Ne-E(L) and Ne-E(H). S-process xenon is only found in 2C1Oc in minor amounts.

It was possible to resolve Cα in 2C1Oc only because the sample was relatively free of isotopically normal polymer or amorphous carbon that could be burnt off below 600 °C. Residue 2C1Oc was in fact only 2 % C (by mass); the majority of the sample being spinel, chromite and hibonite. The same material was used for the ion probe study of Niederer *et al.* (1985) who noted that single aluminium and oxygen-rich grains had the highest ^{13}C (δ^{13}C $\approx$ 2000–4000$\%_0$) abundance; carbon-rich grains were of normal terrestrial isotopic composition. Similarly Zinner & Epstein (1986) only found high δ^{13}C values (up to 2400$\%_0$) by ion probe in a spinel–hibonite–chromite fraction after they had removed isotopically normal carbon using an RF discharge. The isotopic composition measured by gas-source mass spectroscopy for 2C1Oc is clearly only a minimum value for Cα. It is perhaps more appropriate to estimate its ^{12}C/^{13}C as less than 11 based on the most recent unpublished ion-probe data (E. Zinner personal communication) even though this is subject to comparatively large errors.

From the combustion temperature Cα is likely to be some form of disordered graphite that burns at temperatures 100 °C below the spectrographically pure sample used as a standard material. It is not amorphous carbon because 2C1Oc is a coarse fraction although its coarseness could be dictated by the oxide minerals. However, Eberhardt *et al.* (1981), using non-chemical separation techniques, established convincingly that Ne-E in Orgueil does not reside in colloidal material. Given than Ne-E(H) is thought to be trapped in spinel (Lewis *et al.* 1979), it is tempting to postulate that the real carrier of this is Cα enclosed within the oxide mineral, i.e. not available to oxygen during combustion.

Carbon beta

A repeat analysis of 2C1Of that uses a greater number of temperature steps (Carr *et al.* 1983) showed that there were two peaks in the pattern of δ^{13}C values obtained above 600 °C. The first occurs at 850 °C ($+624\%_0$) and the second 100 K higher ($+950\%_0$). The presence of the lower-temperature peak has been confirmed by even higher-resolution studies (D. W. McGarvie *et al.*, unpublished results) and data from Orgeuil (M. M. Grady *et al.*, unpublished results) and the enstatite chondrites Qingzhen and Indarch (Carr *et al.* 1984). D. W. McGarvie (unpublished results) further resolved the high-temperature peak in δ^{13}C values measured for 2C1Of into a doublet, which has subsequently been seen for high density fractions from acid residues of the Murray carbonaceous chondrite (Ming *et al.*, unpublished results). Cβ is clearly not a simple component. It is possible that the peak values in the carbon isotopic composition at different temperatures reflect a single carbon phase being released from separate sites (trapped inside minerals ?). Alternatively there could be several components that have different combustion temperatures because of variation in particle size, degree of crystallinity, etc. If the latter is true, deciding which is the authentic Cβ, host of s-Xe, will require considerable effort, possibly involving physical separation of three markedly similar carbon phases. It will be interesting to establish whether each of the carbon components has a separate noble gas isotopic signature.

A major problem concerning the identity of Cβ is that much of the carbon combusts at temperatures which are higher than those found with standard materials such as graphite or diamond. This would seem to favour the hypothesis that the carbon is shielded by a mineral phase and combustion temperature is a site-dependent phenomenon. There are forms of elemental carbon that have not yet been the subject of high resolution stepped combustion experiments. Londsdaleite (hexagonal carbon), which is observed in iron meteorites (Frondel & Marvin, 1967), where it is thought to have been shock produced, is one possibility, especially because this has more recently been recognized in stratospheric dust grains (Reitmeijer & MacKinnon 1985). Carbyne, once thought to be implicated as an anomalous noble gas host phase (Whittaker *et al.* 1980) could stand reinvestigation and the recently discovered C60 (Kroto *et al.* 1985) is worthy of consideration. It cannot be discounted that graphite made from a very high proportion of ^{13}C atoms (^{12}C/^{13}C $\approx$ 4) is thermodynamically more stable than terrestrial graphite (^{12}C/^{13}C $\approx$ 89).

Carbon delta

Xenon(HL), the anomalous gas associated with Cδ is ubiquitous in primitive meteorites having been identified in CI, CM2, CV3, CO3, H, L, LL and E chondrites (Anders 1981). Because Cδ is very fine-grained and known to undergo combustion in the temperature range 400–600 °C, it is instructive to consider δ^{13}C values measured for various samples in that interval (see figure 2). Isotopic data (M. M. Grady *et al.*, unpublished results) for an acid residue from the CI meterorite Orgeuil (Lewis *et al.* 1983) and a bulk sample of the enstatite chondrite South Oman (Grady *et al.* 1986) are indicative that Cδ is isotopically lighter than at first believed (Swart *et al.* 1983*a*). The δ^{15}N of $-72\%_0$ found for nitrogen released from South Oman in the 400–600 °C temperature interval suggests that Cδ is involved. Thus it may be easier to separate Cδ from enstatite chondrites, where the bulk of the normal carbon is graphite, than it was in the case of Allende, where the major phase was amorphous carbon of δ^{13}C $= -17\%_0$ (Swart *et al.* 1982).

Isotope studies on the Bencubbin meteorite

It is possible that not all the interstellar dust trapped in meteorites is carbonaceous in nature. Recently, interest in the Bencubbin meteorite was aroused by measurements made on gram-sized specimens, which suggested that the bulk sample is enriched in the heavy isotope of nitrogen by a factor of two (Prombo & Clayton 1985). This was the first recognition of an isotopic anomaly involving a light element in a meteorite that was not a primitive chondrite (Bencubbin is a polymict breccia consisting of recrystallized kamacite and silicates together with a variety of chondritic xenoliths welded in a shock melted metal–silicate matrix (Newsom & Drake 1979)). Almost all the nitrogen is uniformly enriched in ^{15}N; it is not confined to the matrix or the xenoliths. In fact, the latter have the lowest δ^{15}N values and may have acquired their nitrogen in a secondary event. Stepped combustion experiments performed on material separated from a sample of metal clasts and matrix (Franchi *et al.* 1986) suggest that the heavy

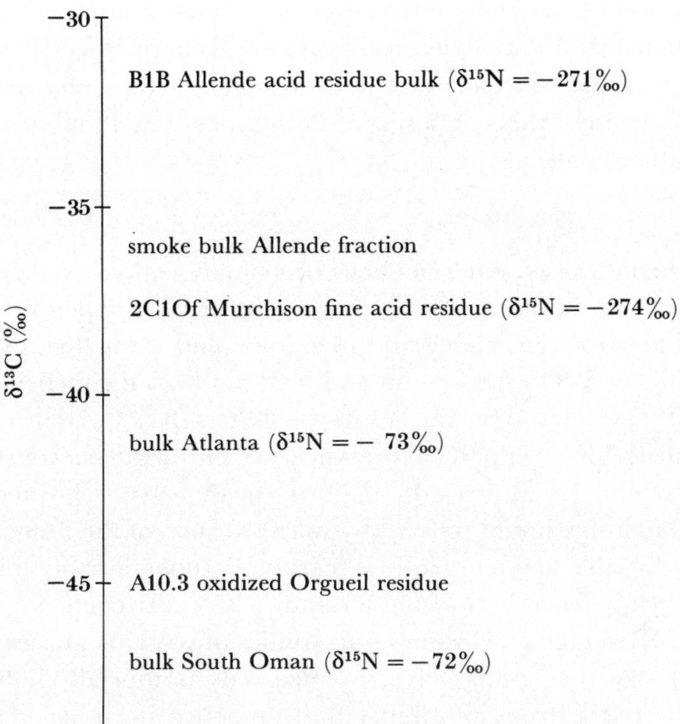

FIGURE 2. Isotopic composition of the combustion of carbon in the temperature range 400–600 °C.

nitrogen (max. δ^{15}N = +1033‰) is located in two components both of which are insoluble in 6 M HCl but only one of which survives treatment with 12 M HCl or 28 M HF. It is very unlikely that the component degraded by strong acid is carbonaceous. Carbon analyses suggest that if the more stable one is carbonaceous then its C:N ratio is unusually low (less than 4). Furthermore, the carbon is not distinguished by an unduly anomalous isotopic composition (δ^{13}C $\approx$ +25‰). A novel chromium-rich sulphide recognized during petrological examination (Newsom & Drake 1979) cannot be ruled out as a possible nitrogen host; the 6 M HCl residue contains 1.7% sulphur (by mass). If sulphides are implicated they could have been produced in a supernova evolved to the silicon-burning state (Clayton & Ramadurai 1977).

Studies of cosmic dust grains

The high δ^{13}C values measured at combustion temperatures above 600 °C for fractions isolated from various primitive meteorites provide compelling evidence for the survival of interstellar grains in the Solar System. With the benefit of hindsight it has proven possible to detect heavy carbon in bulk samples (Halbout *et al.* 1986) or those subject to a minimum of processing (McGarvie *et al.* 1985, 1987). It would be of great significance for our understanding of comets to recognize similar isotopic signatures for materials thought to have derived from cometary sources. Although all micrometeorites are heated to some extent during deceleration in the Earth's atmosphere, such a process should only serve to remove low-temperature organic matter (analogous to organic contamination or meteorite polymer). If any carbon survives, it ought to be the carbon that combusts at high temperature, therefore, stepped combustion experiments of materials of possible cometary origin could be highly informative. A significant fraction of deep-sea spherules may be melted cometary material although a large proportion are more likely to be meteorite ablation debris (Murray & Renard 1883), however, there seems to be unanimity that grains collected in the stratosphere (Brownlee particles) are from cometary sources (D. Brownlee, this symposium). Both types of material have been the subject of carbon isotope measurements.

Deep-sea spherules

A total of six individual deep-sea spherules have been analysed by stepped combustion (Carr *et al.* 1984; Carr 1985). Four of these had clearly been heated to temperatures high enough to cause some melting but two were chondritic in nature and unmelted. Although both the unmelted grains had higher $\delta^{13}C$ values (-16 and $-21\%_0$) than the melted samples, fractions between 600 and 1200 °C were isotopically lighter (-30 to $-34\%_0$). None of the particles can be said to exhibit evidence of a high ^{13}C abundance, which might betray the existence of an interstellar dust component. In all cases the majority of the carbon burned at temperatures below 600 °C with up to a maximum of 6% (by mass) for one of the unmelted samples. The $\delta^{13}C$ value ($-18\%_0$) for the low-temperature carbon is uncommonly high for terrestrial contamination, in our experience between -25 and $-35\%_0$. It could be contamination by marine organisms that have high $\delta^{13}C$ values but, studies of volcanic glasses of deep sea origin (Mattey *et al.* 1984; Exley *et al.* 1986) have revealed only isotopically light organics at low temperature. Thus, there is a strong possibility that unmelted deep-sea spherules have some indigenous carbon which needs to be further explored. It is noteworthy that the isotopic value of $-18\%_0$ is close to that measured for macromolecular material in carbonaceous chondrites (Smith & Kaplan 1970).

Stratospheric particles

Because of their size and friable nature, stratospheric particles present experimental difficulties not encountered with deep-sea spherules. Procedures involving silicone oil are used to ensure that a stratospheric grain adheres to a suitable carrier (platinum foil) for transfer to the combustion system. Unfortunately the silicone oil burns over a wider temperature range than normal laboratory contamination but nevertheless its distinctive light isotopic signature (*ca.* $-40\%_0$) would allow it to be distinguished from heavy carbon of interstellar origin. Isotopic measurements on refractory components in carbonaceous chondrites show that anomalous (high $\delta^{13}C$) compositions may be detected successfully on samples as small as 150 pg. However, when five stratospheric grains, *ca.* 10 µm in diameter with a calculated carbon content of 2 ng, were subjected to stepped combustion, there was no indication of a $\delta^{13}C$ in excess of $-30\%_0$ at temperatures above 600 °C (Carr *et al.* 1986*b*). The bulk isotopic value measured for the grains (plus the silicone oil on platinum foil) was $-37.57\%_0$, marginally heavier than a blank analysis performed in duplicate ($-39.15\%_0$, $-39.20\%_0$). Thus, within the limits of detection of the experiment, it was impossible to recognize the existence of isotopically heavy carbon in stratospheric particles. The results are in keeping with ion-probe measurements reported by McKeegan *et al.* (1985) whose carbon isotopic data for fragments from three individual particles are given in table 2. Although ion-probe studies for carbon have not been refined to a high level of precision, the extreme difference between the particles named Skywalker and Mosquito is believed to be real. None of the values obtained is remotely near the high ^{13}C determined by ion probe for meteorite residues (Niederer *et al.* 1985; Zinner & Epstein 1986).

TABLE 2. $\delta^{13}C$ AND δD MEASUREMENTS FOR STRATOSPHERIC PARTICLES FROM McKEEGAN ET AL. (1985)

(Average errors on an individual measurement: $\delta^{13}C \pm 17.0\text{‰}$, $\delta D \pm 125$ to 520‰.)

particle	$\delta^{13}C$ (‰) (number of measurements)	δD (‰) (number of measurements)
Mosquito	-0.2 (5)	$+125$ to $+2534$ (8)
Skywalker	-38.9 (9)	$+51$ to $+696$ (26)
Calrissian	-14.1 (5)	$+373$ to $+2196$ (7)

In summary, isotopic measurements made on material that could derive from comets seem to agree with the spectroscopic data rather than match values obtained for refractory components from primitive meteorites.

I thank Dr R. H. Carr, Dr M. M. Grady, Dr D. W. McGarvie and Dr I. P. Wright for access to their unpublished data, and Professor E. Anders, Dr R. S. Lewis and Mr T. Ming for permission to use unpublished results. This work was supported by the Science and Engineering Research Council.

REFERENCES

Alaerts, L., Lewis, R. S. & Anders, E. 1980 *Geochim. cosmochim. Acta* **44**, 189–209.
Anders, E. & Heymann, D. 1969 *Science, Wash.* **164**, 821–824.
Anders, E. 1981 *Proc. R. Soc. Lond.* A **374**, 207–238.
Bobrovinikoff, N. T. 1930 *Publ. astr. Soc. Pacif.* **42**, 117–124.
Bergman, I. 1972 *Fuel, Lond.* **51**, 116–119.
Black, D. C. & Pepin, R. O. 1969 *Earth planet. Sci. Lett.* **6**, 395–405.
Briggs, M. H. & Ketto, J. 1962 *Nature, Lond.* **193**, 1126–1127.
Carr, R. H. 1985 Ph.D. thesis, University of Cambridge.
Carr, R. H., Wright, I. P., Pillinger, C. T., Lewis, R. S. & Anders, E. 1983 *Meteoritics* **18**, 277.
Carr, R. H., Carr, L. P., Wright, I. P., Pillinger, C. T. & Crabb, J. 1984 *Lunar planet. Sci.* **15**, 133–134.
Carr, R. H., Wright, I. P., Joines, A. W. & Pillinger, C. T. 1986a *J. Phys. E.* **19**, 798–808.
Carr, R. H., Gibson, E. K. Jr, Rietmeijer, F. J. M., Grady, M. M., Wright, I. P. & Pillinger, C. T. 1986b *Meteoritics.*
Clayton, D. D. & Ward, R. 1978 *Astrophys. J.* **224**, 192–193.
Clayton, D. D. & Ramadurai, S. 1977 *Nature, Lond.* **265**, 427–428.
Clayton, R. N., Grossman, L. & Mayeda, T. K. 1972 *Science, Wash.* **182**, 485–488.
DesMarais, D. J. 1978 In *Proc. 9th Lunar Planet. Sci. Conf.* pp. 2451–2467. Pergamon.
Eberhardt, P. 1974 *Earth planet. Sci. Lett.* **24**, 182–187.
Eberhardt, P., Jungck, M. H. A., Meier, F. O. & Niederer, F. R. 1981 *Geochim. cosmochim. Acta* **45**, 1515–1528.
Exley, R. A., Mattey, D. P., Clague, D. A. & Pillinger, C. T. 1986 *Earth planet. Sci. Lett.* **78**, 189–199.
Franchi, I. A., Wright, I. P. & Pillinger, C. T. 1986 *Nature, Lond.* **323**, 138.
Frondel, C. & Marvin, U. B. 1967 *Nature, Lond.* **214**, 587–589.
Gilmour, I. & Pillinger, C. T. 1985 *Org. Geochem.* **8**, 421–426.
Grady, M. M., Wright, I. P., Swart, P. K. & Pillinger, C. T. 1985 *Geochim. cosmochim. Acta* **49**, 903–915.
Grady, M. M., Wright, I. P., Carr, L. P. & Pillinger, C. T. 1986 *Geochim. cosmochim. Acta* **50**, 2799–2813.
Halbout, J., Mayeda, T. K. & Clayton, R. N. 1986 *Earth planet. Sci. Lett.* **80**, 1–18.
Kerridge, J. F. 1983 *Earth planet. Sci. Lett.* **64**, 186–200.
King, A. S. & Birge, R. T. 1929 *Nature, Lond.* **124**, 127.
Kroto, H. W., Heath, J. R., O'Brien, S. C., Curl, R. F. & Smalley, R. E. 1985 *Nature, Lond.* **318**, 162.
Lewis, R. S., Srinivasan, B. & Anders, E. 1975 *Science, Wash.* **190**, 1251–1262.
Lewis, R. S., Alaerts, L., Matsuda, J. & Anders, E. 1979 *Astrophys. J.* **234**, L165–L168.
Lewis, R. S., Grady, M. M., Wright, I. P., Pillinger, C. T. & Fallick, A. E. 1983 *Meteoritics* **18**, 340.
Mattey, D. P., Carr, R. H., Wright, I. P. & Pillinger, C. T. 1984 *Earth planet. Sci. Lett.* **70**, 196–206.
McGarvie, D. W., Grady, M. M., Gibson, E. K. Jr & Pillinger, C. T. 1985 *Meteoritics* **20**, 708.

McGarvie, D. W., Wright, I. P., Grady, M. M., Pillinger, C. T. & Gibson, E. K. Jr 1987 In *Proc. 11th Symposium on Antarctic Meteorites, Japan.*
McKeegan, K. D., Walker, R. M. & Zinner, E. 1985 *Geochim. cosmochim. Acta* **49**, 1971–1987.
Menzel, D. H. 1930 *Publ. astr. Soc. Pacif.* **42**, 34–36.
Murray, J. & Renard, A. F. 1891 *Challenger report for 1872–1876.* H.M. Stationery Office.
Newsom, H. E. & Drake, M. J. 1979 *Geochim. cosmochim. Acta* **43**, 689–707.
Niederer, F. R., Eberhardt, P. & Geiss, J. 1985 *Meteoritics* **20**, 716–717.
Ott, U., Mack, R. & Chang, S. 1981 *Geochim. cosmochim. Acta* **45**, 1751–1788.
Otting, W. & Zahringer, J. 1967 *Geochim. cosmochim. Acta* **31**, 1949–1960.
Penzias, A. A. 1980 *Science, Wash.* **208**, 663.
Prombo, C. A. & Clayton, R. N. 1985 *Science, Wash.* **230**, 935–937.
Reitmeijer, F. J. M. & MacKinnon, I. D. R. 1985 *Nature, Lond.* **315**, 733.
Reynolds, J. H. & Turner, G. 1964 *J. geophys. Res.* **69**, 3263–3281.
Sandford, R. F. 1929 *Publ. Astr. soc. Pacif.* **41**, 271–272.
Smith, J. W. & Kaplan, I. R. 1970 *Science, Wash.* **167**, 1367–1370.
Srinivasan, B. & Anders, E. 1978 *Science, Wash.* **201**, 51–56.
Swart, P. K., Grady, M. M. & Pillinger, C. T. 1982 *J. geophys. Res.* **87**, A283–A288.
Swart, P. K., Grady, M. M., Pillinger, Lewis, R. S. & Anders, E. 1983*a Science, Wash.* **220**, 406–410.
Swart, P. K., Grady, M. M. & Pillinger, C. T. 1983*b Meteoritics* **18**, 137–154.
Thiemens, M. H. & Heidenreich, J. E. 1983 *Science, Wash.* **219**, 1073–1075.
Trofimov, A. V. 1950 *Dokl. Akad. Nauk SSSR* **72**, 663–666.
Vanysek, V. 1971 In *Comets, asteroids and meterorites* (ed. A. H. Delsemne), 499–503. University of Toledo.
Vanysek, V. & Rahe, J. 1978 *Moon Planets* **18**, 441–446.
Wannier, P. G. 1980 *A. Rev. Astr. Astrophys.* **18**, 399–437.
Whittaker, A. G., Watts, E. J., Lewis, R. S. & Anders, E. 1980 *Science, Wash.* **209**, 1512–1514.
Yang, J. & Epstein, S. 1984 *Nature, Lond.* **311**, 544–547.
Zinner, E., Fahey, A. & McKeegan, K. 1986 *Terracognita* p. 129.
Zinner, E. & Epstein, S. 1986 *Lunar planet. Sci.* **17**, 967–968.

Discussion

SIR WILLIAM McCREA, F.R.S. (*Astronomy Centre, University of Sussex, Brighton, U.K.*). Dr Pillinger speaks of material of extra-Solar System composition, which apparently he finds in very small inclusions in the meteoritic matter he has described. One asks (*a*) how might the material have reached the sites where it is found, (*b*) whether any particles have been found that are composed entirely of such material and that are not inclusions within bodies having otherwise Solar-System composition.

C. T. PILLINGER. Our interpretation of the heavy carbon we find in primitive meteorites is that it is intersteller dust condensed from the expanding shells of evolved stars. However, we have yet to isolate a pure sample; all the material we study is contaminated by phases which were produced in the solar nebula. We could therefore conceive that the interstellar grains are acting and 'seed' crystals. The only source of the isotopically heavy carbon known at present is meteorites but obviously a very logical place to look would be comets, where it could easily be present in greater abundance and variety.

M. ARNOULD (*Institut d'Astronomie, Université Libre de Bruxelles, Brussels, Belgium*). The differential irradiation of portions of the pristine Solar System matter by an intense flux of energetic particles from the young Sun has sometimes been advocated to account for at least some of the observed anomalies. An interesting test of that model would be provided by the search for lithium isotopic anomalies in meterorites. Could such a search be conducted again with the very powerful experimental techniques available today?

C. T. PILLINGER. It should be possible to search for lithium isotopes with an ion probe.

Phil. Trans. R. Soc. Lond. A **323**, 323–337 (1987)

Printed in Great Britain

323

Dynamical relations between asteroids, meteorites and Apollo–Amor objects

By G. W. Wetherill

Department of Terrestrial Magnetism, Carnegie Institution of Washington, Washington, D.C. 20015, U.S.A.

A Monte-Carlo technique has been used to investigate the orbital evolution of asteroidal collision debris produced interior to 2.6 AU. It is found that there are two regions primarily responsible for production of Earth-crossing meteoritic material and Apollo objects. The region adjacent to the 3:1 jovian commensurability resonance (2.5 AU) is unique in providing material in the required quantity and orbital distribution of the ordinary chondrites. This region should also supply a comparable preatmospheric flux of carbonaceous meteorites. The innermost asteroid belt (2.17–2.25 AU), via the v_6 secular resonance, provides a flux *ca.* 9 % that of the ordinary chondrites, and appears to be the strongest candidate for the basaltic achondrite source region. It is unlikely that a significant number of meteorites originate beyond 2.6 AU. It is speculated that enstatite achondrites are derived from the Hungaria region, interior to the main belt, and that iron and stony-iron meteorites originate from many main-belt sources interior to 2.6 AU.

1. Introduction

About 2500 years ago an unpleasant Greek named Heraclitus irritated his erudite contemporaries in a number of ways (Russell 1945). One of these was by telling them that their search for the eternal was a waste of effort, because in fact everything was changing all the time. This dialogue concerning the fundamental priority of substance against process has continued ever since. Those of us who try to understand the history of the Universe, Solar System, and the Earth are the intellectual descendants of Heraclitus. Although the Solar System and its inhabitants will not pass this way again, we still want to know where we have been.

Those of us who serve on too many NASA advisory committees have spent much time providing stirring prose to support the view that 'primitive bodies', the real comets and asteroids of our Solar System, will provide the clues to understanding its earliest days. Actually, this is true. It will be worthwhile not to always think of meteorites as samples of hypothetical 'parent bodies' in the solar nebula, but rather simply as rocks broken from outcrops on real asteroids. Increasingly, we have become aware of the dramatic difference between samples of rocks from planetary bodies, even those as small as the moon, and those from asteroidal objects. The hallmark of a planetary rock is its complex geochemical history, extending to times late in Solar System history. Even though the 4500 Ma age of the Solar System can be inferred from terrestrial and lunar rocks, and the probably martian SNC meteorites, this event is heavily veiled by subsequent events. In contrast, as expected from calculations of thermal evolution of small bodies, the radiometric date of *ca.* 4500 Ma dominates the chronology of all other meteorites, even when these objects have undergone igneous differentiation (basaltic achondrites) or late shock metamorphism (hypersthene chondrites).

[75]

This paper will describe recent work that seeks to identify distinct portions of the asteroid belt as source regions for the most abundant type of stony meteorite (ordinary chondrites) and the majority of differentiated stony meteorites. Many of the Earth-approaching Apollo–Amor objects of 1 km in diameter are members of the same 'collision hierarchy' of asteroidal debris produced in these same regions. For this reason the distinction between meteorites produced in the asteroid belt and those produced as collision fragments of Apollo objects is not a fundamental one. The question of spectrophotometric identification of belt asteroids with their meteoritic and Apollo-asteroid collision fragments is at present unresolved, however.

With significantly more uncertainty, asteroidal source regions of other types of meteorites can be proposed. The identification of at least most meteor showers with periodic comets is well established. It also seems likely that a large number of the Apollo–Amor objects are related to comets rather than to asteroids. It is much less certain whether any survivable meteorites are of cometary origin, but fireball data suggest that this possibility should be considered. Micrometeorites of both cometary and asteroidal origin must have been collected, but their relative proportions are at present unclear.

2. Production of ordinary chondrites in the vicinity of the 3:1 Kirkwood gap

^{39}Ar–^{40}Ar and Rb–Sr ages of ordinary chondrites provide clear evidence for collisional shocks throughout Solar System history (Bogard 1979). This indicates an asteroidal, rather than a cometary, source region. If this is accepted, one can then turn to the question of identifying the specific region in the asteroid belt from which these meteorites are derived. As discussed in detail elsewhere (Wetherill 1985), the distribution of ordinary chondrite orbits, as inferred from fireball data, is strongly peaked toward bodies with perihelia near 1 AU. As a consequence, it is also observed that their geocentric radiants are concentrated toward the antapex of Earth's heliocentric motion, and about twice as many ordinary chondrites fall in the afternoon as compared with the morning. It has been known for some time (Wetherill 1968) that concentration of meteorite perihelia near 1 AU is associated with meteorite-size bodies that first become Earth-crossing while in orbits with $a \sim 2.5$ AU, $c \sim 0.6$ and moderate (not more than 15°) inclination. For small bodies in initial orbits of this kind, vulnerability to collisional destruction while near aphelion in the asteroid belt and/or perturbation to trans-jovian aphelia preclude extensive evolution into smaller orbits prior to Earth impact. Until recently, however, a quantitatively adequate mechanism for providing sufficient bodies with this initial Earth-crossing orbital distribution was lacking.

This deficiency has been removed by the work of Wisdom (1983, 1985) who showed that most asteroidal collision debris with orbital periods near one third of that of Jupiter ($a = 2.48$–2.52 AU) are in 'chaotic' orbits, i.e. their eccentricity increases in a manner resembling random walk up to values of 0.6 or greater. Previous work (Scholl & Froeschlé 1977) suggested that the effects of resonance with Jupiter's motion were more limited, i.e. eccentricities increased only to $ca.$ 0.3 and the long-term orbital evolution was quasi-periodic. Wisdom's discovery provides the long-sought answer to the near-absence of asteroids with these periods (3:1 Kirkwood gap, figure 1), as well as providing a source of meteorites with the required initial Earth-crossing orbital distribution.

Fundamentally the mechanism works as follows. Asteroidal collision debris produced in the vicinity (i.e. within $ca. \pm 0.05$ AU) of the edges of the Kirkwood gap, collisionally ejected at moderate velocities of 50 to 200 m s^{-1}, will have a reasonably high probability of being placed in the chaotic resonant region. On a time scale of not more than 1 Ma, resonant acceleration will lead to excursions in eccentricity, eventually reaching $e \sim 0.6$ (figure 2). The close proximity of the fragment's perihelion to Earth's orbit will lead to close gravitational encounters with Earth, and the resulting perturbation will remove the semimajor axis from the resonant interval.

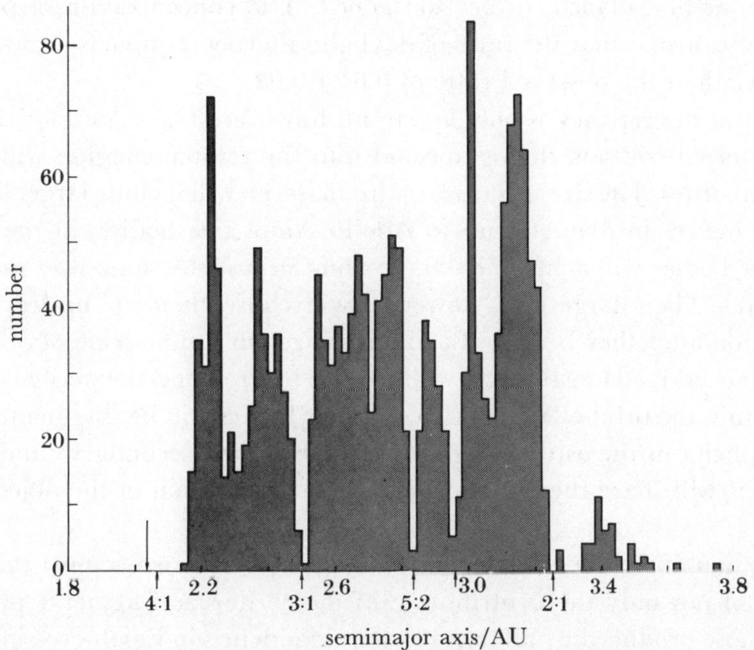

FIGURE 1. Number of catalogued asteroids against semimajor axis (exclusive of Apollo–Amor objects). With the small histogram interval chosen, the nearly complete depletion of the 3:1 resonance is apparent.

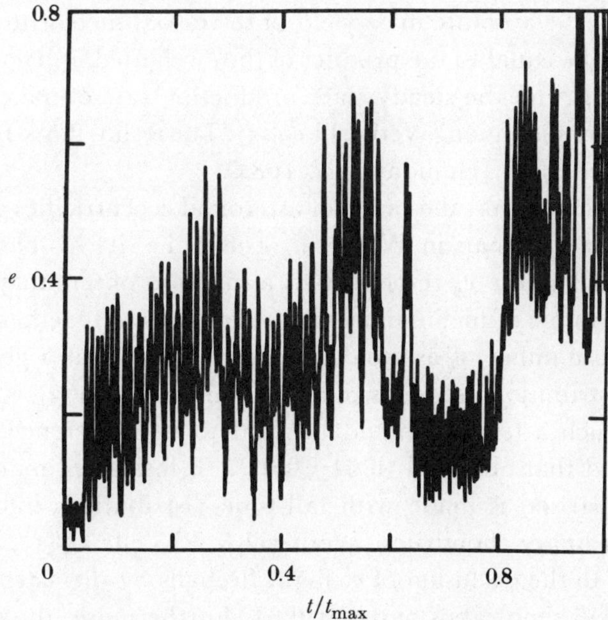

FIGURE 2. Evolution of eccentricity of an asteroid collision fragment in the 3:1 commensurability, as calculated by Wisdom (1983). On this time-scale ($T_{\max} = 2.4$ Ma) the eccentricity becomes large enough (0.6) to cross the orbit of Earth.

The fragment will thus be 'stranded' in an orbit of nearly constant eccentricity *ca.* 0.6 and with perihelion near 1 AU.

The natural unfolding of this situation, described qualitatively above, has been quantitatively studied by a Monte-Carlo technique (Wetherill 1985). Meteorite size (up to 1 m) fragments produced in the vicinity of the 3:1 gap will impact Earth within a few million years, in accordance with observed cosmic ray exposure ages. Quantitatively, the calculated distribution of orbits turns out, at first glance, to be 'too good'. The concentration of perihelia of the impacting bodies is so strong that the ratio of daylight afternoon impacts – total daylight falls is 0.76, even greater than the observed ratio of 0.64 ± 0.02.

At least most of this discrepancy is only apparent, however. The reason for this is that only a portion of the asteroid collision debris injected into the resonant region will be already in the size range of meteorites. The size spectrum of this material will include larger bodies, ranging from those several metres in diameter up to Apollo–Amor size bodies, in the kilometre-size range. These larger bodies will achieve Earth-crossing in just the same way as meteorite-size asteroidal fragments. Their large size, however, will cause them to be less vulnerable to collisional destruction after they become Earth-crossing. On a time scale of *ca.* 10^7 years, the concentration of their perihelia near 1 AU will become blurred and the perihelion distribution will spread well within the orbit of Venus. These larger bodies will be fragmented by collisions while near their aphelia in the asteroid belt, and the resulting secondary (and tertiary, etc.) meteorite-size debris will share the diffused perihelion distribution of the objects from which they are derived.

A proper calculation of the expected distribution of meteorite orbits must therefore include the weighted sum of not only the contribution of meteorite-size fragments produced in the asteroid belt, but those produced by all sizes of asteroidal debris in Earth-crossing orbit as well. A new set of calculations is reported here (table 1), including some improvements over those reported earlier (Wetherill 1985). By combination of steady-state concepts with the observed distribution of small asteroids (Wetherill 1985), determined by the Palomar–Leyden Survey (Van Houten *et al.* 1970), the absolute mass yield of terrestrial meteorite impacts can also be calculated. This yield will be equal to the product of the 'weighted impact yield' for the entire region (last line on table 1) and the steady-state production rate of meteorite size S-asteroid debris (1.97×10^{12} g a^{-1}) in this region (Wetherill 1985). The result, 2.5×10^8 g a^{-1}, agrees with the observed preatmospheric flux (Halliday *et al.* 1984).

In making these new calculations, the range of asteroidal eccentricities and inclinations was chosen in a more realistic way than in Wetherill (1985), i.e. it was based on the observed distribution of asteroids below the v_6 resonance as a function of semimajor axis, as given by Brown *et al.* (1967). The range of inclinations and eccentricities, together with the weighting factor determined by the number of asteroids in the region are also given in table 1. The impact-ejecta velocity distribution used was that of Gault *et al.* (1963).

It is not clear how much attention should be paid to the difference between calculated fall-time ratio (0.688) and that observed (0.64 ± 0.02). If, instead of making use of observed chondrite falls, the comparison is made with fall-time distributions inferred from orbits of fireballs believed to be ordinary chondrites (Wetherill & ReVelle 1981), and consideration is given in the calculations to the exclusion of twilight fireballs by the astrometric network, the calculated and inferred fall-time ratios agree at 0.64. Furthermore, the calculated fall-times were calculated under the assumption of a complete steady state, whereas there is actual

TABLE 1. METEORITES ULTIMATELY DERIVED FROM VICINITY OF 3:1 KIRKWOOD GAP

semimajor axis/AU	asteroidal source regions		
	eccentricity range	inclination range/deg	weighing factor
2.40–2.45	0.05–0.20	2.9–14.9	0.40
2.45–2.48	0.05–0.21	2.9–12.0	0.15
2.52–2.55	0.05–0.30	2.9–14.9	0.14
2.55–2.60	0.05–0.35	2.9–17.8	0.31

source semimajor axis/AU	Earth impacts		
	impact efficiency	weighed impact yield	p.m., total falls
2.40–2.45	2.22×10^{-5}	8.88×10^{-6}	0.699
2.45–2.48	3.73×10^{-4}	5.60×10^{-5}	0.701
2.52–2.55	4.00×10^{-4}	5.60×10^{-5}	0.677
2.55–2.60	3.26×10^{-5}	1.01×10^{-5}	0.675
entire region	—	1.310×10^{-4}	0.688

evidence (e.g. the 6 Ma peak in bronzite chondrite exposure ages (Crabb & Schultz 1981)) that stochastic fluctuations associated with single events are of some importance. Finally, our present understanding of the details of the behaviour of material in the chaotic zone is rudimentary. For example, it was found that Earth was nearly certain to 'pick off' the high-eccentricity fragment during the first set of excursions shown in figure 2. On the other hand, only two such calculations have been presented (Wisdom 1983, 1985) and it is not clear how typical these cases are. For these cases the probability that a Mars perturbation will remove the body from the resonance is 50%. On the other hand, if Earth were less effective in removing the body from resonance, Mars might 'smear' the distributions, and delay Earth-crossing in such a way to favour impacts from larger bodies more. Until these details are better understood, it will not be possible to pursue the question further.

In view of these considerations, it seems best to emphasize the role of the 3:1 resonance as a mechanism for delivering to Earth an adequate number of ordinary chondrites with a pronounced asymmetry in the fall-time distribution. No other asteroidal region shares these properties.

3. SOURCE REGION OF DIFFERENTIATED STONY METEORITES

It has been proposed (Wetherill 1977; Wetherill & Williams 1979) that the most plausible asteroidal source region for differentiated meteorites is the innermost edge of the asteroid belt ($2.17 < a < 2.25$ AU). Many of the asteroids in this 'Flora' region are 'Mars-grazing', i.e. their perihelia will come within ca. 0.05 AU of Mars aphelion for rare, but inevitable, favourable values of the secular variations in the orbital elements of both the asteroid and Mars. Favourably orientated collision ejecta (ca. 100 m s^{-1}) will make even closer approaches to Mars, as a result of their smaller semimajor axes and high values of eccentricity. The latter is a consequence of the amplitude of the secular oscillations in eccentricity becoming larger as the semimajor axis approaches the v_6 resonant value of ca. 2.04 AU (Williams 1969) (figure 3). The smaller perihelia of these ejecta fragments can cause them to be occasional Mars crossers.

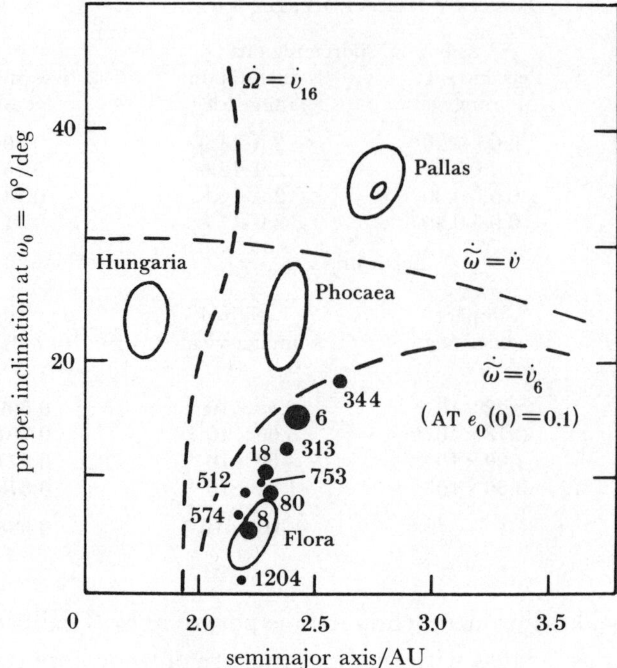

proper inclination at $\omega_0 = 0°/\mathrm{deg}$

semimajor axis/AU

FIGURE 3. Secular resonances and asteroidal regions in the inner asteroid belt, after Williams (1969).

Close encounters to Mars will then lead to a random walk in semimajor axis that will sooner or later cause a to become close enough to the ν_6 resonance to cause the eccentricity to become high enough to permit deep Mars-crossing.

The foregoing orbital evolution will thereby accelerate with time, often leading to Earth-crossing on a time scale of *ca*. 10^8 years. Small, meteorite size, stony fragments will be destroyed by collisions before they become Earth-crossing. In contrast, iron fragments and larger (not less than 10 m) stony fragments can become Earth-crossing. While in Earth-crossing orbit these 'large meteorites' and Apollo objects of various sizes will continue to be fragmented while passing through the asteroid belt and will supply Earth-impacting meteorites. This biasing of the size distribution toward larger initial Earth-crossing masses will reduce the tendency of meteorite perihelia to cluster near 1.0 AU and cause the p.m.:a.m. fall-time distribution to be nearly symmetrical.

This meteorite source mechanism has been studied quantitatively (Wetherill & Williams 1979), verifying the qualitative account given above. In this earlier work, however, it was difficult to estimate very well the absolute flux of meteorites from this region, because of uncertain parameters associated with details of cratering mechanics.

By use of both the approach and results of §2, a much better evaluation of this innermost region of the asteroid belt as a source of stony meteorites has now been made. This improvement is primarily the consequence of several factors.

1. The calculation of the rate of injection of asteroidal debris into the 3:1 gap is now based on the observed distribution of small asteroids and steady-state theory. This is believed to be superior to attempting to estimate the injection rate by use of cratering mechanics. (The result, however, is entirely consistent with cratering mechanical calculations of the injection rate.)

2. The calculated and observed Earth-impact flux of ordinary chondrites from the 3:1 gap agree at *ca.* 10^8 g a^{-1}. Either of these results can then be used to normalize calculations of the Earth-impact rate from other regions of the asteroid belt. In this way, the absolute flux from other regions can be based on the more accurate calculation of the ratio of the yield from a particular region to that from the vicinity of the 3:1 gap.

3. In the earlier work on the source region of differentiated meteorites, the yield from specific asteroids in the Flora region was used to estimate the total yield from this region. A new approach is now used (Wetherill 1985) based on the entire collision hierarchy of asteroidal bodies. This provides a better observational basis for the calculations, and further reduces their sensitivity to details of cratering mechanics.

4. In this work, consideration was taken of the secular perturbations in the orbit of Mars. When this is done, the calculated populated stable region of the innermost asteroid belt agrees well with the observed distribution of asteroids in the vicinity of the ν_6 resonance. If only the present orbit of Mars is considered, the calculated stable region extends well inside that observed. The present correction of this discrepancy should improve the reliability of the calculations. This effect was not included in Wetherill (1985), nor in earlier work.

5. Unlike the results reported by Wetherill & Williams (1979), in this approach the effect of atmospheric ablation of meteorites is explicitly induced in the calculations, and the time of fall distribution explicitly calculated, rather than only estimated.

In this way an entirely new set of calculations has been made of the Earth-impact efficiency of random asteroid ejecta in eight source intervals between the innermost edge of the asteroid belt (2.17 AU) and 2.60 AU and lying below the ν_6 resonance (figure 3). For each of these regions, the source asteroid eccentricity and inclination distributions reported by Brown *et al.*

TABLE 2. METEORITES ULTIMATELY DERIVED FROM THE INNERMOST ASTEROID BELT

asteroidal source regions			
semimajor axis/AU	eccentricity range	inclination range/deg	weighing factor
2.17–2.25	0.10–0.17	2.9–7.0	0.27
2.25–2.30	0.10–0.20	2.9–8.0	0.29
2.30–2.35	0.05–0.20	4.0–9.7	0.22
2.35–2.40	0.05–0.21	2.9–11.5	0.21

Earth impacts			
source semimajor axis/AU	impact efficiency	weighed impact yield	p.m., total falls
2.17–2.25	4.40×10^{-5}	1.19×10^{-5}	0.52
2.25–2.30	3.85×10^{-6}	1.12×10^{-6}	0.55
2.30–2.35	3.18×10^{-7}	7.00×10^{-8}	0.53
2.35–2.40	2.93×10^{-6}	6.15×10^{-7}	0.72

(1967) were approximated (tables 1 and 2). Three values of ejection velocity (50, 150 and 300 m s^{-1}) are shown. These efficiencies include the contribution from meteorite-size fragments produced directly in the asteroid belt from bodies of all sizes, as well as meteorite-size fragments produced from bodies after they have been transferred into Earth-crossing orbit.

It may be seen (figure 4) that in the velocity range that includes *ca.* 98% of the collision

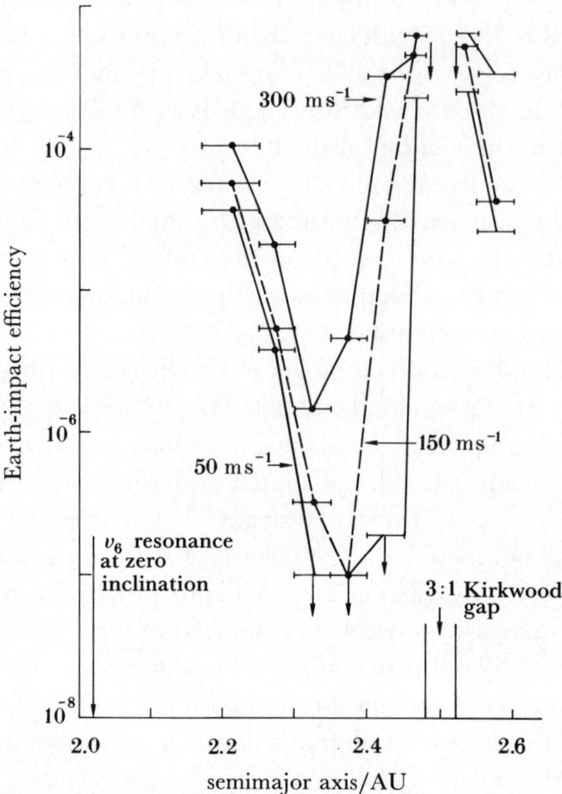

FIGURE 4. Earth impact efficiency for asteroidal collision ejecta as a function of semimajor axis and ejection velocity.

ejecta (Gault *et al.* 1963), the Earth impact efficiency is very dependent on semimajor axis. Efficiencies are very high in the vicinity of the 3:1 Kirkwood gap at 2.5 AU, and moderately high in the innermost edge of the belt. The latter is caused by proximity to the ν_6 resonance. Much higher efficiencies have been calculated for smaller values of semimajor axis in immediate proximity to the ν_6 resonance (and the 4:1 commensurability resonance at 2.06 AU). However, this region is of no relevance to the question of meteorite sources because the number of asteroids in long-lived stable orbits in this region is negligible.

Very low Earth-impact efficiencies are found for material ejected between 2.25 and 2.40 AU in the range of proper inclination and eccentricity considered. The range of proper eccentricities and proper inclinations for each semimajor axis interval encompasses almost all the asteroids in these regions that lie beneath the ν_6 resonance. In addition, there are a few asteroids within higher eccentricities and inclinations that lie closer to the ν_6 resonance. These have higher Earth-impact efficiencies. Insofar as this region of the asteroid belt is a potential source of meteorites, the objects in these orbits are those that should be considered. Calculations for the largest such object, that of the *ca.* 80 km diameter C-asteroid 313 Chaldaea, together with several other specific orbits, are given in table 2. The relatively high efficiency found for this orbit is attributable to its minimum perihelion already being slightly inside the maximum aphelion of Mars (Williams 1979). Therefore, in this sense, this object (and any retinue of smaller collisional debris) is already a Mars-crosser, and some of even its lowest velocity collision ejecta will be even deeper Mars-crossers. Only at higher velocities (*ca.* 1 km s^{-1}) does this yield from more typical asteroids in this region approach that found between 2.17 and 2.25 AU, but only *ca.* 10^{-3} of the ejecta will have velocities this high. Therefore, assuming these calculations to be valid, meteorites from asteroids in this intermediate region (including 4 Vesta at 2.36 AU) should be absent or at most very poorly represented in meteorite collections.

Earth-impact efficiencies have been calculated for the entire range of impact velocities that make a significant contribution for each region of the inner asteroid belt. The yield of each velocity interval was weighted in accordance with the mass against velocity distribution of Gault *et al.* (1963), in accordance with the data used for 'solid rock' by Greenberg & Chapman (1983), and then weighted by the number of asteroids in each semi-major axis interval as given by the Palomar–Leiden survey. The results of this calculation are given in table 2.

TABLE 3. EARTH IMPACT EFFICIENCIES AND FALL TIMES FOR SPECIFIC ASTEROIDAL ORBITS IN THE INNER ASTEROID BELT

asteroid	a	proper elements (Williams 1979) e	i	ejection velocities m s^{-1}	impact efficiency	fall-time ratios
313 Chaldaea	2.38	0.23	12.4°	50	3.9×10^{-5}	0.53
				100	4.8×10^{-5}	0.54
				150	2.7×10^{-5}	0.52
				200	4.6×10^{-5}	0.56
4 Vesta	2.36	0.22	6.4°	400	$< 10^{-7}$	—
				450	3.7×10^{-5}	0.71
				500	1.3×10^{-4}	0.71
8 Flora	2.20	0.14	5°	25	9.3×10^{-6}	0.50
				50	1.4×10^{-5}	0.52
				100	2.2×10^{-5}	0.53
				150	3.9×10^{-5}	0.55
				200	6.0×10^{-5}	0.51
				300	1.2×10^{-4}	0.52
6 Hebe	2.42	0.15	15°	150	$< 10^{-7}$	—
				200	3.9×10^{-5}	0.59
				300	4.6×10^{-4}	0.62

The total yield from the region between 2.17 and 2.25 AU is found to be about 9% that from the vicinity of the 3:1 Kirkwood gap. The time of fall distribution is markedly different, morning and afternoon falls being more nearly equal in number. If both the asteroids in this region as well as those near the 3:1 resonance are of ordinary chondritic composition, admixture of meteorites from this region with those ultimately derived from the 3:1 resonance will reduce the expected afternoon fall ratio of ordinary chondrites only to about 0.67. Therefore there is no strong observational reason for excluding the possibility that the *entire* inner asteroid belt is the source region of ordinary chondrites. In this case, the 3:1 gap would simply be the primary source, because of the greater impact efficiency of asteroidal debris produced in that region.

On the other hand, the achondritic meteorites must come from somewhere, and their observed total p.m./total fall ratio of 0.53 is definitely inconsistent with that expected for the 3:1 resonance region. Because observational bias must discriminate against recovery during night-time a.m. hours, the actual ratio should be even less. Although the data are meagre, from the work of Simonenko (1975) it seems unlikely that achondritic perihelia are as concentrated near 1 AU nearly as strong as those of the ordinary chondrites (figure 5). Achondrites, exclusive of SNCs (probably of Martian origin) and the enstatite achondrites, comprise 8 % of the observed stone meteorite falls. The basaltic achondrites represent 6 % of the observed falls, and have a 24 h fall-time ratio of 0.51. The relatively high proportion of these achondrites among meteorite falls implies that they must be derived from a prolific source. This region of the

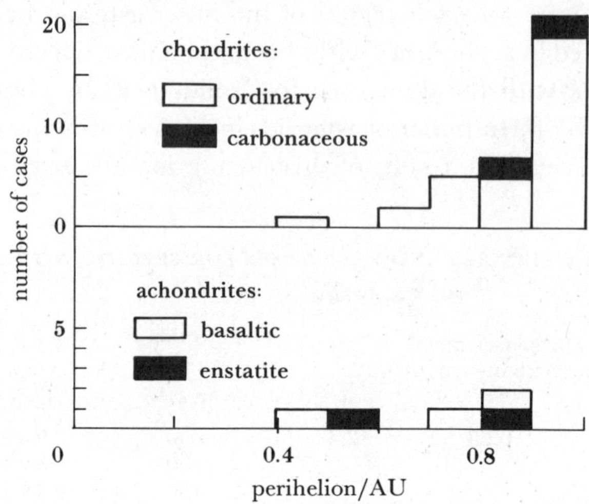

FIGURE 5. Observed distribution of achondrite perihelia, based on Simonenko (1975).

asteroid belt between 2.17 and 2.25 AU is the only candidate known now for a source region that provides the appropriate orbital distribution and impact rate. Very few of these meteorites will be direct collision ejecta from belt asteroids. Rather, they will be secondary (and tertiary etc.) fragments of larger fragments, ranging up to the largest Apollo–Amor objects, such as 433 Eros (20 km diameter). The expected Earth impact rate of these *large objects* is similar to that from the 3:1 region, in agreement with the observation (Kyte & Brownlee 1985) that melt rock data (see, for example, Palme *et al.* 1981) suggest chondritic bodies may not dominate the Earth impact flux of crater-producing bodies.

Because of its basaltic reflectance spectrum, the 400 km diameter asteroid 4 Vesta is often suggested as the primary source of basaltic achondrites (e.g. Drake 1979). The principal problem with this otherwise attractive suggestion is that Vesta ($a = 2.36$ AU) not only lies within the region (2.30–2.40 AU) for which the yield is lowest, but its rather low eccentricity (0.09) and inclination (7°) cause it to be in an unproductive portion of this region. For this orbit, a minimum ejection velocity of *ca.* 450 m s^{-1} is required before a statistically significant Earth-impact efficiency (more than 10^{-7}) is found (figure 4). Even if as much as 10 % of the asteroidal material in this region of the belt were Vesta-derived (for which there is no spectrophotometric evidence), the yield from this 'Vesta Source region' would be only 10^{-3} that of the 3:1 gap. Therefore only *ca.* 10^5 g a^{-1} of this material should impact the Earth, when normalized to the *ca.* 10^8 g a^{-1} flux of ordinary chondrites. If instead of considering the Vesta material to be this entire collision hierarchy, only meteorite-size fragments from Vesta itself are considered, the yield is much lower, *ca.* 1 kg a^{-1}. This is similar to the mass of basaltic achondrites added annually to museum collections, thereby requiring 100 % efficiency in collection of this material, including that falling in the ocean.

4. OTHER SOURCE REGIONS AND OTHER TYPES OF METEORITES

The regions between 2.17 and 2.60 AU below the ν_6 resonance, contain most of the asteroidal mass in the inner asteroid belt. The high inclination Phocaea and Pallas regions (figure 3), above the ν_6 resonance are sparsely populated (*ca.* 4 % as many bodies as the region studied). The Monte-Carlo techniques used in the low-inclination region are probably not applicable in this high-inclination region because of nonlinear mixing of the free and forced oscillations of the secular perturbations. Therefore in this case only qualitative considerations are possible now. In this context, it may be expected that the ν_5 resonance may be expected to behave qualitatively similar to ν_6 in expediting the evolution of Mars-crossing orbits, but because of the smaller mass injected into this region, its effects should be more minor. There is no theoretical requirement that asteroids in this region need to be represented in meteorite collections at all, but a small contribution is possible.

The outer asteroid belt, beyond 2.60 AU, can be expected to contribute a significant number of meteorites only if the outer resonances at the 5:2 (2.82 AU), 7:3 (2.96 AU) and 2:1 (3.28 AU) commensurabilities can avoid excessive transfer of asteroidal debris to the vicinity (i.e. within *ca.* 0.5 AU) of Jupiter before it is Earth-crossing. If these resonances are as effective as the 3:1 resonance in increasing eccentricity, it is likely that Jupiter will remove this material from the Solar System so effectively that the contribution of the outer belt to the terrestrial meteorite flux will be negligible. Murray (1986) has studied chaotic orbits associated with the 2:1 resonance and found that eccentricities as high as 0.8 could be achieved, sufficient to cause the aphelion of the fragment to reach 5.9 AU, i.e. deep Jupiter-crossing. Mars perturbations and/or collisions would then remove the asteroidal fragment from the resonance, destroy the liberation condition, and expose the body to very strong Jupiter perturbations, leading to ejection from the Solar System.

On the other hand, the extent of this chaotic zone appears to be more limited than the 3:1 resonance. Because of this, it is conceivable that the mechanism proposed earlier (Zimmerman & Wetherill 1973) could permit a small amout of material to undergo more modest eccentricity increases and become Mars-crossing before it is Jupiter-crossing. The semimajor axis could then be perturbed inside the resonance, and combined Jupiter and Mars perturbations could random-walk the perihelion toward Earth-crossing. The nearly bimodal nature of the 'maximum-eccentricity' map in Murray (1986) makes this possibility seem very unlikely.

The Hungaria region at 1.95 AU (figure 3) deserves investigation as a possible meteorite source. Most of the asteroids in this region, including the largest (434 Hungaria, 23 km diameter), display the rare, high-albedo, featureless, reddish reflectance spectra of the E-asteroids that are similar to laboratory data on enstatite achondrites. Monte-Carlo techniques available now do not permit quantitative calculations for high-inclination orbits of this kind, particularly considering the proximity of Mars, the ν_{16}, ν_6 and 3:1 resonances. Qualitatively, however, it seems that this source region could be unique by producing fragments with aphelia initially inside the main asteroid belt, and thereby relatively immune to collisional destruction. This could be the reason why enstatite achondrites have remarkably old exposure ages (Schultz & Kruse 1978). Despite the small mass of its largest asteroid, the Palomar–Leyden survey appears to indicate the presence of a surprisingly large number of smaller bodies in this region. If combined with a sufficiently high Earth-impact efficiency, this region would be a strong candidate for the source region of the enstatite achondrites.

Turning to meteorite classes that have not been considered, the region adjacent to the $3:1$ gap contains approximately equal numbers of C-asteroids and S-asteroids. If the usual identification of carbonaceous meteorites with C-asteroids is made, then the expected pre-atmospheric terrestrial flux of carbonaceous meteorites should be comparable to that of ordinary chondrites. The conventional explanation of why the observed fall rate is only 6% that of the ordinary chondrites, greater ablation and fragmentation during passage through the atmosphere, is likely to be correct. The fall-time ratio of these meteorites (*ca.* 0.8) is probably consistent with a $3:1$ resonance source, but of little statistical significance. There is also evidence that some Earth-impacting meteoroids from cometary orbits could be recoverable (Wetherill & ReVelle 1982). If so, these are likely to be carbonaceous. For a cometary source (either live comets or cometary Apollo objects), no fall-time asymmetry is expected. The important possibility that large samples of cometary material may be available in the form of meteorites should be pursued, but there is no strong reason to believe they are contained in our present meteorite collections.

A similar treatment of iron and stony-iron sources would be difficult, because of large uncertainties in fracturing and cratering mechanics, asteroidal identification, and fall statistics for meteorites of these kinds. Based on fall rates, a total source at least as productive as that of the achondrites, and possibly that of the ordinary chondrites, is required. Qualitatively, an inner asteroid belt source region (less than 2.60 AU) appears very probably, simply because of the apparent absence of productive regions in the outer belt. It is quite possible, however, that iron fragments can be accelerated to higher velocities without destruction. Therefore proximity to the $3:1$ or ν_6 resonance may be less critical, and a broader sample of metallic sources in the inner asteroid belt may be present in meteorite collections. Although fall-time statistics are inadequate to be useful, the fact that all five hexahedrite falls occurred in the morning is curious and, if not a fluke, would be difficult to explain.

The matter of identification of S-asteroid spectra in the $3:1$ resonance and the Flora region with the most abundant classes of chondrites and achondrites is still unresolved. Gaffey (1984) presents cogent reasons for believing that 8 Flora is a differentiated asteroid, rather than being chondritic, and at least tentatively extends these arguments to S-asteroids as a whole. Dermott *et al.* (1985) present UBV data indicating a difference in spectroflectance of S-asteroids interior to 2.4 AU relative to those with larger semimajor axes. Qualitatively, a difference would be expected if the achondrite sources are S-asteroids in the Flora region and the ordinary chondrite sources are S-asteroids near the $3:1$ resonance. Feierberg *et al.* (1982) present data and arguments to support their position that typical S-asteroids are of ordinary chondritic composition and mineralogy.

One strong conclusion from quantitative studies of asteroid dynamics is that the ordinary chondrites require a major asteroidal source region, such as half the asteroids within ± 0.05 AU of the $3:1$ resonance. Although it is true that meteorite collections represent a biased sample of the asteroid population, the extent of these selective effects is limited. The natural length scale on which selection can operate is that imposed by collision ejection velocities of several hundred metres per second, i.e. $\pm$ *ca.* 0.05 AU. This precludes the exquisitely precise sampling that would be required to deliver to Earth only the material from a nearly unobservable population of asteroids. Stochastic effects associated with single large collisional events are undoubtedly present, but completely fail as a quantitative explanation of the dominance of the meteorite flux by ordinary chondrites.

One possibility that merits further attention is the relation between asteroidal size and spectral reflectance. The meteorite-size fragments in our collections are nearly the end members of a collisional hierarchy that extends through 'Apollo-size' asteroids of 0.1–10 km up to much larger bodies. The immediate parent bodies of most meteorites are concentrated toward the small mass end of this hierarchy. It is conceivable that for some presently unknown reason the detailed mineralogical composition of smaller asteroids differs from the large bodies on which the most precise spectral reflectance data is available, perhaps because these large bodies are collisional residues, or because of some winnowing process that occurs as material passes down the collisional hierarchy. The tendency of small Earth-approaching objects to more often resemble meteorites (McFadden *et al.* 1985) may be relevant to this question. If the relation between chondrite-like Apollo objects and ordinary chondrites is more than a coincidence there must also be an appropriate asteroidal source for these Apollo objects. For both, the quest for an ordinary chondritic source leads back to at least intermediate size (*ca.* 50 km diameter) bodies in the main asteroid belt.

5. SUMMARY AND CONCLUSIONS

1. Combination of observational data on meteorite orbits and dynamical theory strongly suggest that the ultimate source region of ordinary chondrites and Apollo–Amor objects with ordinary chondrite mineralogy is that portion of the main asteroid belt adjacent to the 3:1 commensurability resonance at 2.50 AU. To agree with the observed terrestrial fall rate of these meteorites, the expected contribution of the entire size spectrum of S asteroids in this region is required.

2. The expected meteorite flux from the innermost asteroid belt (2.17–2.25 AU) via the v_6 resonance is found to be 9 % that of the 3:1 resonance ordinary chondrite source. In contrast with the 3:1 resonance source, no significant asymmetry between morning and afternoon falls is expected for meteorites ultimately derived from this region. It is proposed that this region is the source of basaltic achondrites, exclusive of SNC meteorites, and possibly some other achondrites as well. Stony meteorites ultimately from this region usually arrive on Earth through the intermediary of larger bodies, including those observed as Apollo–Amor objects.

3. The preatmospheric terrestrial flux of carbonaceous meteorites from the 3:1 resonance region should be comparable to that of the ordinary chondrites, i.e. *ca.* 10^8 g a^{-1}. This is in addition to carbonaceous material of cometary origin, possibly recoverable only as micrometeorites.

4. The source regions of other meteorite types is uncertain. It is speculated that the Hungaria region, interior to the main belt at 1.95 AU, may be a source region of enstatite achondrites, and that iron and stony-iron meteorites may be derived from the entire asteroid belt interior to 2.6 AU.

5. There are many opportunities for further advances in our understanding the planetological context of meteoritic material. These include better theoretical understanding of the dynamics of secular and commensurability resonances, and development of techniques for calculating the orbital evolution of bodies coupled to more than one of these resonances (particularly bodies with inclinations greater than 20°). We also need to be able to interpret remote sensing data in a less ambiguous way. It may be hoped that *in situ* study by missions to asteroids will provide the 'ground truth' that may be necessary to accomplish this.

References

Bogard, D. D. 1979 Chronology of asteroid collisions as recorded in meteorites. In *Asteroids* (ed. T. Gehrels) (558–578 pages). Tucson: University of Arizona Press.

Brown, H., Goddard, I. & Kane, J. 1967 Qualitative aspects of asteroid statistics. *Astrophys. J. Suppl.* **14**, 57–124.

Crab, J. & Schultz, L. 1981 Cosmic ray exposure ages of the ordinary chondrites and their significance for parent body stratigraphy. *Geochim. cosmochim. Acta* **45**, 2151–2160.

Dermott, S. F., Gradie, J. & Murray, C. D. 1985 Variation of the UBV colors of S-class asteroids with semi-major axis and diameter. *Icarus* **62**, 289–297.

Drake, M. J. 1979 Geochemical evolution of the eucrite parent body: possible nature and evolution of asteroid 4 Vesta? In *Asteroids* (ed. T. Gehrels) (765–782 pages). Tucson: University of Arizona Press.

Feierberg, M. A., Larson, H. P. & Chapman, C. R. 1982 Spectroscopic evidence for undifferentiated S-type asteroids. *Astrophys. J.* **257**, 361–372.

Gaffey, M. J. 1984 Rotational spectral variations of asteroid (8) Flora: implications for the nature of the S-type asteroids and for the parent bodies of ordinary chondrites. *Icarus* **60**, 83–114.

Gault, D. E., Shoemaker, E. M. & Moore, H. J. 1963 Spray ejected from the lunar surface by meteoroid impact. *NASA TN 1767* (39 pages).

Greenberg, R. & Chapman, C. R. 1983 Asteroids and meteorites: parent bodies and delivered samples. *Icarus* **55**, 455–481.

Halliday, I., Blackwell, A. T. & Griffin, A. A. 1984 The frequency of meteorite falls on the Earth. *Science, Wash.* **223**, 1405–1407.

van Houten, C. J., Van Houten-Groeneveld, I, Herget, P. & Gehrels, T. 1970 The Palomar-Leiden survey of faint minor planets. *Astron. Astrophys. Suppl.* **2**, 339–348.

Kyte, F. T. & Brownlee, D. E. 1985 Unmelted meteoritic debris in the late Pliocene iridium anomaly: evidence for the ocean impact of a nonchondritic asteroid. *Geochim. cosmochim. Acta* **49**, 1095–1108.

McFadden, L. A., Gaffey, M. J. & McCord, T. B. 1985 Near-Earth asteroids: possible sources from reflectance spectroscopy. *Science* **229**, 193–195.

Murray, C. D. 1986 Structure of the 2:1 and 3:2 Jovian resonances. *Icarus* **65**, 70–82.

Palme, H., Grieve, A. F. & Wolf, R. 1981 Identification of the projectile at the Brent crater, and further consideration of projectile types at terrestrial craters. *Geochim. cosmochim. Acta* **45**, 2417–2424.

Russell, B. 1985 *A History of Western Philosophy*, book one, part I, chapter IV. New York: Simon and Schuster.

Scholl, H. & Froeschlé, 1977 The Kirkwood gaps as an asteroidal source of meteorites. In *Comets, asteroids, meteorites* (ed. A. H. Delsemme) (587 pages). University of Toledo Press.

Shulz, L. & Kruse H. 1978 *Nuclear track detection* **2**, 65–103.

Simonenko, A. N. 1975 Orbital elements of 45 meteorites. *Atlas.* (67 pages). Moscow: Nauka.

Wetherill, G. W. 1968 Time of fall and origin of stone meteorites. *Science, Wash.* **159**, 79–82.

Wetherill, G. W. 1977 Fragmentation of asteroids and delivery of fragments to earth. In *Relationships between comets, minor planets and meteorites* (Proc. I.A.U. Colloquium 39, Lyon) (ed. A. H. Delsemme) pp. 283–292. University of Toledo.

Wetherill, G. W. 1985 Asteroidal source of ordinary chondrites. *Meteoritics* **20**, 1–22.

Wetherill, G. W. & ReVelle, D. O. 1981 Which fireballs are meteorites? A study of the Prairie Network photographic meteor data. *Icarus* **48**, 308–328.

Wetherill, G. W. & ReVelle, D. O. 1982 Relationship between comets, large meteors and meteorites. In *Comets* (ed. L. Wilkening), pp. 297–319. Tucson: University of Arizona Press.

Wetherill, G. W. & Williams, J. G. 1979 Origin of differentiated meteorites. In *Origin and distribution of the elements* (ed. L. H. Ahrens), pp. 19–31. Oxford: Pergamon Press.

Williams, J. G. 1969 Secular perturbations in the solar system. Ph.D. dissertation, University of California, Los Angeles.

Williams, J. G. 1979 Proper elements and family memberships of the asteroids. In *Asteroids* (ed. T. Gehrels), pp. 1040–1963. Tucson: University of Arizona Press.

Wisdom, J. 1983 Chaotic behavior and the origin of the 3/1 Kirkwood gap. *Icarus* **56**, 51–74.

Wisdom, J. 1985 Meteorites may follow a chaotic route to Earth. *Nature, Lond.* **315**, 731–733.

Zimmerman, P. D. & Wetherill, G. W. 1973 Asteroidal source of meteorites. *Science, Wash.* **182**, 51–53.

Discussion

G. Turner (*Department of Physics, University of Sheffield, U.K.*). How do meteorite cosmic ray exposure ages fit into the dynamical model presented? In particular to what extent do they reflect fragmentation when the object is in Earth crossing orbit, or in its original orbit in the asteroid belt? To what extent if any do they reflect the dynamical origin?

G. W. Wetherill. An advantage of the dynamical model is that it avoids uncertain assumptions regarding cratering mechanics by treating the entire system, both the asteroid belt and Earth-approaching population, as a single steady-state entity. A price that one pays for this is that distinctions of the kind asked about tend to be suppressed in the calculation, but still something can be said. Cosmic ray ages record the production of unshielded (less than about 2 m diameter) bodies from larger bodies as the ultimate, penultimate (or antepenultimate, etc.) stage of the global fragmentation hierarchy leading to the production of meteorites.

For the principal asteroidal source region, the 3:1 gap, I can estimate that about 10% of the Earth-impacting meteorite-size (100 g–1 t) fragments are produced by collisions in the asteroid belt. A somewhat greater fraction enter the broader unshielded size range while still in the asteroid belt. Meteorite-size fragments of all these larger unshielded bodies will have a multiple exposure history. The remaining cosmic ray exposures will be initiated by fragmentation of non-belt objects disrupted or cratered by collisions when their perihelia are inside the asteroid belt, including many in Earth-crossing orbits. These bodies include astronomically observable Apollo–Amor objects, but primarily intermediate size (e.g. 1–100 m diameter) bodies. Of course, all the bodies under consideration were ultimately fragments of belt asteroids.

For objects originally in the innermost belt, brought into Earth-crossing orbit by the ν_6 resonance, the fraction of cosmic ray ages initiated in the asteroid belt would be very small.

I would expect that the different orbital evolution of objects originally in these distinct asteroidal source regions would be reflected in their cosmic-ray age distribution, but only in a subtle way. I have not studied this question. Because I expect the difference would be slight, I would give higher priority to a serious quantitative test of the predictions this model makes regarding multiple exposure ages, taking into account stochastic fluctuations from the steady state. This could lead to a fine-tuning of the model (or even disclose a terrible problem with it), that would help our understanding of questions such as the origin of the 6 Ma peak in the bronzite chondrite exposure-age distribution.

F. L. Whipple (*Smithsonian Astrophysical Observatory, Cambridge, Massachusetts, U.S.A.*). Would Professor Wetherill care to comment on the SNC meteorites? Do they really come from Mars?

G. W. Wetherill. From consideration of their geochemistry, petrology, and radiogenic isotopes, these meteorites bear a strong signature of planetary, rather than asteroidal origin. Mars is the best candidate for being this planet. There is still a problem in understanding the mechanism by which an adequate quantity of material is accelerated to the martian escape velocity by an impact consistent with the martian cratering record and the young age of these meteorites.

Phil. Trans. R. Soc. Lond. A **323**, 339–347 (1987)

Printed in Great Britain

A review of cometary sciences

By F. L. Whipple

Smithsonian Astrophysical Observatory, 60 Garden Street, Cambridge, Massachusetts 02138, U.S.A.

This paper presents an elementary description of comets and their nature. It deals with the contribution from comets to the solid particles that produce the Zodiacal Light and discusses the possibility that some comets in short-period orbits may degenerate into asteroids. The evidence suggests that this may well have happened for the Trojan and other of the outer asteroids but that the near-Earth asteroids and the meteorites are largely not cometary in origin. The last section of the paper deals with the possible nature of comets and their origin such that some might become superficially indistinguishable from asteroids. Space missions are clearly needed, to answer some of these basic questions.

Introduction

This introduction is intended to present a broad elementary description of the phenomena and nature of comets. The reader who wishes to pursue the general subject in more detail is referred to books edited by Delsemme (1977), McDonnell (1978) and Wilkening (1982).

Emphasis centres on the relation of comets to the interplanetary complex and possibly to asteroids and planets, on the nature of the cometary nucleus and on its possible modes of evolution. Cometary phenomena such as ion tails will largely be ignored as will orbital characteristics, which are treated by Wetherill (this symposium).

At great solar distances, comets appear as pointlike sources of reflected sunlight, observationally indistinguishable from stars except for their motion across the stellar background. As comets approach the Sun, usually well within Jupiter's distance, they develop a hazy coma, sometimes with an apparently starlike central condensation. Because of the limited resolving power of telescopes, this condensation may be several hundred to thousands of kilometres in diameter. More active comets may develop diffuse comas in several tens of thousands of kilometres in diameter and also tails, up to 10^8 km or more in length, directed generally away from the Sun and lagging a few degrees behind the orbital motion about the Sun.

This activity arises entirely from the heart of the comet, its nucleus, an intimate mixture of ices and dust ranging in dimension from less than 1 km to the order of 10 km. When the Sun's radiation becomes adequate to produce significant sublimation of the ices, the resultant vapour escapes from sunlit areas on the nucleus carrying with it embedded dust and meteoroidal material. The dust becomes observable as it scatters and reflects sunlight whereas the gas shines by fluorescence. In this process, the gaseous atom or molecule absorbs a quantum of sunlight and then reradiates the energy usually in two or more quanta of lower energy, and therefore in redder light than that of the absorbed quantum. No evidence suggests that comets contribute any of the radiation observed, although the warming of extremely cold ices may release a certain amount of energy stored in the form of imperfect crystalline structure in *amorphous* ices. The Sun is overwhelmingly the prime source of cometary activity and observed radiation.

The fine dust of dimensions in the micrometre range frequently shows the silicate signature in its infrared reflection spectrum near a wavelength of 10 µm. This dust is forced away from the Sun by light pressure with accelerations rarely exceeding that of solar gravity. Thus, with the proper geometry, we often see such dust tails as relatively stubby, curving at conspicuous angles behind the orbital motion of the comet.

The composition of cometary comas and tails as observed by optical, infrared, ultraviolet and radio sensors is listed in table 1. The preliminary results from the missions to Halley's comet reported in subsequent papers of this symposium should add materially to the entries in this table and to their physical interpretation. Suffice it to say that gas-phase chemistry near the nucleus of a comet in the presence of solar radiation can make and destroy compounds, primarily of oxygen, carbon, nitrogen and hydrogen in such a way as to make detailed deductions about the ices in comets from the data of table 1 a quite impossible task. From the study of many comets, however, the observed lines and bands of H, O and OH lead to the sum of their abundances as roughly H_2O, indicating that water ice is the major icy component of comets. Carbon compounds, although numerous, constitute altogether only perhaps 2 % of the total mass of the ices, carbon being deficient with respect to oxygen and nitrogen when compared with solar abundances. The observable dust component varies strikingly from comet to comet, some comets displaying almost none by reflected sunlight in their spectra. Perhaps a third of the mass of an average comet consists of dust and earthy particles.

TABLE 1. SPECIES OBSERVED IN COMETS

coma, head	H, C, C_2, $^{12}C^{13}C$, C_3,
	CH, CN, CO, CO_2, CS, HCN, HCO, CH_3CN,
	NH, NH_2, NH_3,
	O, OH, H_2O, S, S_2
(near sun)	Na, K, Ca, Cr, Mn,
	Fe, Co, Ni, Cu, V(?), Ti(?)
ions (tail)	C^+, CH^+, CN^+, CO^+, CO_2^+,
	N_2^+, OH^+, H_2O^+, Ca^+, H_2S^+
dust	silicates and hydrocarbons, mostly dielectrics
	(see later papers in this symposium for additions
	from Halley Comet missions).

The molecules in space generally have lifetimes of a number of hours to a few days against the dissociating and ionizing effects of sunlight. Because the gas leaves the surface of the nucleus with a velocity of the order of 0.5 km s^{-1} at the Earth's distance from the Sun, the coma may, therefore, attain an observed diameter of 30 000 to 100 000 km ($2 \times 0.5 \times 86\,400$) km. The diameter of the coma may tend to decrease as the comet approaches the Sun because the lifetimes of the species decrease proportionally to the solar radiation while their velocity increases only as its square root.

Positively charged ions that are formed when solar radiation removes electrons from the various species receive especial treatment in the cometary coma. They become subject to the action of the *solar wind*. The Sun's high atmosphere blows off about 10^6 t s^{-1} of extremely hot gas at a speed of some 400 km s^{-1}. The gas of this solar wind is thoroughly ionized at temperature of the order of 10^6 K. Hence it becomes a *plasma*, meaning that the ions and electrons that compose it carry with them strong electrical currents and magnetic fields comprising total energies comparable to the total kinetic energy of motion of the ions and electrons. In the rare gases of a comet's coma, these magnetic fields of the solar wind can selectively entrap any electrically charged ions present and carry them along, away from the Sun. In this fashion the solar wind forms the huge nearly straight tails of comets, seen only by fluorescence of the ions present, particularly of carbon monoxide, which by chance is conspicuous in visual light. The main components of the solar wind, hydrogen and helium, constituting 98% by mass, are quite invisible because these gases are too hot to radiate. Their electrons have been removed by ionization leaving them radiationally impotent.

The solar wind largely ignores the neutral species in a comet's coma although some cometary atoms or molecules are ionized by charge exchange and then carried into the ion tail. The ionized gases in a comet's tail are so tenuous that the acceleration by the solar wind sometimes exceeds the Sun's gravitational acceleration by as much as a thousand times. These rapid motions and high accelerations in ion tails were mysterious, indeed, until 1950 when the late Ludwig Biermann recognized the function of the solar wind. The solar wind itself was observed directly in the early days of the space age. The plasma theory developed by Hannes Alfvén in 1957 is extremely difficult to apply. Hence the direct observations of fields and charged particles in the ion tails of comet P/Giacobini-Zinner by the ICE spacecraft and of Halley's Comet by the various space probes are substance for new theoretical developments in plasma physics.

Although comets are being continuously depleted by loss of matter and by perturbation of their orbits by the planets, the supply of comets appears to be maintained by comets disturbed from extremely long orbits because of the attraction of passing stars and the galactic centre. This cloud of comets postulated by Öpik (1932) and Oort (1950) is known as the Öpik–Oort Cloud and extends to about a thousand times Pluto's distance from the Sun. Comets attain short-period orbits by the attractions of the planets.

The interplanetary complex and asteroids

The major contribution of comets to the Solar System is the maintenance of small particulate matter in the interplanetary medium providing the material for the Zodiacal Light and the Gegenschein. The annual meteor streams of the Leonids and Perseids were first associated specifically with their parent comets in the 1860s. Some 15 such associations are now recognized and Elsson-Steel (1986) has recently produced evidence that essentially all sporadic meteors are of cometary origin.

Several lines of evidence point to the ejection of moderate sized masses from the surfaces of comets. Not only do many comets split but active comets such as Halley's for which near-nucleus observations are possible, show sharp condensations that must involve ejected coherent pieces. Comet P/Encke shows almost no continuum in its spectrum but its associated Taurid meteors sometimes include fairly bright fireballs. Radar observations of comet IRAS-Araki-Alcock, 1983 VII, in its near-Earth passage indicated an accompanying but detached cloud of particles large enough to be conspicuous at a wavelength of 13 cm (Campbell *et al.* 1983). Cometary antitails, seen when brighter comets are observed as the Earth crosses their orbit planes, arise from sizeable particles lying very close to the planes of the orbits. The difficulties of the Halley Comet missions as they crossed the orbit plane of the comet attest to the existence of these sizeable particles.

In view of the various methods now available to study cometary debris including infrared observation from spacecraft such as IRAS, perhaps it is time to re-evaluate the total contribution of comets to particulate material in the interplanetary complex. Several investigators including Delsemme (1976), Roser (1976), Kresàk (1980) and Mukai *et al.* (1983) doubt that comets can supply the few tonnes of material per second (Whipple, 1967) required to maintain the zodiacal particles, which are largely destroyed by collisions.

There remains a question as to whether the asteroid Phaethon (no. 3200, 1983 TB), apparently the parent body for the Geminid meteor stream, may be an old comet nucleus. The Geminid stream has a small aphelion distance (just beyond Mars's orbit) and a near-record small perihelion distance of 0.14 AU. The observed colour of Phaethon has been in doubt. Tholen (1985) reports that broad-band photoelectric photometry at five wavelengths from 0.3 to 0.9 μm show Phaethon to be slightly bluer than the Sun, implying a rare F classification, whereas Cochran & Barker (1984) and Belton *et al.* (1985) find it to be of S class on the basis of spectroscopic observations. The S classification would place Phaethon colourwise among typical asteroids whereas the F classification would mean it is possibly cometary.

The Geminid meteoroids themselves are very dense relative to cometary stream meteoroids. Verniani (1967, 1969) finds their density to be *ca.* 1.0 g cm^{-3}, about three times the average for those in streams. Whether this high density represents simply the survival of the toughest bodies in short-period orbits so near to the Sun, or whether it represents basically a meteoritic density of carbonaeceous chondritic nature remains an open question.

The general question as to whether ageing comet nuclei may become indistinguishable from rocky asteroids became an obvious issue with the introduction of the icy conglomerate model for the cometary nucleus. Substantive evidence to settle the question remained elusive until recent years. An important related question concerns the source of the near-Earth asteroids, the Aten–Apollo–Amor groups, in orbits somewhat resembling those of old comets. These bodies have quite finite lifetimes against planetary collisions, measured in tens of millions of years. Ageing comets seemed to be a likely renewable source for such kilometre-sized bodies whereas the asteroid belt seemed impotent to renew the supply.

Infrared spectroscopy and photometry have now been applied to a large number of asteroids and also to a few inactive comet nuclei at great solar distances. Photometry alone provides a comparison of the spectral reflectivities of the bodies in question, and the addition of diameter measures, either directly or via temperature measures, adds a knowledge of the albedos.

A very thorough study of the superficial appearances of comets compared with the various classes of asteroids has been made by Hartmann *et al.* (1987). Their compilation of cometary albedos (geometric reduced to visual wavelengths) from 13 measures or averages (17 comets in all) leads to a mean value of 0.051 ± 0.010. The values range from 0.01 to 0.13 with σ for one determination of ± 0.037. The scatter may well arise largely from measuring errors and from dusty atmospheres. The mean value is in excellent agreement with the values determined for the nucleus of Halley's Comet by the *Vega* and *Giotto* space probes, 0.04. The Moon's geometric albedo is 0.115, with a Bond albedo of 0.065. If the typical cometary nucleus geometrically scatters and reflects light like the Moon, the Bond albedo of comets would average about 0.02, extremely black indeed!

The distribution of infrared photometric colour parameters for comets are restricted on the V–J against J–K diagram and on the J–H against H–K diagram (see Hartmann *et al.* 1985) to regions occupied by asteroids of colour classes C, P, and D, particularly class D (see Gradie & Tedesco 1982, for definition). The C-, P- and D-class asteroids have extremely low albedos and are somewhat reddish. They occupy the outer regions of the asteroid belt, including the Trojans, which move near the lagrangian points in Jupiter's orbit. Hidalgo, long recognized as an asteroid having an orbit like that of a short-period comet, is, for example, in colour class D as are the tiny 1983 SA and 1984 BC, with aphelia also exceeding 5.3 AU. The three comets P/Neujmin 1, P/Arend-Rigaux and P/Schwassmann-Wachmann, 1, are also of colour class D, (Hartmann *et al.* 1987). The mysterious Chiron, in a 'chaotic' orbit between Saturn and Uranus, has a similar colour, a subset of class C. Thus Hartmann *et al.* (1986) find that 11 asteroids with orbits suggesting a possibly cometary origin fall in the colour classes of D(5), P(1), C(1), and C-like (4).

On the other hand, among 13 Aten–Apollo–Amor objects Hartmann *et al.* (1986) find only one in the C class (an Aten) and the others in more typical asteroidal colour classes. Because meteorites are probably mostly fragments from asteroids in near-Earth orbits, the apparent asteroidal character of these bodies is consistent with the chemical nature of meteorites as compared with Brownlee particles. In his accompanying paper Wetherill (this symposium) elaborates on Wisdom's (1983) theory concerning the chaotic perturbations of the Aten–Apollo–Amor asteroids from the heart of the asteroid belt.

Because of the similarities of the apparent surfaces of the outer asteroids to those of comets and also to some of the icy satellites of the giant planets, Hartmann *et al.* (1987) support the thesis that ices formed near Jupiter's orbit and beyond during the formation of the Solar System. In their picture, Jupiter comets, Saturn comets, etc., accreted as building blocks of the outer planets and also contributed to the satellite systems. Because of violent collisional losses near the giants, Jupiter and Saturn, the comets of the Öpik–Oort Cloud may have been derived primarily from the comets near Uranus and Neptune.

COMETS TO ASTEROIDS: HOW?

There are at least three obvious circumstances whereby comets could develop into bodies superficially like asteroids and yet another process that might lead to similar results

(*a*) In their growth, comets may have first accreted from rocky material and later added a dusty-ice envelope. Such a comet in a short-period orbit would eventually sublime away its icy envelope and become an inactive asteroidal body.

(*b*) Large comets with radii greater than perhaps 20 km may have accreted from dusty-ice particles but have been heated by radioactivity until the volatiles were removed from their cores. When finally exposed, these cores might appear asteroidal. This subject has been discussed by Whipple & Stefanik (1966). Only if the accumulation of comets occurred in a time less than a few million years could radioactive ^{26}Al, if present, have heated the cores of small comets. Otherwise the usual radioactive elements of ^{40}K, etc. could have, but only for rather large comets.

(*c*) The coarser meteoritic material in active comets may fall back to the surface, insulate and finally choke off cometary activity even though an icy core remains. Only H_2O ice would probably remain in the core as its temperature would probably have risen to a level that would sublimate more volatile ices or amorphous ices.

(*d*) More speculatively, collisions among comets during their accretion period or even later may have volitilized much of the ice and left large volumes of meteoritic material throughout the cometary bodies. The destructiveness of such collisions has not been studied in detail, being barely suggested by Donn (1963). Possibly, though, such collisions may have returned most of the materials of the comets to the solar (or primitive) nebula. The kinetic energy at a velocity of 2.3 km s^{-1} equals the latent heat of vaporization of extremely cold H_2O ice (49000 J mol^{-1}). If, for example, comets of the Öpik–Oort Cloud were the primary building block of Uranus and Neptune, those we have recovered must have been removed early in the evolution of those planets before large velocities were built up by the growing protoplants. Or perhaps these have luckily escaped head-on collisions. The nucleus of Halley's Comet, incidentally, appears to be a badly battered body.

The observations of comets produce indirect evidence that the active comets we observe may not have large meteoritic cores, thus weighing against circumstances (a) and (b). Sekanina (1977), in his study of split comets, finds that the smaller pieces broken from comets move away from their primaries by the action of differential, non-gravitational jet forces radial to the Sun. Their survival times correlate with their non-gravitational forces according to about the same logarithmic relation independent of the survival time or identity of the piece. Thus eighteen pieces broken off from a dozen different comets are somewhat similar in structure. No evidence to date suggests that split comets are inherently different from other comets but, of course, rocky cores may not have split off.

The various members of the Kreutz Sun-grazing family of comets appear to be much alike. They almost certainly are pieces recently broken tidally from a very large comet. For this, however, asteroidal inclusions, if present, probably would not have been observed.

Several comets appear to have disintegrated and disappeared while under observation. Sekanina (1984) has described the process for a half dozen, including the most famous, P/Westphal, 1852 IV, that faded out on the way to perihelion as 1913 VI, never to be found again. His description is: 'When discovered, they typically display a prominent, star-like central condensation, but a fading sets in very suddenly and the central condensation disappears usually in a matter of days. At the same time the coma is often (but not always) expanding gradually and becomes progressively elongated. Its surface brightness is decreasing (sometimes with erratic light variations superimposed) until the comet's whole head completely vanishes. The tail can become the brightest part of the object and survive the head.'

If these descriptions are truly to be accepted as recording the death throes of small comets, it is interesting that Sekanina usually finds the remnant dust tails to consist of particles greater than 50–100 µm. In contrast, the end of the Sun-grazer 1887 I, undoubtedly a broken piece, involved micrometre and submicrometre particles, typical of most ordinary comets. Were the initial grains at the very cores of comets typically greater than 100 µm in dimension? In any case, these disintegrating comets also tend to support the thesis that active comet cores do not consist of huge meteoritic or rock aggregates.

Half a dozen or so other short-period comets have not been rediscovered even after thorough searches with improved telescopic equipment. Probably they were active for a while and then lay dormant. Thus a few comets seem to have died passively, ceasing to show hazy or central condensations even near perihelion. If discovered now, their stellar appearance would cause them to be called asteroids in short-period cometary orbits. Marsden (1970) lists two such examples and discusses some of the orbital considerations. All of these comets probably have thick layers of meteoritic debris that insulate the ices of their interiors from rapid solar heating. One wonders how many of the Trojan asteroids or those of D, P or C colour classes would develop comas and appear cometary if they should collide with sizeable interplanetary bodies.

The high temperatures observed on the inactive surface of the nucleus of Halley's Comet demands that this surface be covered with a good insulating material, presumably dust. The evidence that Encke's comet (Whipple & Sekanina 1979) has one hemisphere that is now mostly inactive supports the hypothesis that accumulated particulate debris covers underlying ices and curtails comet activity.

Many of the larger particles raised by sublimation must fall back to the nucleus. The largest ones may never rise from the surface. Many must survive breakage from wasting on slopes and mesa-like elevations, from material falling back, and from thermal stresses of night to day on a rotating nucleus. In fact it seems easier to imagine processes to choke off comet activity than to promote it.

All of this evidence is consistent with comets having been formed by the accretion of interstellar-type dust at extremely low temperatures, the basic material being of the nature described and studied in the laboratory by Greenberg (1984).

Space missions to comets and asteroids are clearly needed to lead to an understanding of the nature and origin of these fundamental building blocks of the Solar System.

CONCLUSIONS

Comets contribute most of the particles that produce the Zodiacal Light and intercept the Earth's atmosphere as meteors. The colorimetric and reflective characteristics of comets and the outermost asteroids are so similar as to suggest that some short-period comets finally become inactive, indistinguishable from some asteroids. But this is probably not true for the near-Earth asteroids. The cometary evidence suggests that ageing comets may become inactive because they are choked by overlying particulate material that prevents solar heat from sublimating the underlying ices. No strong evidence suggests that pristine cometary cores are intrinsically rocky. Space missions to comets and asteroids are urgently needed.

This study has been supported by the Planetary Geology Program of the U.S. National Aeronautics and Space Administration.

REFERENCES

Belton, M. S. J., Spinrad, H., Wehinger, P. A. & Wychoff, S. 1985 International Astronomical Union circular no. 4029.

Campbell, D. B., Harmon, J. K. Hine, A. A., Shapiro, I. I. & Marsden, B. G. 1983 *Bull. Am. astr. Soc.* **15**, 800 (abstract).

Cochran, A. L. & Barker, E. S. 1984 *Icarus* **59**, 296–300.

Delsemme, A. H. 1976 In *Interplanetary dust and Zodiacal Light* (Proc. IAU Colloq. no. 31) (ed. H. Elsasser & H. Fechtig), vol. 48, pp. 314–318.

Delsemme, A. H. 1977 *Comets asteroids meteorites*. University of Toledo Press.

Donn, B. 1963 *Icarus* **2**, 396.

Elsson-Steel, D. 1986 *Mon Not. R. astr. Soc.* **219**, 47–74.

Gradie, J. & Tedesco, E. 1982 *Science, Wash.* **216**, 1405–1407.

Greenberg, J. M. 1984 *Adv. Space Res.* **4** (9), 211–212.

Hartmann, W. K., Cruikshank, D. P. & Tholen, D. J. 1985 *Ices in the Solar System* (ed. J. Klinger, D. Benest, A. Dollfus & R. Smoluchowski), pp. 169–181. Dordrecht: D. Reidel.

Hartmann, W. K., Tholen, D. J. & Cruikshank, D. P. 1987 *Icarus* **69**, 33–50.

Kresàk, L. 1980 *Solid particles in the Solar System* (ed. E. Halliday & B. A. McIntosh) pp. 211–222. Dordrecht: D. Reidel.

Marsden, B. G. 1970 *Astr. J.* **75**, 206–217.

McDonnell, J. A. M. 1978 *Cosmic dust*. Chichester: John Wiley and Sons.

Mukai, T., Mukai, S., Schwehm, G. H. & Giese, R. H. 1983 *Cometary exploration.* (ed. T. I. Gombosi) Hungarian Acad. Sci., vol. 11, pp. 135–141. Hungarian Academy of Science.
Oort, J. 1950 *Bull. Inst. Netherlands* **11**, 259.
Öpik, E. J. 1932 *Proc. Am. Acad. Arts Sci.* **67**, 169.
Roser, S. 1976 In *Interplanetary dust and Zodiacal Light (Proc. IAU Colloq. No. 31)* (ed. H. Elsasser & H. Fechtig), vol. 48, pp. 319–322. New York: Springer-Verlag.
Sekanina, Z. 1977 *Icarus* **30**, 574–594.
Sekanina, Z. 1984 *Icarus* **58**, 81–100.
Tholen, D. J. 1985 *International Astronomical Union circular* no. 4034.
Verniani, F. 1967 *Smithson. Contr. Astrophys.* **10**, 181–195.
Verniani, F. 1969 *Space Sci. Rev.* **10**, 230–261.
Whipple, F. L. 1967 *The Zodiacal Light and the interplanetary medium* (ed. J. L. Weinberg), pp. 409–426. NASA SP-150.
Whipple, F. L. & Sekanina, Z. 1979 *Astr. J.* **84**, 1794–1909.
Whipple, F. L. & Stefanik, R. P. 1966 Nature et origine des cometes. *Soc. Roy. Liège.* (5) **12**, 33–52.
Wilkening, L. L. 1982 *Comets*, University of Arizona Press.
Wisdom, J. 1983 *Meteoritics* **18**, 422–423.

Discussion

G. TURNER (*Department of Physics, University of Sheffield, U.K.*). To what extent is the surface roughness observed by Giotto thought to be the product of impacts or internal action such as that giving rise to the jets? With regard to impacts has anyone calculated the likely history of collisions during Halley's presumably brief time in the inner Solar System?

J. A. M. McDonnell (*Unit for Space Sciences, University of Kent at Canterbury, U.K.*). Regarding the question of the origin of surface features: a figure for the erosion rate from solar heating of 1 m per perihelion passage has been estimated before these spacecraft encounters. Considering the probable age of Halley of 20000 years in the inner Solar System, and the erosion rate that can be estimated now by the *in-situ* measurements of Halley, we would estimate some metres to be lost from the surface every 76 years. The surface is therefore fresh and its outer surface would not be expected to bear resemblence to its original state in the Oort cloud or carry the effects of impact cratering. We have to explain Halley's surface, therefore, solely in terms of ablation and erosion under solar heating.

M. K. WALLIS (*Department of Applied Mathematics and Astronomy, University College, Cardiff, U.K.*). Professor Whipple was billed to give a review of cometary science; what I have heard amounts to cometary geography with perhaps a little cometary geology. It in no way justifed the preposterous claim that his icy-conglomerate model has been 'nicely substantiated by the space missions'. Indeed, the dust analysers on both missions have shown not a conglomerate mixture of ices with mineral dust, but largely organic grains with little metallic or silicon components. The black surface covering most of the nucleus is presumed carbonaceous, not the favoured mineral dust. There are two or three large ougrassing regions, not the small-scale 'weathering' inhomogeneities that he depicted. He has in recent years adopted the variation of 'exotic ices' following processing by cosmic rays, that give complex organic molecules in 'wild' comets arriving after some millions of years in the outer Solar System (Whipple 1986). Yet Halley's comet showed its complex organics after 76 years away from us. He has modelled a sublimating rotating nucleus as uniformly heated over the surface (Whipple & Huebner 1976). The Halley-environment modellers did not adopt that extreme of isotropic outgassing, but were still a striking failure with 35% of the total emission coming from the nightside in their model (Devine 1981). To many of us, the icy-conglomerate reference model, characterized by sublimating ices with inert dust, has turned out to be misleadingly inflexible. It is time for us to drop its restrictive outlook and develop models that can explain the wealth of phenomena detected by the spacecraft and ground-based studies.

References

Devine, N. 1981 ESA SP-174, pp. 25–30.
Whipple, F. L. 1986 *The mystery of comets*, ch. 20. Smithsonian Institution Press.
Whipple, F. L. & Huebner, W. F. 1976 *A. Rev. Astr. Astrophys.* **14**, 143–172.

F. L. WHIPPLE. I find only strong support for my 1950 icy-conglomerate model of comets in the space-probe observations of Halley's Comet. My theory always involved the ejection of material on the sunlit side of the nucleus. I never supported theories that involved heating of the side away from the Sun. The composition of the ices except for the predominate water ice still remains in question as does the mass fraction of 'earthy' material. Regarding Halley's Comet, I wrote in 1951 (Whipple 1951), 'Suppose the radius to be 10 km (probably a generous estimate)…'.

Reference

Whipple, F. L. 1951 *Astrophys. J.* **113**, 464–474.

Phil. Trans. R. Soc. Lond. A **323**, 349–367 (1987)

Printed in Great Britain

The history of Halley's Comet

By D. W. Hughes

Department of Physics, University of Sheffield, Sheffield S3 7RH, U.K.

The history of Halley's Comet can be approached in three ways. First we can start with the origin of the comet and follow it through its sojourn in the Oort cloud, its transition from a long period orbit into a relatively short-period orbit until we finally arrive at its present position as a typical middle-aged comet with a large associated meteoroid stream. Secondly we can retreat from four and a half thousand million years of history to a mere two thousand years and we can start our history with the first known record made of Halley's Comet by mankind. In our third approach we can wait until the comet actually received its name, a christening for which we must thank French mathematical astronomers, and then we can chart its highlights as the first periodic comet, the first predicted cometary return and more recently as the spur to considerable scientific endeavour and government spending. This paper will review all three approaches.

INTRODUCTION

Halley's Comet is justifiably famous. It belongs to that group of comets whose return to the Sun can be predicted and thus prepared for. At present this group only contains about 120 members and P/Halley (the prefix P indicating that its orbital period is well known) is the most active, one of the largest and is intrinsically the brightest. It has been seen with the naked eye at each of its previous thirty apparitions and has been brighter than the brightest star in the northern sky in A.D. 374, 607, 837 and 1066. It's brightness near perihelion is such that, in any random selection of historic comets, P/Halley appears at the frequency of about 1 in 8.

The fact that the orbital period of P/Halley is about 76 years, close to our allotted time span of 'three score years and ten', makes its perihelion passage a 'once in a lifetime' phenomenon for most of us. It is not commonplace like Comet Encke, which returns every 3.3 years. Nor is it like Comet Kohoutek speeding past in 1973 not to return again for around 70000 years when it will be long forgotten and probably unrecognizable. P/Halley punctuates history regularly and remorselessly, a celestial exclamation mark every 76 years.

A third reason for the scientific and popular excitement engendered by the return of this comet is associated with the man after whom it was named. The large majority of comets are named after their discoverers. P/Halley is one of the few exceptions. Dr Edmond Halley, M.A., LL.D., D.C.L.(Oxon), R.N., F.S.A., F.R.S., was, in my opinion, England's second greatest scientist (see Hughes 1985 *a*) and the man who, in June and July 1696, reported to the Royal Society that the comets of 1607 and 1682 had similar orbits and were probably returns of the same Solar-System minor body. Halley was the first man to calculate the orbit of this comet (see Hughes 1985 *b*), the first to prove its periodicity and the first to predict its (and any comets) return. Edmond Halley and his comet stand at the watershed of cometary science. Before his time, comets were regarded as evil omens, the precursors of doom, disaster, disease and the death of kings. Their physical nature, chemical composition and orbits were unknown. Isaac

Newton showed the world how to calculate the orbit of a comet; Edmond Halley applied this technique to the data of twenty-four comets. He proved that one comet was periodic and as such was predictable. He also surmised (erroneously as it turns out) that the comets of 1661 and 1680 were also periodic (129 a and 575 a respectively) and regarded all comets as being members of the Solar System (as opposed to interstellar).

Comets are usually divided into two groups. Long-period comets have periods in general between 10 ka and 10 Ma with aphelia a considerable distance from the planetary system, in that region of astronomical deep freeze between the Sun and its neighbouring stars. These comets have orbits which are inclined randomly to the plane of the Solar System. Short-period comets have been captured from the long-period group usually by the gravitational field of Jupiter. The mean short-period comet has an orbital inclination to the ecliptic plane of $10°$, a period of 6.5 years, an aphelion distance of 5.5 AU, a perihelion distance of 1.5 AU and a dirty-snowball nucleus of diameter less than 0.8 km (see Hughes 1982, 1985c). The mass loss suffered by a comet during perihelion passage is a function of its size and the perihelion distance of its orbit. Typically a known short-period comet will lose between 1 and 0.1 % of its mass at each apparition. So short-period comets are decaying quickly and have a finite lifetime.

P/Halley is a non-typical short-period comet. Its nucleus is considerably larger than average and its period of 76 years puts it into an intermediate class between the mean short-period and long-period comets. The orbit has a higher than average eccentricity, which, at aphelion, finds the comet out beyond the orbit of Neptune and 10 AU below the ecliptic. The most unusual feature is the inclination, $162°$. Halley's Comet has a retrograde orbit moving around the Sun in the opposite direction to the planets and the vast majority of short-period comets. The nucleus of P/Halley is only active when the comet is closer to the Sun than about 6 AU. For three years, centred on perihelion passage, the nucleus surrounds itself with a gaseous dusty coma; for two months before and five months after perihelion passage the coma is accompanied by a tail pointing in the antisolar direction. For the remainder of its period the comet is 'dead'.

THE TOTAL HISTORY

The origin of comets is still somewhat of a mystery. The subject has been reviewed recently by Bailey *et al.* (1986). Two main classes of theories are prevalent now. In one, comets were formed just before the planet-building that took place in the nebula of gas and dust that surrounded the early Sun. We can go one step further and suggest that comets are the remnants of a disk of planetesimals, some of which accreted to form the major planets Saturn, Uranus and Neptune.

As the solar nebula evolved, dust, gas and ice became concentrated in an equatorial plane. The temperature gradient across this disc was such that between Saturn and Neptune the main constituents were dust and snow, the former having a composition similar to carbonaceous C1 chondrites and the latter being mainly H_2O snow but containing many impurities such as CO_2, NH_3, CH_4, these probably being trapped in the H_2O matrix as clathrates. The dense equatorial disc consisted of dirty snow particles orbiting the Sun on circular, direct coplanar trajectories. Interparticle velocities were low and collisions led in the main to accretion. This process is summarized in table 1. The loss of mass during this planetary aggregation is considerable and this mass loss is all in the form of dirty-snowball planetesimals having sizes

TABLE 1

(The accretion of Uranus and Neptune took place in a dust ring centred on the Sun, inner radius *ca.* 15 AU, outer radius *ca.* 40 AU, thickness *ca.* 0.1 AU; the growth of these planets is illustrated below; the mass loss during this process was in the form of cometary nuclei.)

mass of largest planetesimal/g	number of bodies	timescale of aggregation/a	total mass of ring region/g
10^6	10^{25}	10–100	10^{31}
10^{14}	10^{16}	1000	10^{30}
5×10^{21}	4×10^7	10^6–10^7	2×10^{29}
4×10^{27}	50		2×10^{29}
10^{29}	2 (i.e. Uranus and Neptune)	3×10^8	2×10^{29}

ranging up to around 10^{22} g. These planetesimals are cometary nuclei. Orbital perturbations induced by close encounters with growing protoplanets led to three general results. The new comets could be perturbed into short period orbits of low perihelion distance. Here they decayed quickly. The comets could be perturbed into parabolic orbits, in essence being thrown out of the Solar System and condemned to a life wandering through the galactic disc in and out of the gravitational potential wells of other stars. A small percentage of the comets would be given slightly less than parabolic energies. Their orbits would have perihelia in the Saturn–Neptune region and aphelia between 2×10^4 and 2×10^5 AU from the Sun. These comets would return to the Sun with periods of between 1 and 10 Ma. As their perihelia were beyond 10 AU they would be inactive having no coma and no mass loss. This cloud of comets surrounding the Sun is named after its main proponent Oort (1950). Their numbers would be depleted either by comets being pulled away from the Sun by passing stars or by being perturbed into lower-perihelion-distance orbits in which they started to lose mass when close to the Sun and were in danger of being captured by Jupiter.

Jovian-capture is not a speedy process and usually takes tens of close planetary passages. During this time the eccentricity is lowered until the comet is finally perturbed into a short period orbit which has an aphelion at Jupiter's orbit as opposed to a perihelion.

The second class of origin theories has comets formed in the interstellar medium and specifically in giant molecular clouds close to regions of star formation. McCrea (1975) suggested that, under certain circumstances, the gas and dust in these regions could decouple from one another and the dust (with associated ice) could then satisfy the Jeans criterion. This being so, the gravitational energy of the region would exceed the thermal energy, leading to this specific region of the interstellar medium collapsing and forming a dirty-snowball aggregate; a cometary nucleus. These new comets are then captured by the Solar System during its passage through the giant molecular cloud. The capture process becomes more efficient if the relative velocity between the Sun and the cloud is low and the spatial density of comets in the cloud is high. As the Sun orbits the galactic nucleus it can thus augment its family of comets each time it passes through a giant molecular cloud. If, however, all the comets were formed at the dawn of the Solar System, the total number will have been decreasing ever since.

Both theories have the comets originating in giant molecular clouds, the first being the parent of our Sun and planets, the second being younger.

The next major highlight in the career of P/Halley was its capture by the planets from a long period, near parabolic orbit into its present day, near stable, inner Solar System orbit. As soon as the perihelion distance of the comet was reduced to bring it into the inner Solar System it started to decay profusely at each close passage of the Sun. Gas and small dust particles (mass up to 10^{-9} g) lost by the comet leave the Solar System and feed the interstellar medium. Large dust particles ($m > 10^{-9}$ g) leave the comet at lower relative velocities and move in orbits that have parameters very similar to those of the parent comet. These particles slowly gain on or fall behind the comet until, after a few thousand years in P/Halley's case, they form an annulus of dust around the comet's orbit. Twice a year the Earth intersects the stream of dust produced by the decay of P/Halley, once in late April and again in mid-October. The geometry is shown in figure 1. As indicated, the annulus of dust is broad, and the comet and mean stream orbits are separated by a few degrees.

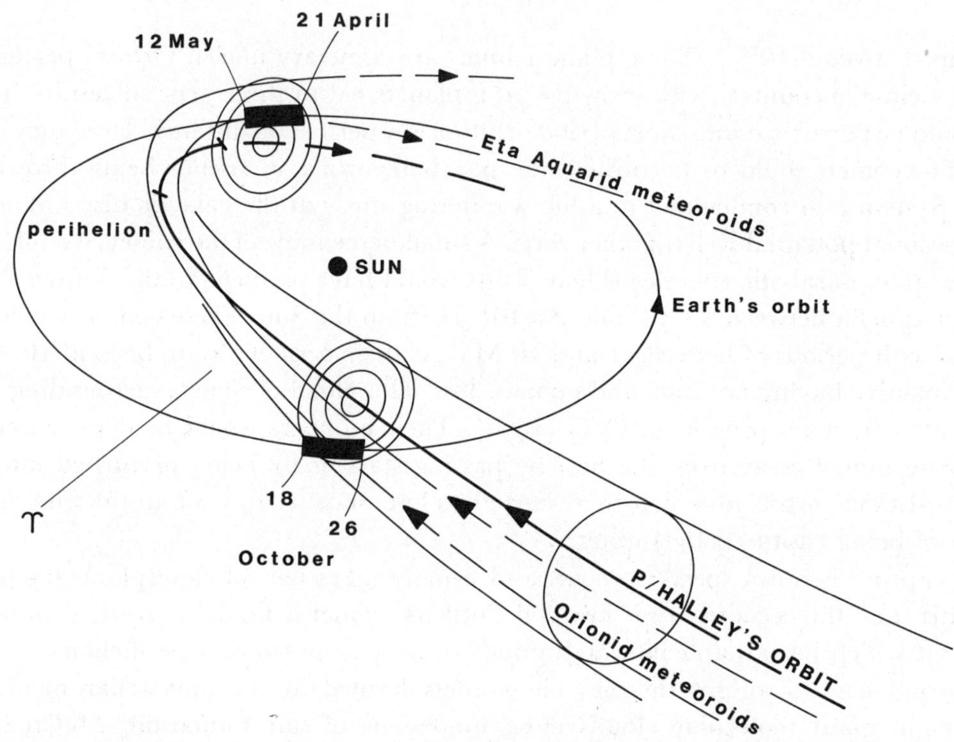

FIGURE 1. An annulus of dust surrounds the orbit of P/Halley and has been produced by the decay of the comet at previous apparitions. Earth intersects the annulus in October, when Halley dust is 'seen' as the agent responsible for the Orionid meteor shower and in April and early May as the Eta Aquarid shower. (The dashed portions of the orbits are below the ecliptic.) In May the Earth passes within 0.065 AU of the comet orbit and in October within 0.154 AU.

The overall history of P/Halley can be assessed by measuring four quantities, the total mass of dust in the dust annulus surrounding the orbit of the comet, the size of the dirty-snowball nucleus now, the density of that nucleus and the mean mass loss of the comet over recent apparitions (see Hughes 1985 d).

McIntosh & Hajduk (1983), using meteoroid shower observations, estimated that the spatial density in the annulus in the region that produced the maximum of the Eta Aquarid shower was 3×10^{-24} g cm^{-3}. In October the Orionid maximum spatial density was about 10^{-24} g cm^{-3}. By extrapolation the authors concluded that the density near the cometary orbit was about 5×10^{-24} g cm^{-3}. Integration of these dust densities around the whole dust annulus gave a total dust mass of the stream of about 5×10^{17} g. During cometary decay low-mass dust particles and gas are also lost by the nucleus. The assumption that there is twice as much mass in this form as there is present in large dust particles leads to the conclusion that in the past P/Halley has lost about 1.5×10^{18} g.

Observations of the nucleus of P/Halley by the multicolour camera on the *Giotto* fly-by mission (Keller *et al.* 1986) show that the mean effective cross-sectional area of the nucleus is 100 km^2, which leads to the radius of the equivalent spherical nucleus being 5.6 km. (The nucleus is irregular in shape being at least 15 km long, *ca.* 10 km wide and having an albedo of about 0.04.)

The density of the nucleus is not known. Comet modellists (see, for example, Newburn & Reinhard 1981) usually assume a density of 1.0 g cm^{-3}. Wallis & Macpherson (1981) concluded that the observed non-gravitational forces can only in general be reconciled with H_2O outgassing if the mean density of the dirty-snowball nucleus is below 0.7 g cm^{-3}. If the density of the nucleus of P/Halley is assumed to be 0.5 g cm^{-3} (a compromise between the density of rock 2.8 g cm^{-3}; ice 0.9 g cm^{-3} and snow 0.08 g cm^{-3}) this, coupled with a radius of 5.6 km, yields a mass of 3.7×10^{17} g. Adding this to the mass that P/Halley has lost in the past indicates that the initial mass of the comet, before its capture into the short period orbit, was of the order of 1.9×10^{18} g and its radius was 9.6 km.

It can be easily calculated from the figures given by Newburn (1981) that P/Halley at its 1910 apparition lost around 5.1×10^{36} molecules. Spectra of the comet taken in 1910 have been interpreted (Newburn & Reinhard 1981) as indicating that the gas emitted by the comet is 83.4 % H_2O, the remainder being molecules of a mean molecular mass 44 u. Therefore the mean molecule mass of all the molecules leaving the comet is 22.3 u, i.e. 3.7×10^{-23} g. Thus the total gaseous mass loss during the 1910 apparition was about 1.9×10^{14} g. Following our previous assumption this is equivalent to a total gas and dust loss of 2.8×10^{14} g.

If the cometary nucleus has a radius of 5.6 km and a mean density of 0.5 g cm^{-3} this is equivalent to it losing a layer of thickness 140 cm from its surface during the 1910 apparition. Most models of cometary decay (see Hughes 1983a) indicate that a comet with an orbit of constant perihelion distance will lose a layer of dirty snow of constant thickness each time it passes perihelion. So for P/Halley to slim from a radius of 9.6 km to its present radius of 5.4 km would take about 3000 perihelion passages. Likewise to decay completely from its present size would take a further 4000 or so close approaches to the Sun. Assuming that the orbital period remains reasonably constant at about 76 years it can be concluded that the comet was captured by Jupiter into its present orbit some 200000 years ago and will, some 300000 years hence, disappear completely, having all decayed into the associated meteoroid stream. In cometary evolution terms P/Halley is typically middle-aged, well behaved and of steady and predictable activity.

Each time P/Halley passes perihelion it loses mass, the nucleus decreases in size and the absolute brightness decreases. The apparent magnitude of a comet follows an equation of the form

$$m = H_0 + 5 \lg \Delta + 2.5n \lg r, \tag{1}$$

where H_0 is the absolute magnitude of the comet (the magnitude it would have if seen at zero phase angle when it was both 1 AU, from the Sun and 1 AU from the Earth), Δ is the comet–Earth distance, r the comet–Sun distance and n the activity index (the molecular efflux from a comet near perihelion is proportional to r^{-n}). In the main, the index n is 4.0 and when this is used in (1), H_0 is replaced by H_{10}. If one assumes that the brightness of a comet is proportional to the surface area of its nucleus it can easily be shown that

$$\Delta H_{10} \approx -0.724 \, \Delta M/M, \tag{2}$$

where $\Delta M/M$ is the fractional mass loss per apparition and ΔH_{10} is the change in absolute magnitude per apparition (see Hughes & Daniels 1983). From the figures given above, $\Delta M/M$ for P/Halley in 1910 is 7.6×10^{-4}. So ΔH_{10} is 5.5×10^{-4} per apparition.

Morris & Green (1982) found that P/Halley in 1910 had a preperihelion absolute magnitude of 5.49 ± 0.07 and a postperihelion absolute magnitude of 5.44 ± 0.05. The logarithmic mean of these values is 5.465. With brightness proportional to surface area it is found that the radius R of a cometary nucleus is proportional to $10^{-0.2H_{10}}$ and its mass M to $10^{-0.6H_{10}}$. The use of P/Halley as a typical comet gives the general relations

$$\lg M = 21.166 + \lg \rho - 0.6 H_{10}$$

and

$$\lg R = 1.842 - 0.2 H_{10}$$

(where M is in grams and R in kilometres and ρ is the density in grams per cubic centimetre). So P/Halley when it was first captured by Jupiter, 200000 years ago, had an absolute magnitude of $+4.30$ making it intrinsically only three times brighter then than it is now.

THE RECORDED HISTORY

Until now, the earliest identifiable record of P/Halley is in the Annals of the Shih-chi. In the seventh year of Emperor Chin Shih-huang, 240 B.C., 'a broom star first appeared at the eastern direction; it was then seen at the northern direction. During the fifth month (24 May–23 June) it was again seen at the western direction' (see Stephenson & Walker 1985).

Edmond Halley (1705) recognized the comets of August–September 1682, October 1607 and August 1531 as being 'his' comet and he guessed correctly that the June 1456 comet was a previous appearance. Hind (1850) (and Laugier 1843) extrapolated backwards to include the comets of 1378, 1301, 1223, 1145, 1066, 989, 912, 837, 760, 684, 608, 530, 451, 374, 295, 218, 141, 66 and 12 B.C. Unfortunately Hind underestimated the tenacity and skill of later astronomical historians by writing 'Previous to the year 12 B.C. the accounts of comets become so vague that it would be vain to attempt to carry the inquiry into more remote antiquity'. He also made four mistakes, twice he picked the wrong comet in the year (912 and 837) and also P/Halley returned in 1222 and 607 not 1223 and 608. The long-term motion of P/Halley can be followed by using detailed celestial mechanics, including the effects of perturbations induced by close passages to the planets. Yeomans & Kiang (1981) calculate that perihelion passages occurred in, for example, September 315 B.C., September 391 B.C. and July 466 B.C. and continue back (with decreasing precision) to October 1404 B.C. From brightness considerations alone any ancient comet record has a 1 in 8 chance of being P/Halley so there is still hope for those willing to continue Hind's 'vain attempt'.

P/Halley, like all members of the Solar System does not have a fixed orbit. The pushes and pulls exerted by the gravitational fields of the main perturbing planets are continually in evidence. The orbital period, or to be more precise the time between perhelion passages, is shown in figure 2. The peaks near A.D. 374 and A.D. 1145 are due to a 2:13 commensurability

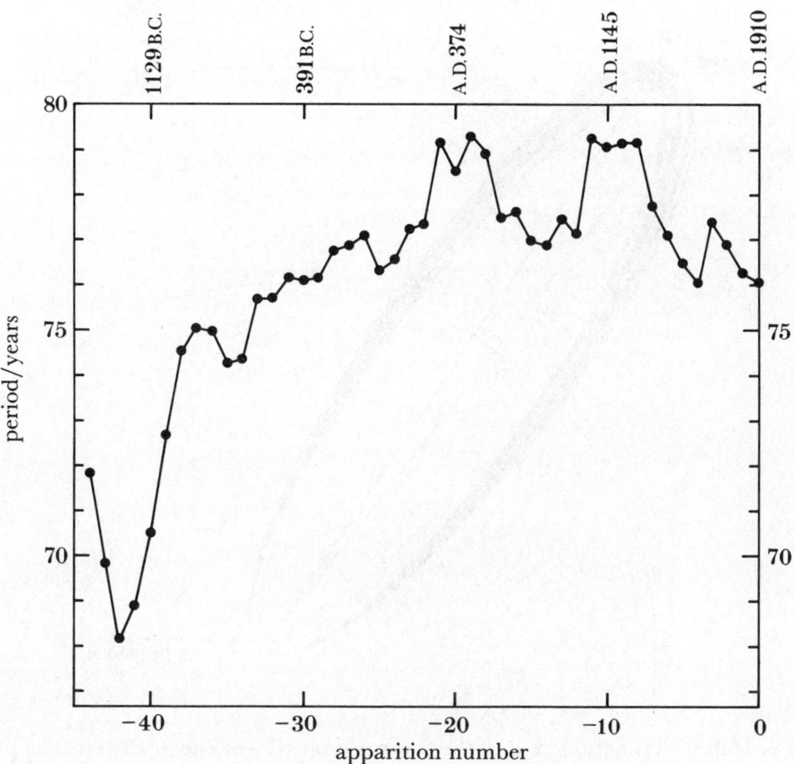

FIGURE 2. The orbital period (time between perihelion passages) of P/Halley for apparitions between 1404 B.C. and A.D. 1910 (data taken from Yeomans & Kiang 1981). So far the earliest identified record of P/Halley is that of June 240 B.C.

between the orbital periods of the comet and Jupiter. Over the past 46 apparitions the mean period has been 76.07 a. The retrograde orbit is precessing slowly and this can be seen from figure 3 in which the orbits between 87 B.C. and A.D. 1910 have been traced out in the mean orbital plane. This precession coupled with the obvious planetary perturbation helps explain why the associated meteoroid stream is so broad.

Hughes (1983b) has investigated the way in which the absolute magnitude of P/Halley has changed during recorded history. This is shown in figure 4. The kindest conclusion is that the absolute magnitude has not changed recognizably during the past 2000 years and has an average value of about 5.40. If, as calculated above, ΔH is 5.5×10^{-4} per apparition, the change in absolute magnitude over the 25 apparitions shown in figure 4 would be $+0.014$, a value entirely in keeping with the previous conclusion.

So P/Halley has not become detectably less luminous during the past few thousand years. It's brightness as recorded by earthbound observers is entirely due to the relative position of our planet at the time of the comet's maximum activity. This is illustrated in figure 5, a diagram which is drawn in a coordinate system which has the Sun and the perihelion of the orbit of P/Halley stationary over the past 2300 years. When the comet is at perihelion the Earth is at the point shown on its orbit. Dashes interior to the Earth's orbit are roughly at monthly intervals. The comet markers along the cometary orbit are at intervals of 10 days. The tail lengths have been obtained by averaging the observations of the 1759, 1835 and 1910 apparitions. Figures above each comet marker represent the reduced magnitude M_{R} of the comet. The apparent magnitude can be estimated by adding $5 \lg \Delta$ to M_{R}, where Δ is the comet–Earth distance in astronomical units.

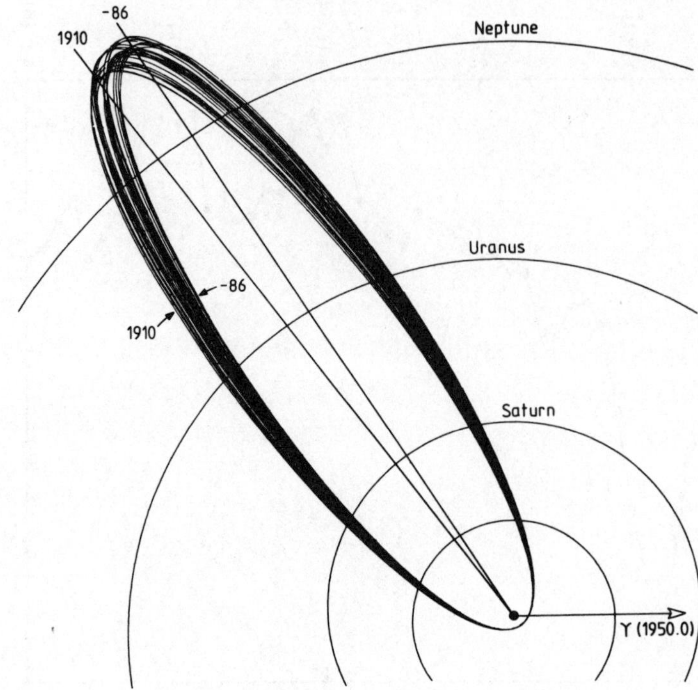

FIGURE 3. The Orbit of Halley's Comet referred to the first point of Aries at epoch 1950.0. The plane of the paper is the plane of the orbit and the curves marked Neptune, Uranus and Saturn have semimajor axes of 30.0, 19.2 and 9.5 AU, respectively, and have been drawn in that plane.

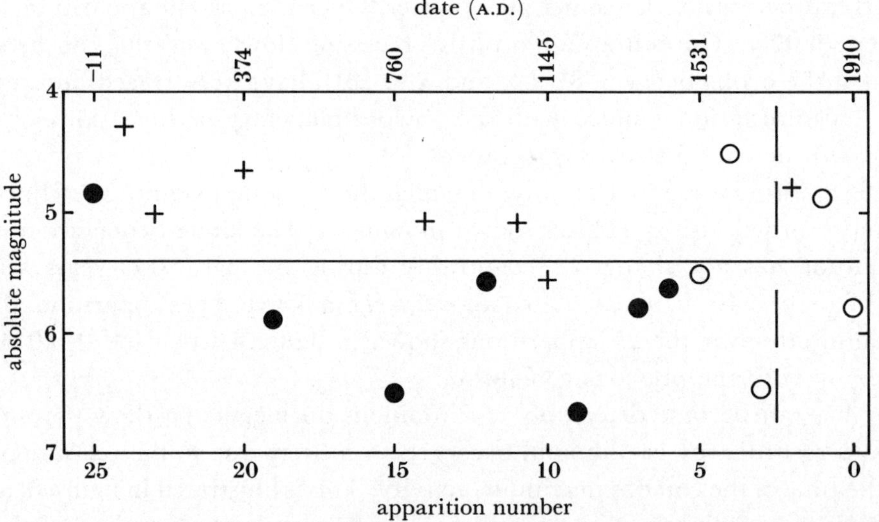

FIGURE 4. The absolute magnitude of P/Halley plotted as a function of the apparition number (1910 ≡ 0) with data from Hughes (1983b). Crosses refer to the postperihelion discoveries, open circles to preperihelion discoveries and filled circles to preperihelion discoveries with favourable Sun and Moon positions. The thin horizontal line represents the mean of the data. The vertical dashed line separates the unexpected from the expected returns.

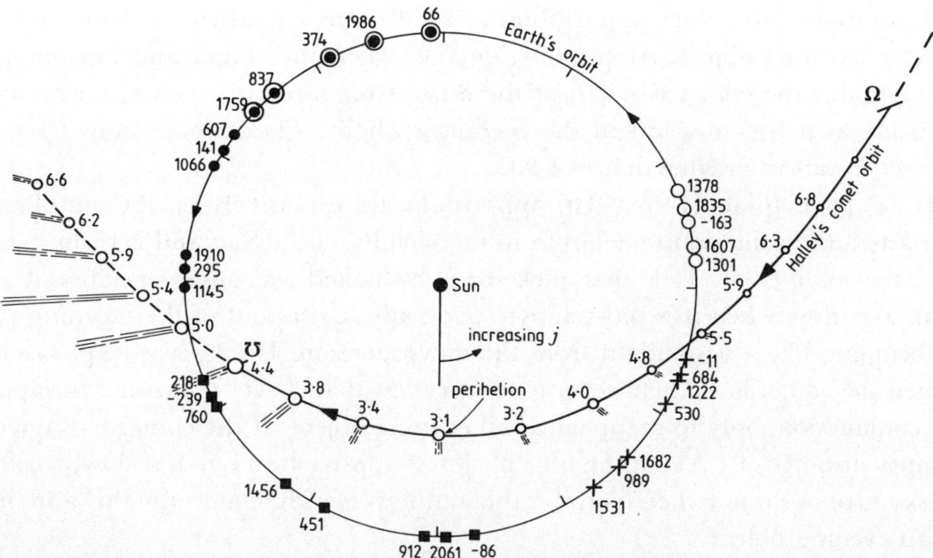

FIGURE 5. The symbols around the Earth's orbit represent the Earth's position at the perihelion passage time of Halley's Comet. Tick marks on the inside of the orbit are at approximately monthly intervals. Apparitions have been divided into five classes A, B, C, D, E. The comet is shown at intervals of 10 days along its path by a series of circles representing the size of the coma. This has a visible diameter of about 200000 km at maximum. The tail of the comet is represented to scale. The numbers above the cometary symbols are the reduced magnitudes. Apparent magnitude can be estimated by adding $5 \lg \Delta$ to this figure, Δ being the comet–Earth distance.

The angle j is the celestial longitude of the Earth minus the celestial longitude of the comet both measured at the time when the comet is at perihelion. The nodal positions given in this figure are for the 1986 apparition (see Hughes 1985 d).

The 'spectacularness' of Halley's Comet is mainly a function of the comet–Earth distance, the rule simply being the smaller the brighter. As the comet coma and tail are more developed after perihelion passage even more spectacular displays occur when comet and Earth meet near the descending node (℧ in figure 5).

Following Bortle & Morris (1984) the apparitions of Halley's Comet can be classified according to the type of visual display. In this paper we increase their three groups to five.

Class A (o) (A.D. 1301, 1378, 1607, 1835) apparitions appear exclusively on the right of figure 5. The smallest value of the comet–Earth distance occurs well before perihelion and is typically between 0.1 and 0.2 AU. The comet is a considerable distance from the Sun in the sky and is seen as a slowly brightening object in the morning sky. Class A returns are specially favourable for observers in the northern hemisphere. Following the comet's closest approach to Earth it moves close to the Sun in the sky thus becoming very difficult to see. It remains in the vicinity of the Sun until it has faded below the naked-eye limit. The comet is usually only seen preperihelion.

Class B (+) (A.D. 1222, 1531, 1682) apparitions have their lowest comet–Earth distance (0.15–0.4 AU) just before perihelion when the comet is interior to the Earth's orbit and in the direction of the Sun. Here Halley's Comet is a morning object. It brightens quickly and is usually seen for a month preperihelion, during which time the brightness changes little. It then disappears into the solar glare shortly after perihelion passage.

Class C (■) (A.D. 218, 451, 760, 1456, 2061) apparitions occur when the comet is at or just past perihelion at the time the comet is closest to Earth. The comet is only seen with the naked

358   D. W. HUGHES

eye for a short time during these apparitions. It usually appears rather suddenly and reasonably brightly as a morning object. At conjunction (i.e. when the comet and Sun have the same ecliptic longitude) the comet is north of the Sun. Afterwards it becomes an evening object, fading quickly as it becomes lost in the evening twilight. These apparitions have minimum Earth–comet distances greater than 0.4 AU.

Class D ($\bullet$) (A.D. 1066, 1145, 1910) apparitions are exclusively on the left of figure 5. As seen from Earth the comet approaches from the vicinity of the Sun and is rushing towards us, hidden in the solar glare. It is first picked up by naked-eye observers after it has passed perihelion, at a time when it is most active. Suddenly it rises out of the morning twilight, its long tail beaming like a searchlight from the dawn horizon. If the closest approach to Earth occurs when the comet has a heliocentric distance less than 1 AU the comet 'disappears' as it moves to conjunction, only to reappear as an evening object. If the comet has moved beyond a heliocentric distance of 1 AU at the time of closest approach to Earth it slowly drifts from the morning sky to opposition (when it crosses the southern meridian at midnight) and finally fades away as an evening object.

Class E ($\odot$) (A.D. 66, 374, 837, 1759, 1986) apparitions are rather a mixed bag, the nature

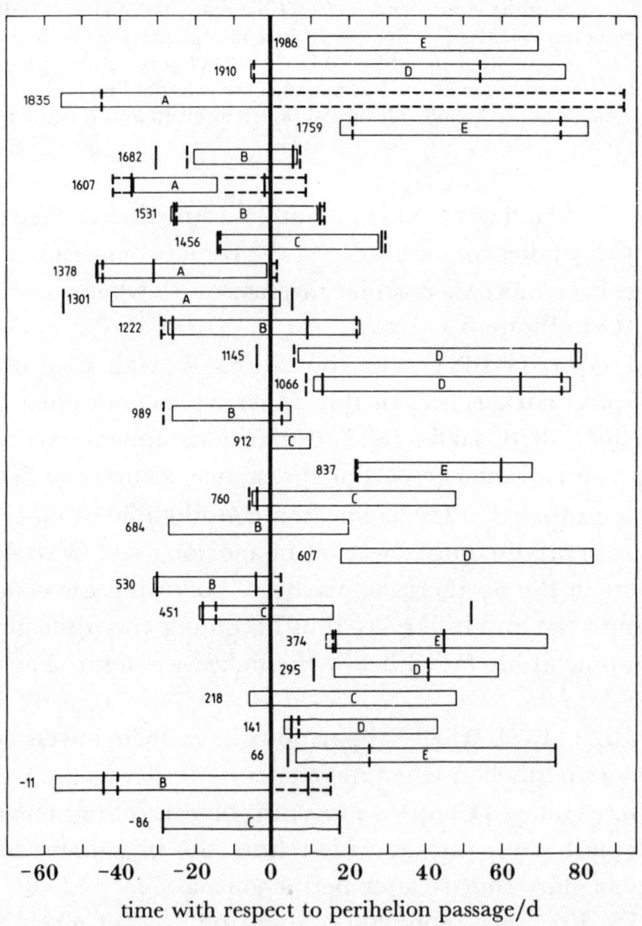

FIGURE 6. The period of naked-eye visibility of Halley's Comet at each apparition is represented by horizontal boxes. The thick central line corresponds to perihelion passage time. The five classes A, B, C, D and E are defined in figure 5. The end times of the full boxes have been taken from Bortle & Morris (1984), the heavy vertical bars from Yeomans & Kiang (1981) and the dashed vertical bars from Broughton (1979).

of the apparition depending on the position of the descending node. The minimum Earth–comet distance varies considerably, being only 0.03 AU for A.D. 837 but being 0.42 AU for the recent 1986 apparition, this later value of Δ occurring on 1986 April 11. The 1986 apparition would have stood a considerable chance of being overlooked by naked-eye observers in pretelescopic days. At its intrinsically most active phase the comet is on the opposite side of the Sun to the Earth making the return most unfavourable.

A survey of the historical records of P/Halley (Broughton 1979; Bortle & Morris 1984; Yeomans & Kiang 1981 and Yeomans *et al.* 1986) have revealed the time periods over which P/Halley has been seen (and recorded) by naked-eye observers. The periods, drawn with respect to perihelion passage time, are shown in figure 6, all data being derived from northern hemisphere observations. The five classes of observations clearly correlate with the position of Earth at the time of perihelion passage. The variation of the apparent magnitude of the comet during its periods of recorded naked-eye visibility are shown in figure 7*a*, *b*. It can be seen that the brightness varies considerably from apparition to apparition. In A.D. 837 the apparent magnitude reached nearly -4 making the comet, albeit only for a very short time, as bright as Venus. The miserly nature of the 1986 apparent magnitude curve is only too clear from figure 7*a*, *b*. The black dots on the curves in these figures indicate the times of first and last recorded sighting. These times vary considerably, mainly as a function of the class of apparition. Very roughly, the first records of classes A, B, C, D and E occurred at apparent magnitudes of 3.3, 2.8, 2.0, 0.6 and 1.8 and the last records at 2.5, 2.3, 1.8, 5.9 and 4.1 respectively. So any broad assumption of a fixed onset magnitude (as in Broughton 1979; Vsekhsvyatskii 1964) is fraught with danger.

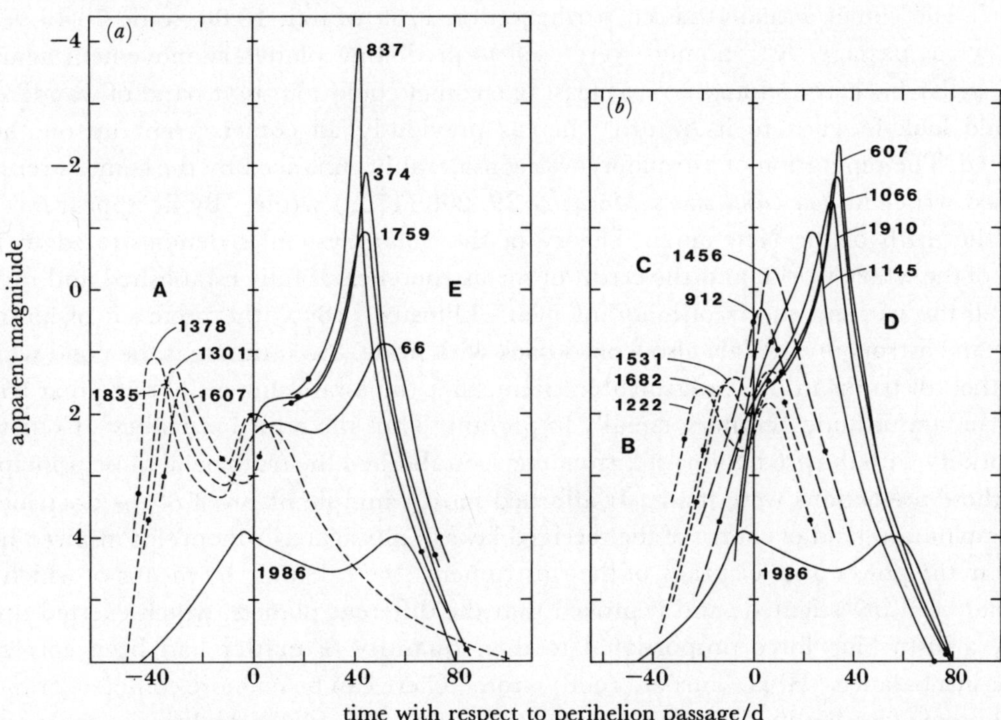

FIGURE 7. The apparent magnitude of P/Halley during its periods of naked-eye visibility plotted as a function of time with respect to perihelion passage, (*a*) for class A and E apparitions (*b*) for class B, C and D apparitions (see figure 5 for a definition) (data from Yeomans *et al.* 1986). The black dots indicate the time and apparent magnitude of the first and last recorded sightings.

The history of the named comet

The name 'Halley's Comet' seems to have been first used in France, the Abbé de la Caille (1759) being the first to suggest such an appellation, in May of that year. Charles Burney (1769) went so far as to state that 'the comet of 1759 is known throughout Europe by the name of Dr Halley's comet' and by the time of its subsequent return in 1835 this nomenclature had become commonplace.

The comet owes its prominence in scientific history to its proven periodicity. It was the *first* periodic comet, Edmond Halley recognizing that the comets of 1531, 1607 an 1682 had similar orbits and were thus probably caused by one and the same parent body returning every 76 years or so (we stress 'or so' because the period between 1531 and 1607 was 76.143 years and that between 1607 and 1682, 74.885 years). Bright periodic comets are rare. They are almost inevitably considerably fainter than the near-parabolic long-period comets. In fact, the ancient cometary records (pre-A.D. 1700) seem to contain only one periodic comet, this being P/Halley which, as we have said previously, seems to crop up at the rate of about 1 in 8. (Anyone with ample time on his hands could have great fun searching for P/Mellish (1917 I), which is the next brightest but has a period of about 145 years.) The list of periodic comets grew painfully slowly, the next additions being P/Lexell (seen only once, in 1770), P/Olbers (1815), P/Encke (recognized in 1819), P/Pons-Brooks, P/Pons Winnecke and P/Beila (recognized in 1826). By 1850 only P/Faye, P/de Vico and P/Brorsen had been added.

P/Halley was the first comet to return to the Sun as *predicted*. Edmond Halley wrote (in a later version of his synopsis, printed in Gregory (1726)) 'It is probable that its return will not be until after the period of 76 years or more, about the end of the year 1758, or the beginning of the next'. The comet actually passed perihelion on 1759 March 13.06, some 76.49 years after its previous passage. Astronomers were used to predicting planetary movement against the 'fixed' stellar background and now at least one comet could join that band of wanderers. People could look forward to its return whereas previously all comets crept up on them unannounced. The reputation of astronomy was considerably enhanced by the comet's return. An unnamed writer in *The Gentleman's Magazine* **29**, 206 (1759) wrote, 'By its appearance at this time, the truth of the Newtonian Theory of the Solar System is demonstrated to the conviction of the whole world, and the credit of the astronomers is fully established and raised far above all the wit and sneers of ignorant men'. Olmsted (1850) the professor of natural philosophy and astronomy at Yale also looked back with pride. 'So intimate is the bond which binds together all truths in one indissolvable chain, that the establishment of one great truth often confirms a multitude of others, equally important. Thus the return of Halley's Comet in exact conformity with the predictions of astronomers, established the truth of all those principles by which those predictions were made. It afforded most triumphant proof of the doctrine of universal gravitation, and of course of the received laws of physical astronomy; it inspired new confidence in the power and accuracy of that instrument (the calculus) by means of which its elements had been investigated; and it proved that the different planets, which exerted upon it severally a disturbing force proportional to their quantity of matter had been correctly weighed, as in a balance.' H. H. Turner (1908) wrote 'There can be no more complete or more sensational proof of a scientific law than to predict events by means of it. Halley was deservedly the first to perform this great office for Newton's Law of Gravitation.' 'Newton's great discovery of the Law of Gravitation, and the story of Halley's comet thus form an integral part of the most important event in the whole history of science.'

[112]

The prediction of the return of P/Halley led to the production of the first comet-finder charts. Figure 8 shows an example and this has been taken from *The Gentleman's Magazine* of September 1756. This monthly magazine contained news, articles of general interest, poems, history, book reviews and scientific pieces. It was published in arrears, the September issue containing September's news and letters, so the finder chart was printed over two and a quarter years

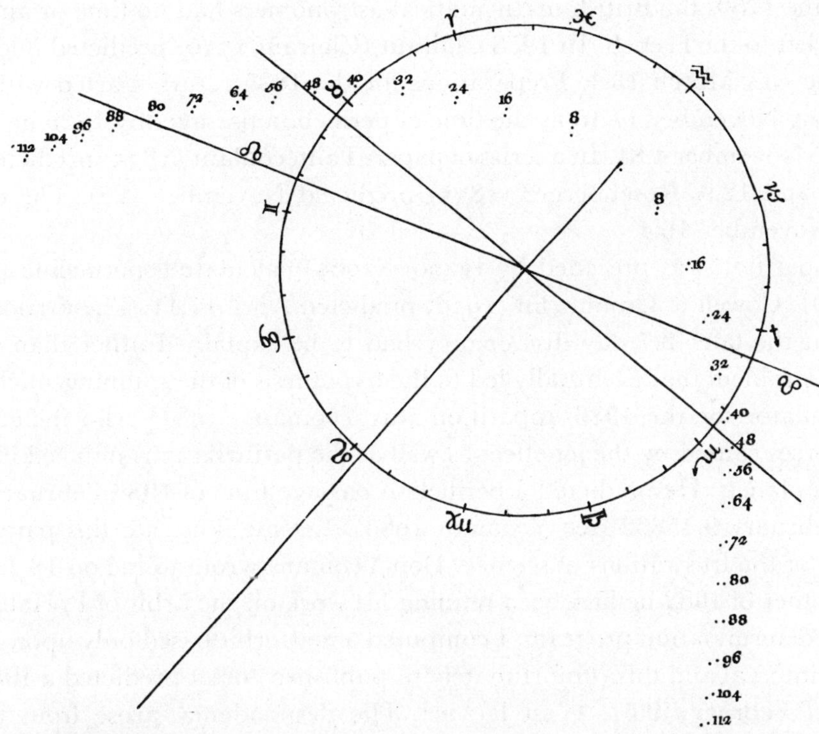

FIGURE 8. One of the first comet-finder charts. The Sun is at the centre of the large circle, which represents the orbit of Earth. The parabolic path is the projection of P/Halley's orbit in the plane of the ecliptic, the path being marked with numbers representing the position of the comet as a function of the number of days before and after perihelion passage. The ascending and descending nodes are represented by ☊ and ☋. Note that the comet moves in the opposite direction to Earth and the angle between the planes of the two orbits is 162°. To quote from *The Gentleman's Magazine* 26, 413: 'To find the comet's place at any time, count how long it is before or after its perihelion, and mark the place in the projection of the parabola: lay one edge of a parallel ruler through that point, and also through the point of the Earth's place in its orbit at that time, and the other edge passing through the Sun, will cut the Earth's orbit at the comet's geocentric place. The tangent of the inclination, taking the perpendicular from the comet's place to the line of the nodes as radius, is the tangent of its apparent latitude, making the curtate distance of the comet from the Earth, the radius.' (The second straight line through the Sun is at *ca.* 18° to the line of nodes and is to help people without tables of tangents do this calculation.)

before the comet was first seen. It had two roles. If perihelion passage time was accurately known the chart gave the rough zodiacal position of the comet over a time interval of 224 days centred on that time. If, on the other hand, a comet had been observed on a specific date and in a specific zodiacal position the chart could be used to predict its movement across the sky assuming that it was Halley's returning comet. The chart could thus be used to check to see if an observed comet actually was P/Halley. The writer of the article also knew that P/Halley had only been seen in the past during a time interval of two to three months each side of perihelion so he used this condition to draw up a table of where the comet may expect to begin to appear in any month of the year (see *The Gentleman's Magazine* 26, 413). People did not know when the comet was due; Edmond Halley's 'about the end of the year 1758 or the beginning of the next' had too easily slipped from the memory. Barker (1755) suggested the end of 1757!

In the mid-eighteenth century we can see P/Halley commence its role as the great spur to the calculations of accurate orbits and ephemerides. All the previous speculation as to its celestial position in 1759 would disappear if the perihelion passage time could be calculated accurately. In the 1750s the British mathematical astronomers had no time or aptitude for the task and it was left to the French. In 1958 Clairaut (Clairaut 1759) predicted mid-April 1759: the correct time was March 13.1. Preparations for the 1835 return started with Damoiseau (1820) predicting November 17.15 as the time of perihelion passage. By 1829 he had changed his prediction to November 4.81. In a series of papers Pontécoulant (1835) predicted November 7.5, 13.1, 10.8 and 12.9, Rosenberger (1835) predicted November 12.0. The comet passed perihelion on November 16.4.

The 1910 apparition was preceded by Ivanov (1909) calculating perihelion passage to be 1910 April 22.91. Cowell & Crommelin (1910) predicted April 17.11. These calculations were so accurate that the later 2.7 day discrepancy had to be explained other than by planetary perturbation, a problem that eventually led to the hypothesis of the spinning-nucleus jet effect. The main calculator for the 1986 apparition was Yeomans (1981) who included the non-gravitational forces caused by the jet effect as well as the perturbations induced in the orbit by the nine known planets. He predicted a perihelion passage time of 1986 February 9.66128. It occurred on February 9.45862 (see Yeomans 1986). Let me illustrate this paragraph by an example of one of the frustrations of science. Don Yeomans wrote to me on 18 January 1983. During the summer of 1982 he had been refining his work on the orbit of P/Halley 'Using an improved orbit determination program, I computed a new orbit based only upon the data over the 1759–1911 interval and this (unfortunately unpublished) orbit predicted a 1986 perihelion passage time of February 9.51. C'est la vie.' The despondency arose from the fact that P/Halley had been recovered on 1982 October 16 and the new observations completely vitiated calculation based only on the 1759–1911 sightings. But still, Yeomans was only 1 h 14 min out.

P/Halley's 1759 return had a negligible effect on our knowledge of the physical and chemical characteristics of comets. *The Gentleman's Magazine* **29**, 206 (1759) was 'disposed to look upon them as containing rather the materials and rudiments of habitable worlds...our earth in particular was once a comet...the Mosaic account of the creation is only an account of its reduction from that state to its present habitable form'.

By 1835 the burgeoning fame of P/Halley ensured that it was widely observed by the great astronomers of the day. F. W. Bessel, J. F. W. Herschel, W. Struve, C. P. Smyth and T. Maclear are typical examples. As is only to be expected it seems that the more they saw, the more complicated the comet appeared to be. 'Struve compared the appearance of the nucleus, about the end of the first week of October, to a fan-shaped flame emanating from a bright point; and subsequently to a red-hot coal of oblong form. On October 12 it appeared like the stream of fire which issues from the mouth of a cannon at a discharge and when the sparks are driven backwards by a strong wind. At moments the flame was thought to be in motion, or exhibiting scintillations similar to those of an aurora Borealis' (see Chambers 1910). Sir John Herschel wrote (1867) 'Although the appearance of this celebrated comet was not such as might be reasonably considered to excite lively sensations of terror, even in superstitious ages, yet, having been an object of the most diligent attention in all parts of the world to astronomers, furnished with telescopes very far surpassing in power those which had been applied to it at its former appearance in 1759, and indeed to any of the greater comets on record, the opportunity thus afforded of studying its physical structure, and the extraordinary phenomena which it presented when so examined have rendered this a memorable epoch in cometic history.'

But what was discovered in 1835, remembering that this was an era before photography and spectroscopy? The answer must be 'very little'. The comet did, however, generate a sunward jet which 'underwent singular and capricious alterations, the different phases succeeding each other with such rapidity that on no two successive nights were the appearances alike' (Herschel 1867, p. 381). These observations encouraged the solid-nucleus hypothesis and it was suggested that reactions to the jet caused the nucleus to alter its direction slightly. Also the tenuous nature of the coma was illustrated by the lack of diminution in star brightness for stars seen very close to the nucleus of P/Halley.

A lot had happened in cometary science between 1835 and the next, 1909, appearance of P/Halley. The comet of 1843 had spewn forth a 100° tail in one day and then, while passing within 100 000 km of the Sun's surface, had whirled this unbroken tail round through 180° in less than two hours. P/Beila had split in two. 1866 I and 1862 II had been accused of being the parents of the Leonid and Perseid meteoroid streams. The spectroscope revealed cometary light to be made up of both reflected sunlight and a hydrocarbon spectrum supplemented by metal lines when close to the Sun. CN, C_3, CH, CO^+ and N_2^+ bands were recognized. Radiation pressure was proposed as the force responsible for tail production. The photographic plate tended to replace the eye at the telescope. 'Comet tails, formerly smooth and ghostly, hardly visible phantoms now appeared on the plates as brilliant torches with rich detail of structure, with bright and faint spots, never before seen or even suspected.'

Much was expected of the third, 1910, predicted return of P/Halley. The comet was not easily seen being unfavourably low in the northern sky during its period of maximum brightness. Barnard (1914) rather gloomily wrote 'It is safe to say that it did not give us any new information concerning these strange bodies'. It was, however, the seed of considerable collaboration between scientists and observatories, the aim being to keep the comet under continuous surveillance. The engendered enthusiasm produced a flood of data, much of which was poorly calibrated. Unfortunately the will to examine this data carefully was, in the main, lacking, and it was twenty years before a major résumé of the results was produced by Bobrovnikoff (1931).

Scientifically, the passing of the cometary nucleus in front of the face of the Sun, the sweeping of the cometary tail over the Earth and the detailed day to day photography produced excitement in cometary circles even if they didn't generate startling results. Solar transits are rare, the only recognized one before P/Halley was that of 1882 II. Nothing was seen, indicating that the cometary nucleus either had a diameter less than 50 km or was a 'bee swarm' of smaller components. The tail of P/Halley swept across Earth during 18 and 19 May 1910. To the public the comet immediately assumed the guise of a gaseous menace. The fact that cyanogen had recently been detected in the tail of Comet Morehouse fuelled the apprehension. Scientists were encouraged to watch out for abnormal atmospheric phenomena, scientific balloons were launched, meteor watches were encouraged; nothing untoward was seen. The analysis of the plethora of P/Halley tail photographs indicated that the knots of plasma were being accelerated away from the Sun with a repulsive force of around 200 times larger than that due to solar gravitational attraction. Studies of the intensity distribution in the tail led to the suggestion that fluorescence was responsible for the line and band spectra.

The 1986 apparition of P/Halley has encouraged immense scientific activity. Remember that 1682 saw the first telescopic observation of the comet and still in 1835 the astronomer could only rely on naked-eye observations and a drawing pad and pencil as the recording medium. 1910 was the era of photographic and spectroscopic endeavour. By 1986 not only had researchers expanded away from the confines of the visual radiation band to encompass ultraviolet, infrared and radio regions, their instruments could now leave the entombing atmosphere of Earth to travel freely into space and across space. Cometary science had not lain dormant between 1911 and 1982. The solid-nucleus model had been reintroduced. Cometary orbital analysis had produced evidence that no comet had entered the Solar System with a hyperbolic velocity thus confining them to membership of the Sun's family. Momentum transfer from a continuously flowing solar wind was found to be a sufficient source of force for ion-tail acceleration. And Oort's discovery of a parsec-diameter cometary cloud surrounding the Sun provided an ample deep freeze for storing comets.

Astronomers banded together in an International Halley Watch (see IHW 1986) an organization designed to coordinate observations and archive data. The goals were to standardize techniques where appropriate, to promote simultaneous observations by many techniques and to organize close temporal sequences of observations during the apparition.

P/Halley's reappearance 25 years into the space age has been the key to loosen the purse strings of some of the space nations. The European Space Agency has funded the *Giotto* Mission, a ten-experiment spacecraft, which flew past P/Halley on 14 March 1986 getting to within 600 km of the nucleus at 00h03 U.T. The Soviet Union funded two spacecraft, *Vega* 1 and *Vega* 2. These had fourteen experiments on board, *Vega* 1 passing within 8890 km of P/Halley on 6 March 1986 (at 07h20 U.T.) and *Vega* 2 passing within 8030 km on 9 March 1986 (at 07h20 U.T. too). Japan launched two spacecraft towards the comet, *Sakigake* and *Suisei*. Three experiments were carried between them and *Sakigake* passed within 6.99×10^6 km of the comet on 1986 March 11 (04.18 U.T.) and *Suisei* to within 151 000 km on 1986 March 8 at 13.06 U.T. The initial plans for all these missions are summarized in Space missions to Halley's Comet, *European Space Agency special publication* no. 1066. The first results from the five space-probes are presented in *Nature*, **321**, 259–366 (1986). They reveal the comet to be essentially as expected. The scientists who for years had produced models of cometary environments to aid the engineers designing space-borne apparatus were not proved to be far out in their surmising. There were very few surprises. The environment was complex just as would be expected around an active source of plasma and dust set in the supersonic flow of the solar wind. At the centre of the coma was a single, avocado-pear-shaped, reasonably smooth black nucleus, again much as predicted. The amount of data returned from the combined missions was immense, a cornucopia of information that will take decades to analyse completely.

March 1986 was a memorable month in the history of cometary science and P/Halley in particular. Before that we had to stay on Earth, or close by, and peer at this three-dimensional object from afar. In essence comets were squashed flat against the background of the sky. Now for the first time spacecraft have been able to dive through an active comet. The third dimension has at last been added.

What comes next? P/Halley will return; and its path across the northern sky in A.D. 2061 can already be traced out. The observers of 1910 could not have imagined a fleet of spacecraft flying by in 1986. Why should it be any easier to imagine what the next 75 years will bring? We can however look closer to hand. We have just had our first 'distant-close' encounter with primordial material. It is certainly within our ability to rendezvous with a comet, to launch a spacecraft that will catch up with a comet and then 'get on board', keeping it company for years as opposed to flying through at around 70 km s^{-1}. Given the necessary funds the scientific space agencies can do the job. The subsequent step is sample return. In 1986 we rushed past Halley's Comet, by 2061 freezer boxes full of cometary material may be regularly being brought back to laboratories for analysis.

REFERENCES

Bailey, M. E., Clube, S. V. M. & Napier, W. M. 1986 *Vistas Astron.* **29**, 53–112.

Barker, T. 1757 *An account of the discoveries concerning comets with the way to find their orbits and some improvements in constructing and calculating their places. For which reason are here added new tables, fitted to those purposes; particularly with regard to that comet which is soon expected to return.* London: J. Whiston and B. White.

Barnard, E. E. 1914 *Astrophys. J.* **39**, 373–404.

Babrovnikoff, N. T. 1931 Halley's Comet in its apparition of 1909–1911. *Univ. Calif. Publs Lick Obs.* **17** (2), 309–482.

Bortle, J. E. & Morris, C. S. 1984 *Sky and Telescope* **67**, 7–12.

Broughton, R. P. 1979 *J. R. astr. Soc. Can.* **73**, 24–36.

Burley, C. 1769 *An essay towards a history of the principal comets that have appeared since the year 1722.* London.

Chambers, G. F. 1910 *The Story of the Comets*, 2nd edn, p. 122. Oxford University Press.

Clairaut, A. C. 1759 *J. Sçavans* **41**, 80–96.

Cowell, P. H. & Crommelin, A. C. D. 1910 Greenwich Observations for the year 1909, Appendix.

de Damoiseau, T. 1820 *Memorie Accad. Sci. Torino* **24**, 1–76.

Gregory, D. 1726 *The Elements of Physical and Geometrical Astronomy*, vol. 2, pp. 881–905. London: D. Midwinter.

Halley, E. 1705 *A synopsis of the astronomy of comets.* London: John Senex. (See also (in Latin) *Phil. Trans. R. Soc. Lond.* **24**, 1882–1899.)

Herschel, J. F. W. 1867 *Outlines of Astronomy*, p. 380. London: Longmans, Green & Co.

Hind, J. R. 1850 *Mon. Not. R. astr. Soc.* **10**, 51–58.

Hughes, D. W. 1982 *Contemp. Phys.* **23**, 257–283.

Hughes, D. W. 1983*a* *Asteroids Comets Meteors. I* (ed. C.-I. Lagerkvist & H. Rickman), pp. 239–257. Uppsala Universitet Reprocentralen HSC.

Hughes, D. W. 1983*b* *Mon. Not. R. astr. Soc.* **204**, 1291–1295.

Hughes, D. W. 1985*a* *J. Br. astr. Ass.* **95**, 193–204.

Hughes, D. W. 1985*b* *Bull. Inst. Maths. Appls* **21**, 146–153.

Hughes, D. W. 1985*c* In *Dynamics of comets; their origin and evolution* (ed. A. Carusi & G. B. Valsecchi), pp. 129–142. Dordrecht: D. Reidel.

Hughes, D. W. 1985*d* *Mon. Not. R. astr. Soc.* **213**; 103–109.

Hughes, D. W. 1985*e* *Q. Jl. R. astr. Soc.* **26**, 513–520.

Hughes, D. W. & Daniels, P. A. 1983 *Icarus* **53**, 444–452.

I. H. W. Staff 1986 Space missions to Halley's comet, *European Space Agency special publication no.* 1066, pp. 161–178.

Ivanov, A. A. 1909 *Astr. Nachr.* **182**, 225–226.

Keller, H. U., Arpigny, C., Barbieri, C., Bonnet, R. M., Cazes, S., Coradini, M., Cosmovici, C. P., Delamere, W. A., Huebner, W. F., Hughes, D. W., Jamar, C., Malaiss, D., Reitsema, H. J., Schmidtt, H. U., Schmidt, W. K. H., Seige, P., Whipple, F. L. & Wilhelm, K. 1986 *Nature, Lond.* **321**, 320–326.

de La Caille, N. L. 1759 *Mémoires de mathématique et de physique, tirés des registres de l'Académie royale de sciences* 522–524.

Laugier, P. A. E. 1843 *C.r. hebd. Séanc, Acad. Sci., Paris* **16**, 1003–1006. (See also *C.r. hebd. Séanc. Acad. Sci., Paris* **23**, 183–189.)

McCrea, W. H. 1975 *Observatory* **95**, 239–255.

Morris, C. S. & Green, D. W. E. 1982 *Astr. J.* **87**, 918–923.

Newburn, R. L. & Reinhard, R. 1981 The Comet Halley Dust and Gas Environment. *European Space Agency special publication no.* 174, pp. 19–24.

Newburn, R. L. 1981 The Comet Halley dust and gas environment. *European Space Agency special publication* no. 174, pp. 3–18.

Olmsted, D. 1850 *The mechanism of the Heavens.* Thomas Nelson: London.

Oort, J. H. 1950 *Bull. astr. Insts Neth.* **11**, 91–110.

Pannekoek, A. 1961 *A history of astronomy*, p. 425. London: George Allen & Unwin Ltd.

de Pontécoulant, P. C. L. 1835 *Mém. prés. div. Sav. Acad. Sci. Inst. Fr.* (2) **6**, 857–947.

Rosenberger, O. A. 1835 *Astr. Nachr* **12**, 187–194.

Stephenson, F. R. & Walker, C. B. F. (eds) 1985 *Halley's comet in history.* British Museum Publications Ltd.

Turner, H. H. 1908 *Halley's Comet, an evening discourse to the British Association, Dublin,* 4 September 1908. Oxford University Press.
Vsekhsvyatskii, S. K. 1964 *Physical Characteristics of comets.* Jerusalem: Israel Program for Scientific Translations.
Wallis, M. K. & Macpherson, A. K. 1981 *Astron. Astrophys.* **98**, 45.
Yeomans, D. K. 1981 *The Comet Halley handbook,* 1st edn, Jet Propulsion Laboratories.
Yeomans, D. K. 1986 In Space Missions to Halley's Comet. *European Space Agency special publication* no. 1066, pp. 179–197.
Yeomans, D. K. & Kiang, T. 1981 *Mon. Not. R. astr. Soc.* **197**, 633–646.
Yeomans, D. K., Rahe, J. & Freitag, R. 1986 Space missions to Halley's Comet. *European Space Agency special publication* no. 1066, pp. 1–24.

Discussion

P. H. FOWLER (*University of Bristol, U.K.*). In his figure Dr Hughes highlighted the apparition of A.D. 837 as being the brightest historical apparition. Can one expect brighter future apparitions, or can Halley even hit the Earth?

D. W. HUGHES. There is absolutely no reason why P/Halley should not get closer to Earth than the A.D. 837 value (see Stephenson & Yau 1985), and the brightness at any specific apparition depends drastically on the minimum geocentric distance. It must be stressed though (see Hughes 1981 for a review) that, size for size, Earth is 46 times more likely to be struck by an asteroid than by a comet. Kresák (1978) concluded that earth would collide with a body (asteroid or comet) of a diameter greater than 1 km every 1.5–2.0 million years. For something as large as P/Halley, impacts would occur about a factor of 30 less frequently.

Additional references

Hughes, D. W. 1981 *Phil. Trans. R. Soc. Lond.* A **303**, 353–368.
Kresák, L. 1978 *Bull. astr. Insts Csl.* **29**, 103–125.
Stephenson, F. R. & Yau, K. K. C. 1985 *J. Br. interplanet Soc.* **38**, 195–216.

SIR BERNARD LOVELL, F.R.S. (*Jodrell Bank, Macclesfield, Cheshire*). It is believed that the η-Aquarid meteor stream which occurs early in May and the Orionid stream of October may be associated with Halley's Comet, and it is possible to estimate the mass of meteoric material in this orbit. How does this compare with the mass loss from the comet computed from the *Giotto* data?

D. W. HUGHES. The comparison is not easy to make because *Giotto* only produced an 'instantaneous' dust-emission value, the instant being 1986 March 14.0 when the comet was postperihelion and at 1.44×10^8 km from Earth and 1.35×10^8 km from the Sun. Also *Giotto*, at minimum, was 605 km from the nucleus. Added to this problem is the fact that P/Halley not only has a steadily varying activity which is a function of its heliocentric distance but also has a fluctuating activity (by a factor of about six) which is a function of its spin and nutation phase. The total dust emission shortly before closest approach has been measured to be 3.1×10^6 g s^{-1}. Thankfully *Giotto* only encountered 26×10^{-3} g of dust during its fly-by, the most massive impacting particle being 1.1×10^{-3} g. If this had hit Earth it would have produced a third-magnitude meteoroid. More massive particles were missed by *Giotto* so the emission result must be taken to be a lower limit.

D. Lynden-Bell, F.R.S. (*University of Cambridge, U.K.*). Some years ago Dr Kiang gave a rather pretty explanation of the periodic variations of the period of Halley's Comet in terms of the inverse pendulum problem. Others had attributed these periods to such things as new planets in the outer parts of the Solar System. In Kiang's theory this was due to gravitational interaction with Jupiter. Is this theory of the period changes still the accepted one?

D. W. Hughes. The variations in the period of P/Halley are explicable simply by invoking the gravitational perturbations due to the known planets plus the non-gravitational forces produced by thermal lag in the thin surface layers of the rotating nucleus. As P. J. Message (Liverpool University) has shown, the main resonance is due to a 13:2 commensurability between the periods of Jupiter and P/Halley. Nothing more exotic is needed.

P. J. Message (*University of Liverpool, U.K.*). The perturbations giving rise to the long-period oscillation in the orbital period (and other orbital parameters) of the comet are those associated with the 13:2 near-commensurability of orbital period with Jupiter. There is a free-amplitude oscillation, or libration, of period about six or seven centuries (depending on the amplitude) arising solely from perturbations by Jupiter, so there is no need to invoke any undiscovered planet! (The comet appears to have moved out of this oscillation now, as is shown by a plot of the intervals between successive perihelion returns.)

J. E. Wilkinson (*Capricornia Institute, Rockhampton, Queensland, U.S.A.*). Dr Hughes suggested that comets may have been formed in the Uranus–Neptune region of the Solar System. Do not the larger amounts of water ice observed, suggest an origin closer to the Sun, say in the Saturn–Jupiter region?

D. W. Hughes. I do not wish to be over specific about the nebula region in question. Comets were probably the 'left overs' of planetary formation throughout the whole Jupiter, Saturn, Uranus, Neptune region and it is highly probable that both the gas to dust mass ratio and to some extent the gas composition varied as a function of heliocentric distance.

Phil. Trans. R. Soc. Lond. A **323**, 369–379 (1987)

Printed in Great Britain

Earth-based observations of Comet Halley dust and gas

By A. J. Meadows

Department of Astronomy, University of Leicester, University Road, Leicester LE1 7RH, U.K.

[Plate 1]

Very extensive ground-based observations of P/Halley have now been made, both to provide a standard cometary archive and to help interpret data from the Halley space probes. A number of new results are reported here, of which the most important is probably the major role played by sporadic activity of the nucleus in the development of the comet.

Introduction

From the viewpoint of the ground-based observer, P/Halley has the virtue that it is a bright comet with a very accurately determined orbit. An observational programme can therefore be planned well beforehand, and can include a more detailed examination of cometary properties than is possible with fainter periodic comets.

P/Halley was first observed at this apparition on 16 October 1982 (which led to its current designation as 1982i), when it was still some 11 AU from the Sun (Jewitt *et al.* 1982). Observations over the following months were consistent with the assumption that sunlight was being reflected from an irregularly shaped, slowly rotating body a few kilometres across.

This estimate of size depended on the value assigned to the albedo, which was highly uncertain. Figures of 0.1–0.2 have often been used for comets, but it has been clear that both higher and lower albedos might be possible. A higher figure would be reasonable if the nucleus had a clean surface made of ice, perhaps reaching the 0.6 achieved by the jovian satellites. However, if loss of volatiles has led to the formation of a 'crust' on the nucleus, the albedo could be very low: less than 0.05 to judge from the values for the darkest asteroids. Interestingly, the observations of the satellites of Uranus by the *Voyager* fly-by mission in January 1986 indicated another reason for expecting a low albedo for cometary nuclei. The lower albedo of these satellites as compared with those of Jupiter and Saturn has been attributed tentatively to the presence in their icy surface of volatiles containing carbon. The influence of solar ultraviolet radiation can convert these volatiles into polymers of low albedo. Such volatiles (e.g. methane) are certainly present in comets.

The problem was that some ground-based observations of P/Halley were taken to imply a higher (0.2–0.3), rather than a lower albedo for its nucleus. Infrared measurements of the nucleus could, in principle, have resolved the uncertainty, but the infrared instruments were too insensitive to detect the comet until it had already become active.

Comets tend to brighten from 6 AU inwards as the solar heat input becomes large enough for water ice to sublime. The first near-infrared measurement was made when P/Halley was at 5.4 AU from the Sun, at which time a small dust coma already appeared to be present (Birkett *et al.* 1985). Measurements in the thermal infrared, initially by the University of Kent and UKIRT staff, came later, when a dust coma was certainly observed. Consequently, direct estimates of the nuclear albedo had to await the *Giotto* fly-by.

OBSERVATIONS OF DUST

Observations of P/Halley have clearly shown that the emission of dust from its nucleus tends to be sporadic and localized, rather than continuous and from the whole surface. The dust often appears to be thrown out in relatively short-lived jets from restricted regions of the nucleus. The forces leading to this emission presumably arise from the rapid sublimation of volatiles, so that the dust is carried along with the outflowing gas. Whereas the gas expands in all directions, the dust tends to continue along its original direction, so forming a definable envelope which, under favourable circumstances, can be observed from the ground. Recent image-enhancement of photographs of P/Halley taken in 1910 has shown that the tracing of such jets is feasible (Larson & Sekanina 1985). The appearance of the envelope produced by a jet depends on the rotational period of the nucleus and the position of the observer relative to the rotational axis. Hence, it is possible, in principle, to deconvolve the observations to provide information on these. The ground-based data from 1910 have suggested a rotational period of 52 h and an obliquity of some 30°. Large numbers of images providing information on dust have been obtained at this apparition. Schmidt plates (exposures of a few seconds) of P/Halley taken for astrometric purposes can apparently provide unexpectedly useful information on jets (see figure 1). Hence, data on this type of activity are currently increasing rapidly.

Jet formation may be limited to specific areas covering less than 10 % of the nuclear surface. There is even a suspicion from a preliminary comparison of 1910 data with some from the present apparition that the surface localities concerned may be similar in both. Activity from

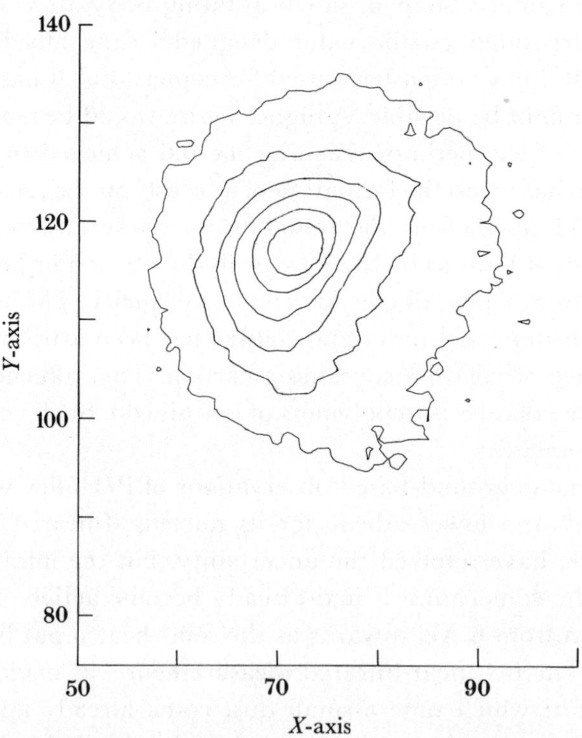

FIGURE 1. Intensity contours of P/Halley obtained from a U.K. Schmidt plate taken for astrometric purposes. The exposure of a few seconds was expected to produce a point image, but the actual image is extended and slightly asymmetrical.

these spots seems to switch on and off at intervals. It might be expected that near-nuclear variations would show some sign of the period of rotation of 52 h. At this apparition, the first signs of activity occurred in 1984 as the comet came within 7 AU from the Sun. Variations in brightness by a factor of up to six were observed over periods of a day or so. These were mainly random in time, but a more detailed analysis has suggested that an underlying two-day period may be present (Morbey 1985). In January 1986, jets were particularly frequent, and were found to have an approximately two-day cycle of activity (Larson 1986).

The dust ejected from the nucleus is acted on by solar radiation, so leading to the formation of an extended dust trail. The first attempt to discuss this theoretically was actually made by Bessel on the basis of observations of P/Halley at its 1835 apparition. Basically, a dust particle takes up an orbit round the Sun that depends on the balance between the solar gravitational attraction (F_G) and the solar radiation pressure (F_R). For a spherical dust particle of radius, a, and density, ρ, at a heliocentric distance, r, we have

$$F_G = (GM_\odot/r^2)\tfrac{4}{3}\pi a^3\rho; \quad F_R = (\phi/c)\,[L_\odot/4\pi r^2]\,\pi a^2,$$

where $M_\odot$ is the mass of the Sun; $L_\odot$ is the solar luminosity; c is the velocity of light; ϕ is the radiation pressure efficiency (i.e. the ratio of the radiation pressure cross section to the geometrical cross section of the particle). Both F_G and F_R are radial and vary as r^{-2}, but act in opposite directions. Hence, the dust particle follows a keplerian orbit under a reduced effective gravitational force, $(1-\beta)F_G$, where β (the acceleration ratio) is given by

$$\beta = F_R/F_G = \text{const.} \times (\phi/\rho a).$$

It is convenient to use this parameter β in considering the formation of a dust tail. Suppose a cometary nucleus continuously releases dust particles with a given value of β. The particles are repelled successively by radiation pressure, and so become strung out along a locus called a syndyne. In reality, cometary dust particles will have a distribution of sizes and may have more than one density. Consequently, particles will have different values of β, and will move along different syndynes. Under these circumstances, it may be more convenient to consider the trajectories followed by dust particles with different β values, all released at the same time. The locus of the corresponding envelope is called a synchrone. Calculation of typical synchrones shows that, for large particles (greater than 0.01 cm), the motion is so slow they can remain in the vicinity of the cometary nucleus for several hundred days. If the particles are released with non-zero velocities, both syndynes and synchrones will be spatially blurred. When cometary dust is produced continuously and isotropically, the resultant dust tail is therefore spread out into a broad fan. We have seen, however, that this has not been the typical form of dust production in P/Halley: one result of its sporadic activity has been the production of a number of distinct dust tails visible simultaneously. It should be added that P/Halley was better placed for observing its dust tails after perihelion, at which time a distinct antitail was noted.

Ground-based infrared observations of P/Halley dust have allowed short-term changes in its amount to be monitored quantitatively and for the changing temperature of the dust to be measured. Dust production after perihelion was up by a factor of 2–4 as compared with amounts before perihelion: this was presumably connected with the increased jet activity seen. There were indications that the dust size distribution also changed with time. One interesting point emerged from radio observations of high spatial resolution of OH emission made in November

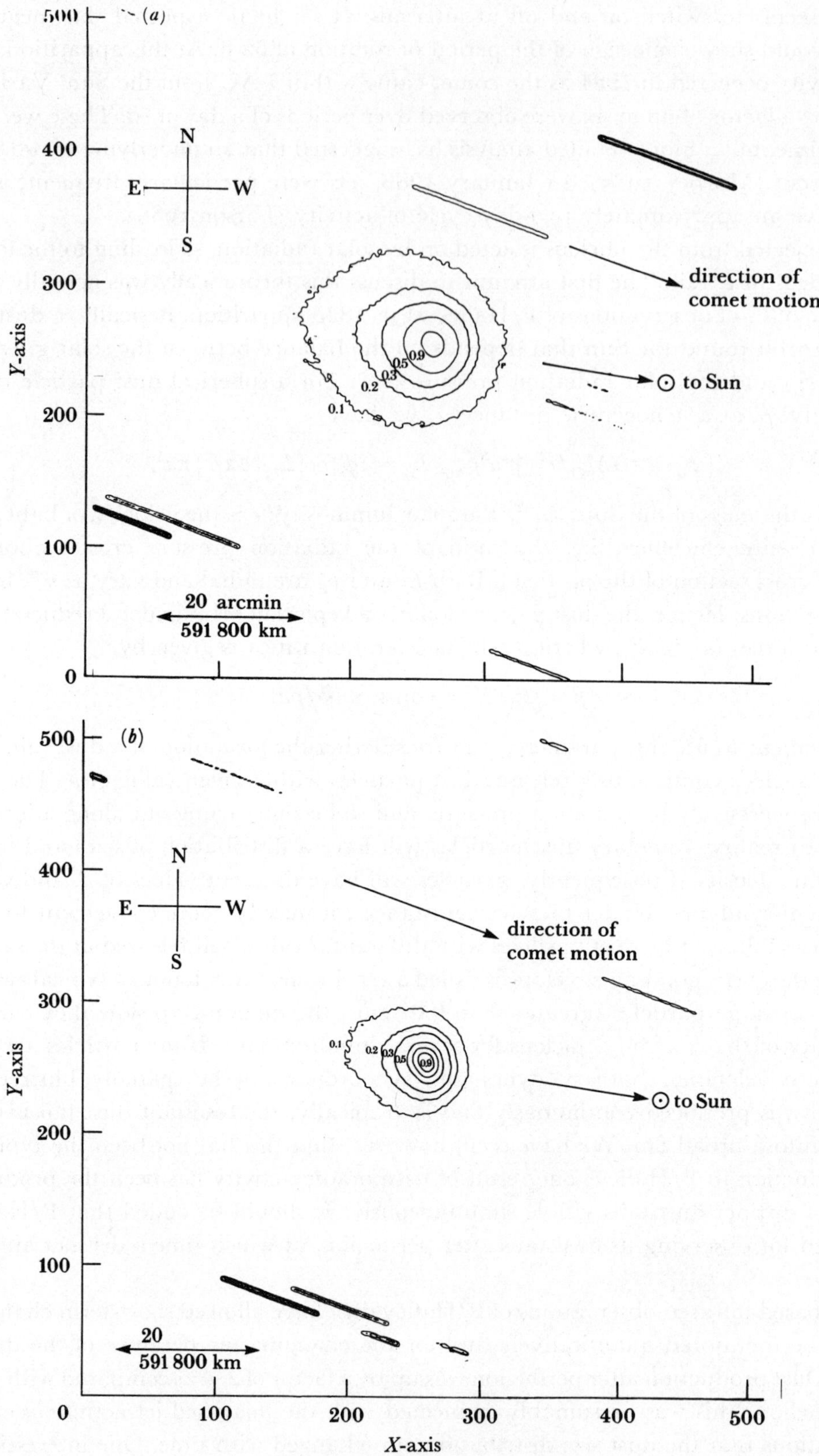

FIGURE 2 *a*, *b*. For description see opposite.

1985. These indicated that OH, instead of decreasing smoothly in amount away from the nucleus, appeared to be concentrated into clumps. A possible explanation (though not the only one) is that large fragments of ice broke off from the nucleus and began to sublime as they separated. An infrared absorption feature due to water ice was observed in November (Bregman *et al.* 1985). Emission features corresponding to the C–H bond and to silicates have also been identified.

It is of interest to compare the parallel changes in the gas and dust content of the coma. The decrease in observed flux as a function of distance from the centre of the coma can be compared for C_2 emission and the continuum (corresponding to reflected light from dust) for two successive months, November and December 1985. The observations clearly indicate that flux from the dust is similar in both months. This flux falls off linearly with distance, implying that the mass of dust present is conserved as the emitted particles expand outwards. The C_2 fluxes do not fall off linearly with distance, and there is some difference between the two months. (Comparison of gas and dust emission is shown in figure 2*a*, *b*.)

OBSERVATIONS OF GAS

Interpretation of observations of gaseous emission from comets is more difficult than interpretation of dust observations, because photoionization and photodissociation leads to a complex mix of interrelated atomic and molecular species. Collisions in the innermost part of the coma can lead to chemical reactions, the products of which can then be acted on by solar ultraviolet radiation. Ground-based observations relate to the end-products of these interactions seen in the outer coma.

The simplest model for interpreting the gas emission is due to Haser (1957). It assumes that, in the first place, molecules evaporating from a cometary nucleus lead to a coma density given by

$$n(r) = (\phi/4\pi r^2) \exp(-r/v\tau),$$

where $n(r)$ is the number density at a distance, r, from the nucleus; ϕ is the production rate; v is the outflow velocity; τ is the mean lifetime of the molecule. The distance to which the molecule can be traced (its scalelength, γ) is equal to $v\tau$.

For daughter molecules produced from this parent, the corresponding density distribution is given by

$$n_d(r) = \frac{\phi_p \gamma_d}{4\pi v_d r^2 (\gamma_p - \gamma_d)} \left[\exp\left(-\frac{r}{\gamma_p}\right) - \exp\left(\frac{-r}{\gamma_d}\right) \right],$$

where p represents parent and d represents daughter. This simple model often gives a fair approximation to the observed distribution, but needs to be replaced by more complex models when detailed studies are undertaken. However, quick-look observations of P/Halley data have typically been expressed in terms of the Haser model.

FIGURE 2. (*a*) This U.K. Schmidt narrow-band image (O I, 40 min exposure) contains gaseous emission from [O I] and NH_2 along with reflected sunlight from dust. The contours represent various light intensity levels. (*b*) This image (IF, 40 min exposure) taken on the same night (5 December 1985), consists almost entirely of reflected sunlight from dust. A comparison of parts (*a*) and (*b*) indicates both the dominance of the gaseous emission and its less rapid fall off with distance from the nucleus. The contours represent various light-intensity levels.

[125]

The development of P/Halley spectra has followed the general lines that were predictable from other cometary spectra, but a number of new, or more detailed, results have also been obtained. No significant emission was observed until February 1985, when weak CN bands were detected (Wyckoff *et al.* 1985). By September 1985, the spectrum showed strong emission from CN, C_2 and C_3. OH had by then already been detected by radio astronomy measurements (Brockelée-Morvan *et al.* 1985) and in the ultraviolet region. A month later, OH was also detected in the visible spectrum. By this time, most of the usual cometary emission features were appearing; along with carbon features (now including CH) were NH, NH_2, H_2O^+ and OI. HCN was detected subsequently (Despois *et al.* 1985), and was found to vary in intensity over periods of a day or so. HCN observations have proved particularly important because this molecule has been considered the likely parent of the common cometary radical CN. Combined observations in the visible and radio regions now show that this cannot be true for P/Halley at least. Direct measurements of the H_2O molecule for the first time (Mumma *et al.* 1985), together with OH, O and H measurements, confirm the importance of H_2O as a constituent of the nucleus of P/Halley. It has been found in previous comets where observations have been possible that radio detection of OH indicates different amounts of the radical present as compared with other methods. This has also been found for P/Halley, but the very detailed observations available offer an excellent chance of resolving the discrepancy.

Significant day-to-day variations in the intensity of OH and H have been observed, along with less obvious fluctuations in other species (CN, C_2 and C_3). These variations suggest that the rotation of the nucleus plays a larger role in the development of the gas coma than had been supposed. A surprising discovery was that the extended hydrogen coma undergoes variations with a similar period. Observations of the hydrogen coma from Ly-α images obtained by the Japanese spacecraft, *Suisei*, during November and December 1985 indicated considerable activity with a period of 2.2 d. NaI emission was detected in January 1986, when it was also found to be strongly variable (Barbieri *et al.* 1986). However, unlike the other emission features, it was found to be asymmetrically distributed relative to the nucleus.

DESCRIPTION OF PLATE 1

FIGURE 3. This print of Comet Halley, showing clearly for the first time the formation of the tail, is a contrast-enhanced photograph. It was made from plate number B10578 exposed for 25 min on the night of 9 December 1985 with the U.K. 1.2 m Schmidt Telescope at Siding Spring Observatory in Australia.

The telescope tracked the comet, which was moving relatively fast against the star background in the constellation of Pisces, resulting in the stars being trailed.

Photography by D. F. Malin from an original plate by the U.K. 1.2 m Schmidt Telescope. Reproduced with permission from the Royal Observatory, Edinburgh.

FIGURE 4. Comet Halley on 10 March 1986. This high-contrast composite positive photograph of Comet Halley is made from two exposures of ten minutes in blue light with the U.K. 1.2 m Schmidt Telescope. In this photograph the comet extends for some ten degrees across the sky covering two original photographic plates. The lower, filamentary ion tails show a disconnection about four degrees from the comet's head where material was blown away from the comet's head by the solar wind. The movement of this disconnection can be seen in comparison with a photograph taken the previous night. The smoother dust tails can be seen curving away from the comet to the north of the photograph.

This photograph complements the images obtained from various spacecraft which made their closest approaches to the comet between 6 March and 19 March.

Original plates J10830 and J10831T taken with the U.K. Schmidt Telescope in Australia. Photography by B. W. Hadley, Royal Observatory, Edinburgh. Reproduced with permission from the Royal Observatory, Edinburgh.

[126]

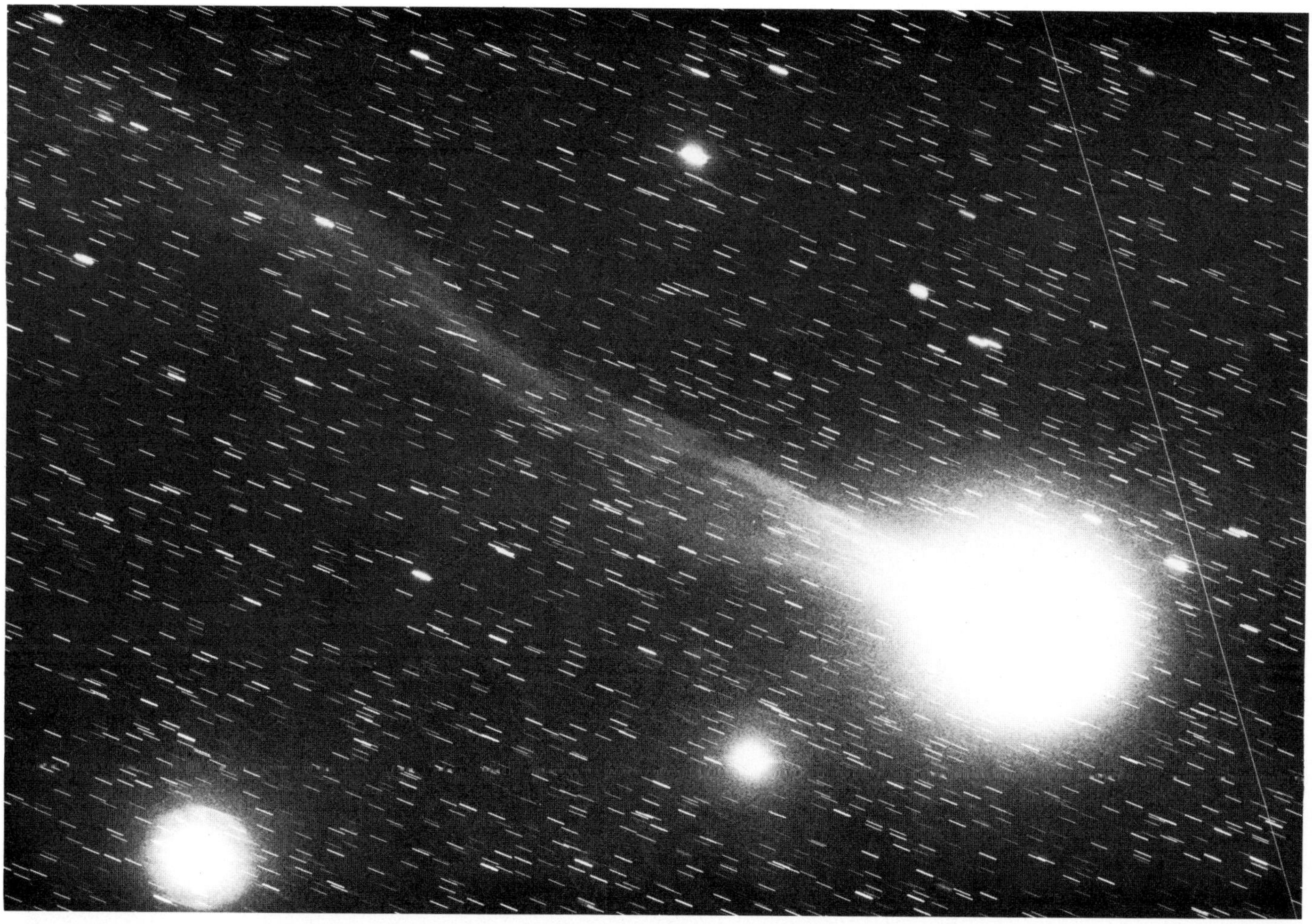

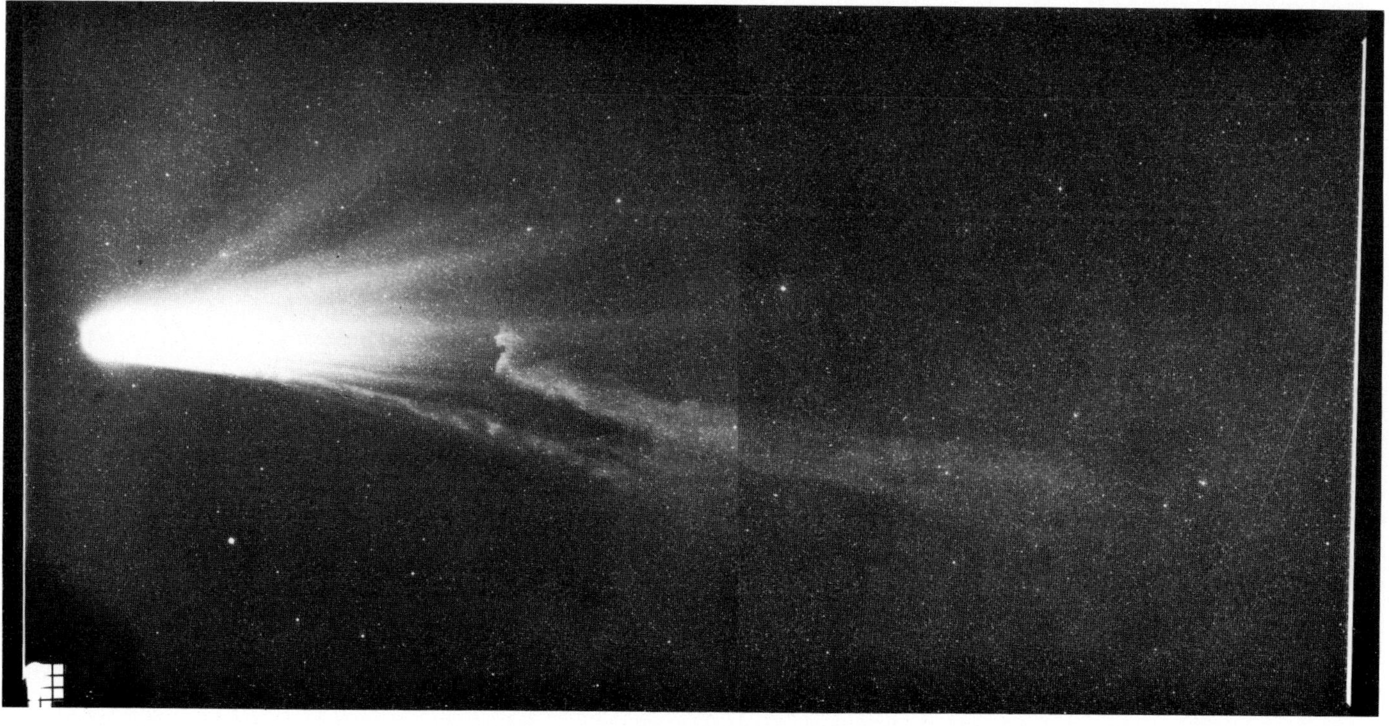

FIGURE 3. For description see opposite.

FIGURE 4. For description see opposite.

(*Facing p.* 374)

Towards the end of 1985, an extended plasma tail began to develop, and the characteristic tail emissions of N_2^+ and CO^+ were soon detected. The tail was reported to be up to 43° long in April 1986. By November 1985, the full range of plasma-tail activities were being observed. These included rays forming and changing in times of less than a day, knots moving down the tail with velocities of up to 50 km s^{-1}, and appreciable brightness changes of the whole tail from night to night. Tail oscillations with a period of 1–2 d were suspected. These might be related to the rotation of the nucleus, but could also be characteristic plasma wave velocities (Jones & Meadows 1969). The most spectacular events observed were tail discontinuities. These occurred in January and March 1986, both coinciding with reversals of the interplanetary magnetic field.

Conclusions

A major reason for the very extensive ground-based observations of P/Halley at this apparition has been to provide a standard, in whose terms observations of other comets can be interpreted, or reinterpreted. From that viewpoint, the much more detailed information now available on features commonly observed in comets will be invaluable. Inevitably, however, attention has tended to concentrate on results that are new. Here, the observations of dust development in P/Halley, and especially the frequency of jets, are notable. The most important observation concerning the gas component is probably the confirmation of the importance of H_2O. Overall, the most striking result has been the realization of the extent to which activity in P/Halley is dominated by emission from a few restricted regions of a slowly rotating nucleus.

I am grateful to members of the U.K. Schmidt Telescope Unit of the Royal Observatory Edinburgh for access to their photographic data, to Dr S. F. Green, University of Kent, for provision of infrared data and to Dr I. P. Williams for information on photometric observations. I am also grateful to Miss C. M. Birkett and Mr A. Fitzsimmons for their assistance in preparing this paper.

References

Barbieri, C., Iijima, T. & Sabbadin, F. 1986 IAU Circ. no. 4169.
Birkett, C., Green, S., Longmore, A. & Zarnecki, J. 1985 IAU Circ. no. 4025.
Bockelée-Morvan, D., Bourgois, G., Crovisier, J. & Gérard, E. 1985 IAU Circ. no. 4106.
Bregman, J. D., Witteborn, F. C., Rank, D. M. & Wooden, D. 1985 IAU Circ. no. 4149.
Despois, D., Forveille, T., Schraml, J., Bockelée-Morvan, D., Crovisier, J. & Gérard, E. 1985 IAU Circ. no. 4142.
Haser, L. 1957 Bull. Acad. r. Belg. Cl. Sci. (5e) **43**, 740.
Jewitt, D. C., Danielson, G. E., Gunn, J. E., Westphal, J. A., Schneider, D. P., Dressler, A., Schmidt, M. & Zimmerman, B. A. 1982 IAU Circ. no. 3737.
Jones, D. R. L. & Meadows, A. J. 1969 Observatory **189**, 184.
Larson, S. M. 1986 IAU Circ. no. 4159.
Larson, S. M. & Sekanina, Z. 1985 Astron. J. **90**, 823.
Morbey, C. 1985 IAU Circ. no. 4034.
Mumma, M. J., Weaver, H. A., Larson, H. P., Davis, D. S. & Williams, M. 1985 IAU Circ. no. 4154.
Wyckoff, S., Foltz, C., Heller, C., Wehinger, P. A., Wagner, M., Schleicher, D. & Festou, M. 1985 IAU Circ. no. 4041.

Discussion

J. D. Shanklin (*British Antarctic Survey, Cambridge, U.K.*). The visual light curve (of Comet Halley) shows an approximate 2 mag brightening between 2.0 and 1.7 AU preperihelion and a corresponding fading postperihelion. Are there any comments?

A. J. Meadows. No; the low value of the nuclear visual albedo should imply that most of the rapid increase in brightness would occur further out.

M. K. Wallis (*Department of Applied Mathematics, University College, Cardiff, U.K.*). Because the emitting area at the base of the jets is rather limited, estimates that may not be conclusive give too little input of solar radiative energy into the source area to sublimate H_2O ice. Does Professor Meadows have ideas on alternative energy sources, or does he conceive that the nucleus has a substantial admixture of material other than H_2O with a much lower latent heat?

A. J. Meadows. Substances more volatile than H_2O may be partly contained in clathrates and partly free. The former may trap the latter below the surface of the jet-emitting area until the nucleus has approached quite closely to the Sun. Evaporation of the overlying clathrate might then lead to explosive production of the volatile. Perhaps rather less plausible is the old idea of 'unstable' molecules; unstable in the sense that they undergo strong exothermic reactions unless kept at a low temperature.

M. E. Bailey (*Department of Astronomy, The University, Manchester, U.K.*). Could Professor Meadows comment on the implication of the very early observed onset for sublimation (*ca.* 6 AU), together with the apparently strong domination of water ice in the nucleus, as inferred from ground-based and satellite observations? In particular, is there yet any information as to what the more volatile component is or was?

A. J. Meadows. The distance from the Sun at which gas first appears in appreciable quantities depends on a number of factors. For example, it can vary with nuclear albedo, and this may be important for P/Halley. Substances more volatile than water may be present in the nucleus incorporated into a clathrate structure, in which case they will tend to come off at the same distance from the Sun as water does. However, some volatiles may have condensed separately, so they can evaporate first. There are several molecules from which to choose, but because CO_2/CO seems commonest next to H_2O, this may be the most likely substance.

G. Fielder (*Department of Environmental Sciences, University of Lancaster, U.K.*). Professor Meadows discussed the 3.4 μm CH feature in the spectrum of Halley. Does he, or anyone else, have a laboratory material that reproduces the features?

A. J. Meadows. Considerable work has been done on this. Perhaps Dr Wickramasinghe would like to comment on this.

D. T. Wickramasinghe (*Department of Mathematics, Faculty of Science, Australian National University*). It is now well established that there is an interstellar feature near 3.4 μm due to CH stretching vibrations in organic material within interstellar dust. The characteristics of this feature are somewhat variable from one source to another depending on the astronomical environment in which the dust is found. The feature corresponding to dust in the diffuse interstellar medium is most clearly seen in observations of the Galactic-centre source IRS7 and has a central wavelength of 3.4 μm. Observations of some molecular cloud sources have shown another feature centred at 3.53 μm that is also likely to be due to organic material in the dust. The possibility that cometary dust may have properties similar to dust in the interstallar medium has been widely discussed in the literature.

On 30 and 31 March, and 1 April 1986 we obtained spectroscopic observations of Comet Halley in the wavelength region 2–4 μm with a view to search for evidence of organic material in cometary dust. The observations were carried out with the 3.8 m Anglo Australian telescope with a circular variable filter that gave a wavelength resolution $\Delta\lambda/\lambda \sim \frac{1}{100}$. Figure 1 shows the results that were obtained with a 5 in × 10 in† rectangular aperture on 31 March 1986 when

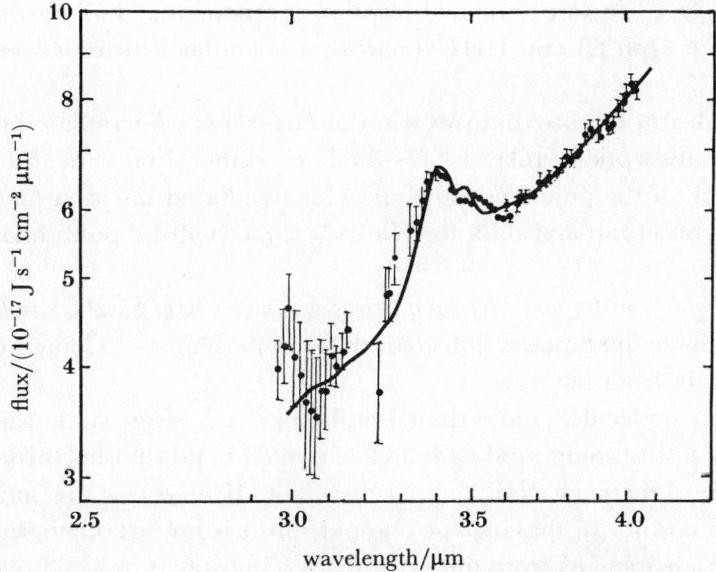

FIGURE D1. The infrared spectrum of Comet Halley on 31 March 1986. The solid curve is a fit of the cometary data with a bacterial grain model at 320 K.

the comet was at a heliocentric distance of 1.17 AU. A more detailed account of the observations will be presented elsewhere. The short wavelength region (2–3 μm) is dominated by scattered solar radiation and is essentially featureless at our resolution. Longward of 3 μm we see a sharply rising dust emission continuum on which is superimposed an emission feature peaking at 3.4 μm. The detection of this feature points to the presence of dust in the inner halo of Comet Halley that is predominantly organic in character. We note that our data do not show any evidence of a water-ice absorption feature that is expected to have a central wavelength of 3.07 μm. The band characteristics of the 3.4 μm emission feature are generally similar to those observed for the diffuse interstellar absorption feature in IRS7, but there are some differences. As in the interstellar case an identification of the 3.4 μm feature with simple organics can be excluded. Possible candidates are the non-volatile residues produced in Urey–Miller type experiments and biochemicals and organics of biological origin. It has been shown previously that the bacterial grain model of Hoyle & Wickramasinghe provides an excellent fit to the IRS7 data and it is accordingly of considerable interest to investigate whether a similar model might also fit the present data.

The solid curve in the figure is the calculated flux representing the sum of the scattered

† 1 in = 2.54 cm.

sunlight component and an optically thin dust emission component, which has been computed assuming a dust temperature of 320 K. The theoretical curve has been normalized to agree with the observational flux at $\lambda = 4$ μm. The observed strength and shape of the central 3.4 μm feature are in good agreement with the bacterial grain model, the two sets of data deviating from one another by no more than 5% at any wavelength. The largest significant deviation occurs near 3.5 μm. The astronomical observations show the presence of a broad secondary-emission peak extending from 3.46–3.55 μm whereas the bacterial grain model shows a sharper peak at 3.48 μm. This may indicate that there is a second independent component that contributes to this feature or that the cometary grain component is subtly different from that in the line of sight to IRS7. We note that an absorption component at 3.53 μm due to interstellar dust has been seen in ρ Oph 29 and HH100 and that a similar feature is seen in emission in HD 97048.

It may also be significant to note that spectrum of *Escherichia coli* irradiated to 1.5 Mrad† at 77 K shows a sharp absorption spike at $\lambda \sim 3.53$ μm rather than an absorption peak at 3.48 μm. A comparison of the predicted flux curve for irradiated bacteria (as would occur in a solar flare) and the observational data for Halley's comet will be published elsewhere.

H. W. KROTO (*School of Chemistry and Molecular Sciences, University of Sussex, Brighton, U.K.*). The previous discussants have juxtaposed infrared data from Halley's Comet with their own laboratory infrared data from bacteria.

As almost all organic molecules (more than 3 million known) show similar features, it is not in fact difficult to find a pure compound with a C–H stretch band that fits this resonant feature as well as do bacteria. One such compound is farnesol. If mixtures are included then the number is essentially infinite. In making the comparison it is important to separate the C–H stretch resonance of *Escherichia coli* from the extraneous, almost featureless, broad background, which comes to a peak near 3.0 μm (1). The main contributor to the background component is likely to be an opacity due to solid-particle scattering which is independent of chemical composition and perhaps also a very broad OH absorption. The separation can be achieved reasonably satisfactorily by passing a smooth curve through the *E. coli* spectrum in the regions below 3.3 μm and above 3.6 μm and interpolating over the intermediate region. The C–H stretch feature of farnesol has, according to the literature, an overall contour very closely matching that of *E. coli* and a listed wavelength maximum at 3.42 μm, together with subsidiary shoulders listed at 3.37 and 3.49 μm; wavelengths that are to be compared with the analogous features of *E. coli*. If the farnesol absorption is scaled according to Hoyle and Wickramasinghe's recipe and added to the background component of *E. coli*, the resulting curve is seen to have the same characteristic shape as *E. coli* and where the curve departs to any significant extent (between 3.30 and 3.42 μm), it matches the astrophysical data significantly better than does *E. coli*. It is to be noted that Hoyle and Wickramasinghe rest their case on the fact that the particular shape shown is unique to living matter. Unfortunately it is not, as myriads of simple compounds and an infinite number of mixtures possesses similar features.

However, the main point of this contribution is rather different and very simple. Because of the large errors that attend the measurements, even if bacteria were responsible for the observed IR flux, a perfect fit to the points cannot be expected on simple statistical grounds. Indeed any

† 1 rad = 10^{-2} Gy = 10^{-2} J kg^{-1}.

curve that passes within error through the observations is as good as any other. This point is so fundamental to the fitting of experimental data that it is usually assumed to be part of basic scientific procedure.

Sir Fred Hoyle, F.R.S. (*Department of Applied Mathematics and Astronomy, University College Cardiff, U.K.*). I disagree. I have not seen a satisfactory fit arising from any abiotic model. Our laboratory spectra were all carefully calibrated to enable a calculation of the opacity function $\tau(\lambda)$ at all wavelengths over a broad band for our particular grain model. Our transmittance $T = 100 \, e^{-\tau}$ was presented on a graph grid from which $\tau = \tau(\lambda)$ could be read out. No such calibrated curve was shown for any of the cases that H. W. Kroto discussed. Any competing abiotic model must have a $\tau(\lambda)$ function that does not depart from the bacterial $\tau(\lambda)$ curve by more than 3% at any wavelength between 3.2 and 4 μm.

Phil. Trans. R. Soc. Lond. A **323**, 381–395 (1987)

Printed in Great Britain

381

Giotto observations of Comet Halley dust

By J. A. M. McDonnell, J. C. Zarnecki, R. E. Olearczyk, S. C. Chakaveh,
G. S. A. Pankiewicz and S. T. Evans

Unit for Space Sciences, University of Kent at Canterbury, Canterbury, Kent CT2 7NR, U.K.

Measurements of the dust environment of Comet P/Halley by the dust experiments
on the *Giotto* and *Vega* spacecraft are discussed, as well as some aspects of the dust
modelling that was carried out before the spacecraft encounters with the comet.
These data have highlighted certain shortcomings in the models. For example,
significant fluxes of grains ($m < 10^{-17}$ kg) smaller than allowed by the models have
been detected, as well as a far higher ratio than predicted of light to dark side emission
from the cometary nucleus. The mass-loss rate is determined and implications for the
past history of the comet are discussed. It is also shown how the dust-impact data
from the spacecraft experiments can be used to derive information on the distribution
of emission over the surface of the nucleus.

1. Introduction

Within a period of eight days in March 1986, five spacecraft successfully achieved a close
encounter with Comet Halley some five weeks after the comet's perihelion passage (on
9 February 1986). The closest approach to the cometary nucleus varied from *ca.* 7×10^6 km (for
Sakigake) to *ca.* 605 km (for *Giotto*). A wide variety of scientific experiments was carried by these
spacecraft involving dust, plasma and imaging instruments. Almost all experiments performed
well and a vast quantity of data of unprecedented quality in terms of cometary research has
been obtained. Very preliminary results have recently been presented (*Nature Suppl.* **321**
pp. 259–366 (1986)).

2. *In-situ* dust measurements

The study of cometary dust was possible hitherto only through ground-based optical and
infrared observations of the light scattered by the dust and of the thermal emission from the
grains. This has led to some progress in the understanding of the nature of the grains but
detailed understanding is possible only through *in-situ* measurements as carried out in particular
by *Giotto* and *Vega* 1 and 2. Table 1 summarizes those experiments that made direct *in-situ*
measurements of the dust. Other instruments provided further information on dust properties
through remote sensing; e.g. on *Giotto*, the OPE (optical probe experiment) and HMC (Halley
multicolour camera) measured the light scattered by the dust while the GRE (*Giotto* radio science
experiment) measured the deceleration produced on the spacecraft by the impinging gas and
dust; on *Vega* 1 and 2, the TVS (television system), IRS (infrared spectrometer) and TKS (three-
channel spectrometer) instruments were all able to measure radiation scattered or emitted by
the dust in various wavelength ranges (see table 2).

[133]

TABLE 1. SPACECRAFT EXPERIMENTS PERFORMING *IN-SITU* DUST MEASUREMENTS

spacecraft	instrument	instrument aims	principle of detection	limiting mass sensitivity/kg	geometric area/m²	instrument mass/kg	coincidence detection
Vega 1 encounter, 6 March 1986 07h 20min 06s; miss distance, 8890 km; encounter velocity, 79.2 km s⁻¹	PUMA	dust particle elemental composition	time-of-flight ion mass spectrometer	3×10^{-19}	7.5×10^{-3}	19	coincidence with impact flash
	SP-1	dust particle flux and mass spectrum	impact plasma detection	10^{-19}	8.1×10^{-3}	2	internal? (positive and negative charge measurements)
	SP-2	dust particle flux and mass spectrum	piezoelectric and impact plasma detection	10^{-19}	5×10^{-2} (piezoelectric) 4×10^{-3} (plasma detection)	4	internal (three piezoelectric sensors) no
Vega 2 encounter, 9 March 1986 07h 20min 00s; miss distance, 8030 km; encounter velocity, 76.8 km s⁻¹	DUCMA	dust particle flux and mass spectrum	PVDF foil detection	1.5×10^{-16}	7.5×10^{-3}	3	no
	FOTON	large dust particle detection	foil perforation				
Giotto encounter, 14 March 1986; miss distance, 605 ± 8 km; encounter velocity, 68.4 km s⁻¹	PIA	dust particle elemental composition	time-of-flight ion mass spectrometer	3×10^{-19}	5×10^{-4} to 10^{-6} (dependent on impact rate)	9.89	coincidence with impact flash
	IPM-P	dust particle flux and mass spectrum	impact plasma detection	10^{-20}	1.19×10^{-2}	2.26 total	internal (positive and negative charge measurement) and possible coincidence with IPM-M
	IPM-M	dust particle flux and mass spectrum	piezoelectric detection	4×10^{-13}	1.19×10^{-2}		coincidence with IPM-P
	MSM	dust particle flux and mass spectrum	piezoelectric detection	4×10^{-12}	2.2		possibly internal (three piezoelectric sensors)
	RSM	dust particle flux and mass spectrum (front-shield penetration)	piezoelectric detection	10^{-9}	2.2		coincidence with front shield detectors
	CIS	dust particle counter	thin foil penetration	3×10^{-13}	0.1		possible coincidence with MSM

TABLE 2. CHARACTERISTICS OF ENCOUNTER: SPACECRAFT INSTRUMENTS PERFORMING REMOTE-SENSING OBSERVATIONS OF THE COMETARY DUST

spacecraft	instrument	type	dust-related aims	reference
Giotto	HMC	CCD telescope	imaging of dust jets and near-nucleus dust distribution through two narrow-band ($\Delta\lambda \sim 20$ nm) filters	Schmidt *et al.* (1986)
	OPE	photopolarimeter	determination of dust spatial density and scattering cross section by observing scattered light in three narrow-band filters ($\Delta\lambda \sim 3$–10 nm) and two polarizations	Levasseur-Regourd *et al.* (1986*a*)
	GRE	radio Doppler and ranging measurements	determination of total mass (dust and gas) impacting on spacecraft by observation of Doppler frequency shift due to atmospheric drag (no special on-board equipment required)	Edenhofer *et al.* (1986*a*)
Vega 1 and 2	TVS	CCD telescope	similar to *Giotto* HMC	Sagdeev *et al.* (1986)
	IKS	infrared spectrometer (2.5–5 μm and 6–12 μm), and imager (7–10 μm and 9–14 μm)	determination of dust spectral features and nuclear dust temperature	Combes *et al.* (1986)
	TKS	ultraviolet, visible and near infrared spectrometer	determination of dust spatial density and size spectrum from near-infrared scattered spectrum	Krasnopolsky *et al.* (1986)

2.1. *Measurements from the dust impact detection system, DIDSY*

The dust impact detection system (DIDSY) consists of five independent subsystems for the determination of the cometary dust efflux mass spectrum (figure 1) (McDonnell *et al.* 1986*a*). Three piezoelectric sensors mounted on the spacecraft-front dust-protection shield register dust grains of mass greater than 4×10^{-12} kg while a similar sensor mounted on the Kevlar rear shield detects the larger dust grains that can perforate the front shield ($m \gtrsim 10^{-9}$ kg). Mounted separately on the dust shield, the IPM (impact plasma and momentum) subsystem comprises a plasma detector (IPM-P) of limiting mass 10^{-20} kg, together with a piezoelectric crystal (IPM-M) fixed to the same target plate (limiting mass 4×10^{-13} kg). The IPM-P sensor consists of two identical halves, one of which is covered by a thin (*ca.* 2.5 μm) foil to provide limited particle density information. Also on the shield, the CIS (capacitor impact sensor), which comprises a thin film capacitor bonded onto the front shield, detects particles whose mass exceeds a threshold value of 3×10^{-13} kg. All these subsystems are controlled by the CDF (central data formatter), performing on-board data processing and formatting.

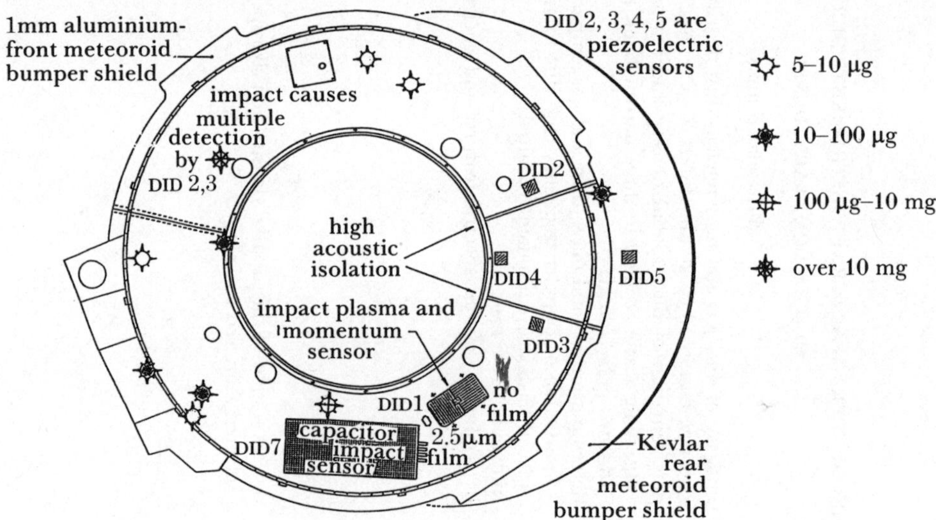

FIGURE 1. DIDSY sensor configuration on the outer dust shield and DID 5 on the rear Kevlar shield. Position of the ten largest particle impacts determined until now are indicated.

The properties of the subsystems and of the approximately 12000 impacts so far identified up to the point of encounter are summarized in table 3 (McDonnell *et al.* 1986*b*). All data in this table refer to the time period from first impact detection; up to -43 s for IPM-P, -5 s for MSM, RSM and IPM-M and -10 s for CIS. Time refers to encounter at 14 March 00h 11min 00s U.T. ground received time.

At -21 min an impact occurred on IPM causing saturation of both IPM-M and IPM-P giving a lower mass limit of some 5×10^{-11} kg. The CIS suffered a short circuit at -16 min which corrected itself at about -1 min. At approximately -44 s the largest mass, some 4×10^{-5} kg, was detected by all three front shield sensors. Around this time the IMP-P became erratic; the mode of operation of two of the MSM sensor (DID2 and DID3) became erratic and then changed to a fixed, non-preferred mode.

Preliminary analysis of the count rates from the various sensors has yielded fluxes of particles; a time profile shows the activity measured in the various mass ranges (figure 2).

TABLE 3. SUMMARY OF PERFORMANCE DATA AND MEASUREMENTS FROM DIDSY

sensor subsystem	limiting mass/kg	geometric area/m^2	effective area/m^2	first detection (time and distance)	total number detected[†]	peak rate/s^{-1}[†]	peak flux/$(m^{-2}\ s^{-1})$[‡]
DID 1 (IPM-M)	4×10^{-13}	1.19×10^{-2}	1.19×10^{-2}	-24 min (9.7×10^4 km)	1027	155 (-8 s)	4.5×10^4
DID 1 (IPM-PA) (sensor open)	10^{-20}	5.96×10^{-3}	5.96×10^{-3}	-62 min (2.5×10^5 km)	1100	31 (-49 and -44 s)	5.2×10^3
DID 1 (IPM-PB) (sensor + foil)	10^{-20} ($+2.5$ µm foil)	5.96×10^{-3}	5.96×10^{-3}	-36 min (1.5×10^5 km)	80	10 (-47 s)	1.6×10^3
DID 2/3 (MSM)	4×10^{-12}	1.94	2.5×10^{-1}§	-65 min (2.7×10^5 km)	1465	63 (-26 s)	1.5×10^3
DID 4 (MSM)	4×10^{-12}	2.2×10^{-1}	1.56×10^{-1}§	-70 min (2.9×10^5 km)	906	39 (-28 s)	5.2×10^2
DID 5 (RSM)	10^{-9}	2.2	~ 0.4 (est.)§	-208 s (1.4×10^4 km)	11	3 (-7 s)	7
DID 7 (CIS)	3×10^{-13}	10^{-1}	10^{-1}	-34 min (1.4×10^5 km)	7551	726 (-28 s)	5.3×10^4

† Not corrected for saturation and dead time effects.
‡ For all masses above the limiting mass.
§ Depends on incident mass spectrum (see text).

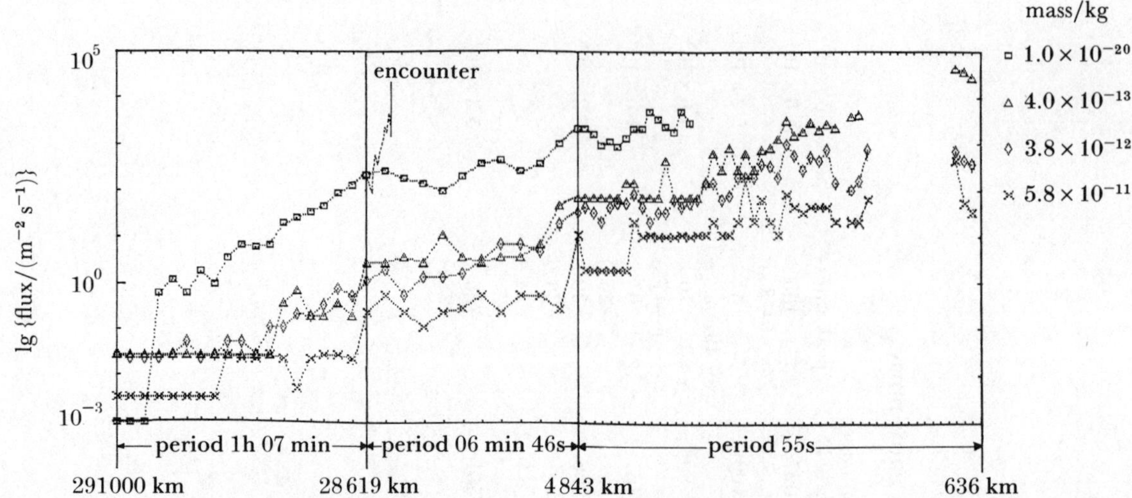

FIGURE 2. Flux rates (cumulative im mass) detected by various of the DIDSY subsystems, for the pre-encounter period, with the mass threshold limit for each profile shown. Note the expansions in the time axis for the rates nearest the time of closest approach. For comparison, the dotted line on the 1.0×10^{-20} kg line indicates, on the same timescale as the first period, the flux up to closest approach. No corrections are made for sensor dead-time effects, but this should be significant only for the last few seconds before closest approach.

3.1. *Cometary gas emission and dust acceleration*

The basis for the dynamical behaviour and distribution of dust is the so-called 'fountain model' based on Whipple's icy-nucleus model (Whipple 1950). This model can account for the motion of particles in the range 10^{-17}–10^{-3} kg; these are ejected with a velocity that depends on particle size, and move (relative to the comet nucleus) under the influence of radiation pressure (Divine 1981). The Halley encounters have shown many parameters assumed in the model to be ill-founded, but its general properties are substantiated.

The cometary nucleus is represented as a sphere of radius R_N of an inhomogeneous mixture of rocks, dust grains and volatile ices (Whipple 1950; 1951). The gas flux, related to the nucleus surface temperature T_N is given by (Delsemme & Swings 1952):

$$Z = P_s \, (2\pi \bar{M} k T_N)^{-\frac{1}{2}} \, \text{kg m}^{-2} \, \text{s}^{-1}, \tag{1}$$

where P_s is sublimation pressure, $\bar{M}$ is mean gas molecular mass, k is Boltzmann's constant and T_N is nucleus surface temperature.

This assumes that ice sublimes to form a gas with a molecular speed of

$$\bar{V} = (8k T_N/\pi \bar{M})^{\frac{1}{2}}. \tag{2}$$

Close to the nucleus surface, the sublimating ice–gas escapes with rate of Q_g kg s^{-1}, thereby accelerating the released dust; loading during expansion may be significant.

The radial flux of gas molecules at the surface of a spherical nucleus is given by

$$N = f(\zeta, \xi) \, (Q_g/R_N^2) \, \text{kg m}^{-2} \, \text{s}^{-1}, \tag{3}$$

where $f(\zeta, \xi)$ represents the source distribution function, and ζ and ξ are the spherical co-ordinates of the emission point on the surface of the nucleus (see §3.3).

Only dust grains that acquire enough force from the gas flow to overcome the nucleus's gravitational attraction can escape from the comet and an upper limit of 3.3×10^{-3} kg is assumed now (Divine *et al.* 1986). The radial flux of such dust grains is described by a mass distribution $g(m)$ given by (Divine 1981)

$$dN = \mu \bar{M} Q_g / R_N^2 \, f(\zeta, \xi) \, g(m) \, dm \, \text{m}^{-2} \, \text{s}^{-1}, \tag{4}$$

where μ is the dust:gas mass ratio.

We have established a parametric form for the terminal velocity V_T of the liberated grains as a function of gas velocity V and particle radius, a, which encompasses the data of T. I. Gombosi (personal communication), namely

$$V_T = \bar{V} \frac{(a_g/a)^{\frac{1}{2}}}{1 + (a_g/a)^{\frac{1}{2}}}, \tag{5}$$

where $a_g = a_p \, (0.59/r_\odot)^{2.2}$ is the radius of the particle accelerated to $\frac{1}{2}V$ (which varies with heliocentric distance). For Halley near encounter we take

$$a_p = 2 \times 10^{-6} \, \text{m},$$

$$r_\odot = \text{heliocentric distance in AU}.$$

Alternative and more complete forms have been given (see, for example, Delsemme & Miller 1971) but are less tractable in fitting observational data. Observational data from the Halley *Giotto* NMS instrument (Krankowsky *et al.* 1986) suggests a gas velocity of near 900 ms^{-1}, although pre-encounter modelling had assumed a value of *ca.* 660 ms^{-1}.

The motion of dust particles under radiation pressure results only in zero, one or two solutions at a point for dust trajectories from the nucleus, corresponding to locations beyond, on or within the envelope respectively. The 'two solution region' corresponds to trajectories originating directly from the nucleus or 'reflected' from the outer envelope. On the envelope these two solutions converge to produce an enhancement in local dust density.

A parameter η permits the specification of times of flight from the nucleus; the parameter is given by

$$\eta = \frac{R}{(2E+S) + [(2E+S)^2 - R^2)]^{\frac{1}{2}}}, \tag{6}$$

where R is the distance from the comet nucleus, S is the distance from the nucleus along the Sun–comet axis and E corresponds to the apex distance, which is the maximum distance any particle can travel in the sunward direction. The time of flight for travel from the nucleus for the two solutions is given by

$$t = (4ER\eta^{\pm 1})^{\frac{1}{2}}/V, \tag{7}$$

where η^{-1} and η^{+1} represent direct and reflected particles respectively. The ratio of the density of direct to reflected particles can be expressed as

$$\delta = \frac{f(\zeta_a, \xi_a) \, [(2E+S) + (2E+S)^2 - R^2]^{\frac{1}{2}}]}{f(\zeta_b, \xi_b) \, (R^2)}, \tag{8}$$

where $f(\zeta_a, \xi_a)$ and $f(\zeta_b, \xi_b)$ represent the source distribution functions for direct and reflected particles, respectively. Predicted values for *Giotto*'s trajectory show that δ varies from five at the edge of the outer envelope ($R = 10^5$ km) to 5×10^4 at the point of closest approach.

3.2. *Comparison with other models*

One shortcoming of the 'fountain model' is that it does not consider realistically the behaviour of large particles. As studied by Fertig *et al.* (1984), the larger particles have lower terminal velocities, which, because of the changing heliocentric distance, results in their confinement to a limited and asymmetric area around the nucleus–Sun line. This in turn causes the shape of the overall particle cloud to deviate greatly from the paraboloid of rotation typical of smaller particles.

In addition the 'fountain model' does not cater for the more general case where the nucleus moves around the Sun on a keplerian orbit. Models recently developed at ESOC (Massone *et al.* 1985) simulate such behaviour. The motion of dust particles in such cases can be expressed by

$$r = \phi V + \psi \, \Delta\kappa, \tag{9}$$

where r is the position vector of the dust particle, $\Delta\kappa$ represents the dependency of radiation pressure, V is the initial velocity vector of dust grains; and ϕ, ψ depend only on the orbit of the nucleus and the time of arrival at the required location.

It is essential to note that the pre-encounter mass distributions used to develop a framework for all modelling were determined explicitly from infrared observations of dust in comets. This was studied by several researchers such as Mukai (1977), Krishna Swamy & Donn (1979) and Hanner (1980).

Grain temperatures as a function of particle radius and heliocentric distance can be derived from theoretical modelling of thermal emission (Hanner 1982). These are obtained by equating the amount of solar energy absorbed by the grains to the thermal energy emitted by them. Because the emission is inefficient at wavelengths large compared with grain size, this relation shows that grains of a given temperature cannot radiate efficiently at wavelengths greater than ten times their radius and implies an observational limit for detection by infrared techniques of about $0.1\ \mu\mathrm{m}$ for the radius of emitting grains. Considering a typical density of about $1\ \mathrm{g\ cm^{-3}}$ for such particles, a mass cutoff of 10^{-16}–10^{-17} kg is implied. However, the data from *Giotto* and *Vega in-situ* dust experiments now demonstrate the existence of a high spatial density of grains with masses up to three magnitudes smaller than this limit (i.e. 10^{-20} kg) (McDonnell *et al.* 1986*b*).

3.3. *Nucleus emission model*

It is important to use a realistic function that describes the pattern of dust and gas emission from the nucleus. The Comet Halley Working Group uses the form (Divine 1981)

$$f(\zeta, \xi) = (0.0159)\ (5 - 3\cos\theta), \tag{10}$$

where θ is angular distance from antisubsolar point. The function has a maximum emission from the subsolar point and minimum from the antisubsolar point. This function accounts for emission due to both solar radiation and heating from the neighbouring coma and leads to a light–dark peak emission ratio of four. A new set of functions has now to be considered to account for the unexpected flux measurements from the spacecraft encounter with Halley.

Measurements show a strong concentration in forward ejecting jets, and hence a more concentrated function was tested, namely

$$f(\zeta, \xi) = (3/4\pi) \cos^2 \theta, \; \theta > 90° \quad \text{and} \quad f(\zeta, \xi) = 0 \quad \text{for} \quad \theta < 90°. \tag{11}$$

Figure 3 shows the result of fitting to the cumulative mass obtained from DIDSY at a distance of 15 600 km from the nucleus. The mass distribution, $g(m)$, and source function, $f(\zeta, \xi)$ of Divine (1981) have been used. The poor fit, as shown by discrepancies in the slope, flux and cut-off are indicative of the shortcomings inherent in much of the pre-encounter modelling. Better agreement between the *in-situ* data and theoretical models is being sought by using modified mass and source functions. It is already found that, for example, more concentrated source functions, of the type described above, do give an improved fit to the experimental data.

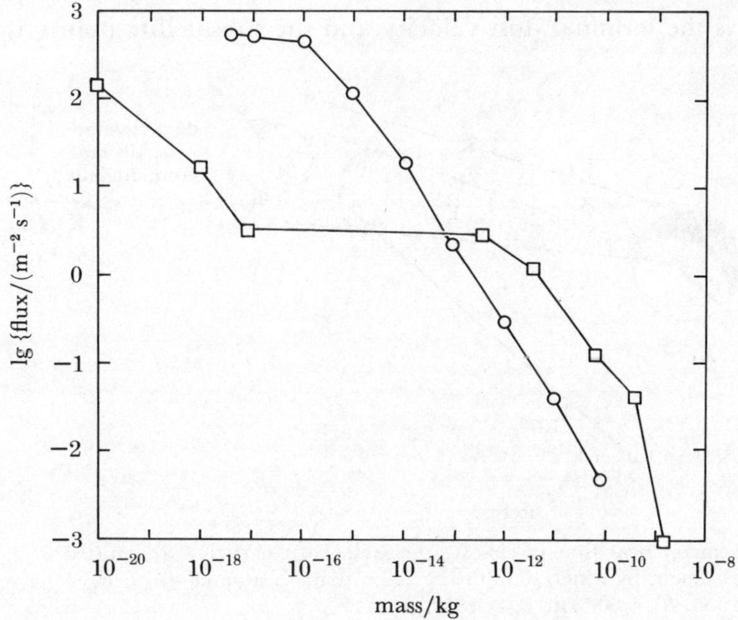

FIGURE 3. Comparison of actual and predicted cumulative mass spectrum at a distance of 15 600 km from the nucleus. □, Measured DIDSY data over an integration period of 23 s; ○, predicted cumulative spectrum based on the dust model of Divine (1981). Discrepancies in slope, cut-off and flux level are clearly illustrated.

3.4. *Nucleus source activity mapping from direct fly-by measurements*

Recording the dust flux on the shield of *Giotto* has, of course, enabled the flux of particles to be measured along the spacecraft's trajectory. Because the typical apex distances of such particle trajectories are of the order of 10^5 km (Divine 1981), and the parabolae are approximately described by straight lines near the source, we can assume the dust particles to have travelled in a straight line between the emission point and the interception point at *Giotto*. In this way, the dust flux can be transformed back to a given position on the nucleus of the comet, according to the mass of particle arriving at the spacecraft, because the acceleration (occurring within several nuclear radii, namely *ca.* 15 km) is radial with respect to the comet nucleus (Divine *et al.* 1985).

Such a technique can be used to map active areas of the nucleus, where comparatively large dust emissions have occurred. By making (θ_0, ϕ_0) the equatorial coordinates of the subsatellite point at closest approach, the subsatellite track can be drawn; a term resulting from the rotation of the nucleus is included to give the coordinates of dust emission θ_t, ϕ_t (see figure 4).

$$\theta_t = \theta \ (\text{subsatellite}) + \theta \ (\text{emission point rotation})$$

$$= \left\{ -\arctan\left[\frac{V_s t \cos \epsilon}{R_0}\right] - \frac{t}{P_c} \times 360° + \theta_0 \right\} - \left\{ \frac{R}{V_T} \times \frac{360°}{P_c} \right\}, \tag{12}$$

$$\phi_t = \arcsin\frac{V_s t \sin \epsilon}{R} + \phi_0,$$

where V_s is the spacecraft velocity, t is time after encounter, R_0 is the distance of closest approach, ϵ is the angle subtended by the nuclear equator and ecliptic along the trajectory, P_c is the rotation period, V_T is the terminal dust velocity and the subsatellite points are θ_0 and ϕ_0.

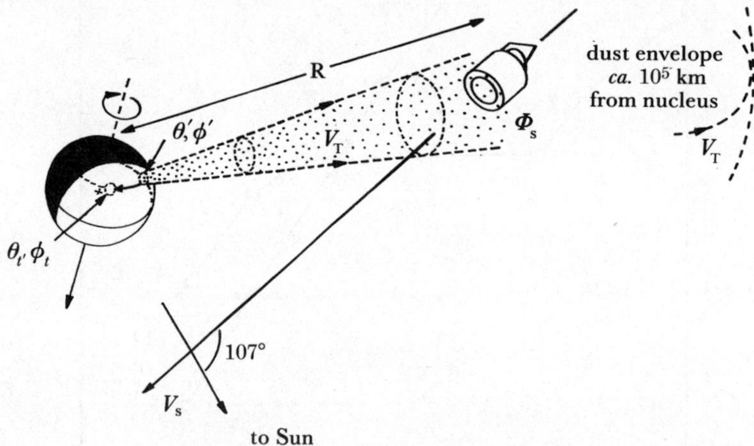

FIGURE 4. Encounter geometry near time of closest approach. Dust particles are emitted from point (θ', ϕ') and detected by *Giotto* at time t, by which time the emission point is situated at (θ_t, ϕ_t) relative to the spacecraft, due to nucleus rotation. $R_0 = 600$ km, $\theta_0 = 185°$, $\phi_0 = 25°$.

By looking at fluxes detected within different mass ranges, as distinct from threshold sensor measurements, whose terminal velocities are different, we have a dispersion of particle emission points around the nucleus. In this way, a picture of the nucleus can be built up by following back different particles to their appropriate emission points.

3.5. *Nucleus source mapping applied to* DIDSY *in-situ data*

A preliminary study of the potential for nucleus source mapping has been made with dust fluxes measured by the DID sensors. A total mass range is available from 10^{-20} kg to the maximum mass of 40 mg (largest particle); here, we show the tracks of emission points corresponding to $m_i = 10^{-11}$, 10^{-9} and 10^{-4} kg in figure 5. The subsatellite longitude and latitude of *Giotto* at closest approach are taken to be $\theta_0 = 185°$, $\phi_0 = 25°$ with $\epsilon = 58°$.

In figure 5 we also show the flux rate for the DID1M sensor, corresponding to particles of about 4×10^{-13} kg. This activity can then be related to the emission track of such masses (see §5). In this way, analysis of DIDSY data will reveal more detail of the nuclear surface.

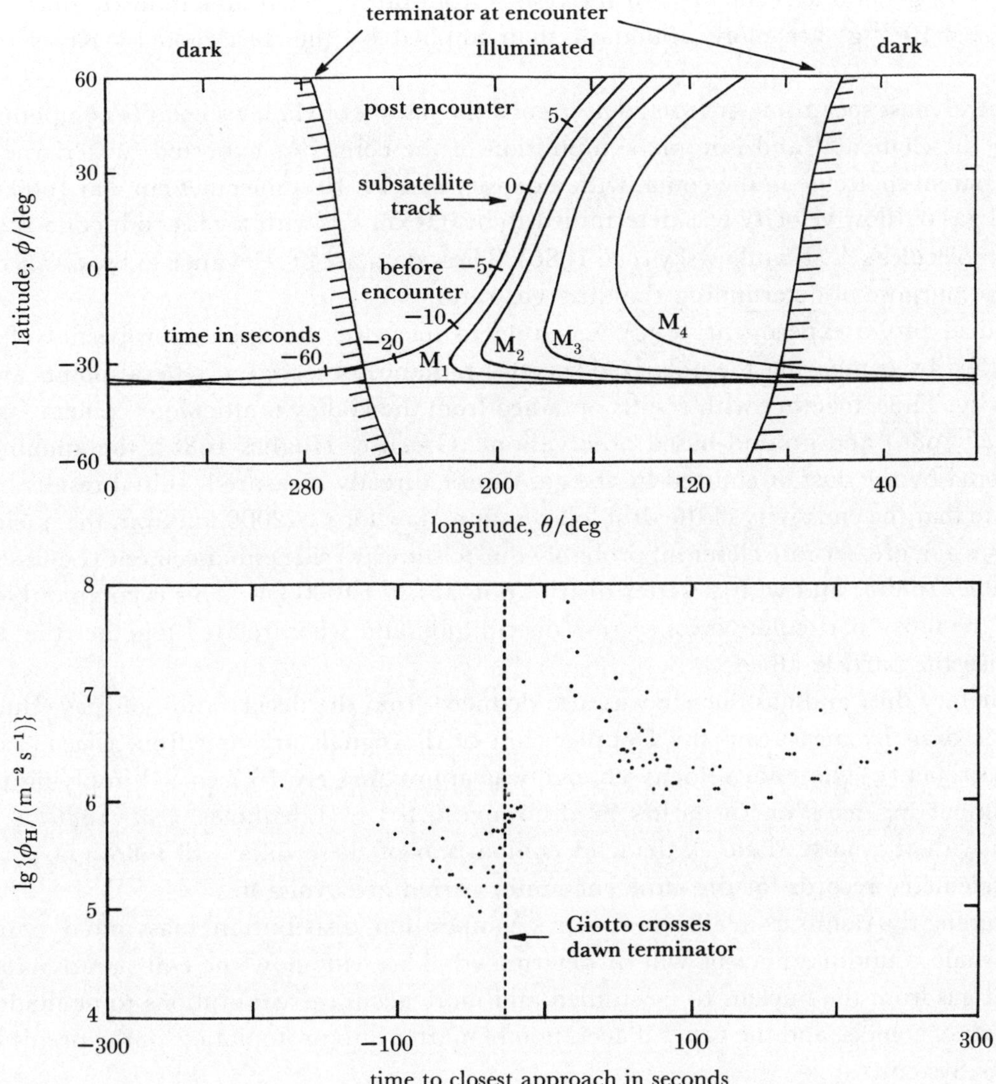

FIGURE 5. A latitude–longitude map showing emission tracks for particles of four different masses around Halley's nucleus. The track marked M_1 corresponds to masses of 4×10^{-13} kg, and a plot of the nucleus dust flux against time to closest approach is shown below. Tracks marked M_2, M_3 and M_4 correspond to masses of 10^{-10}, 10^{-8} and 10^{-6} kg, respectively. The subsatellite track is also shown, with tick marks indicating, in seconds, time to closest approach. We also show the flux at Halley, ϕ_H, normalized to an R^{-2} dependence for particles of mass 4×10^{-3} kg.

4. *Synergistic measurements related to* DIDSY

In addition to DIDSY, dust detection on the *Giotto* and *Vega* spacecraft is by the particulate impact analyser PIA and PUMA (Kissel 1986), which provides information on the composition of the individual dust particles and the average composition and mass distribution of cometary matter released by Halley. Preliminary results indicate the presence of H, C, N, O, Na, Mg, Si and Fe. One group significantly comprises particles that are rich in H, C, N and O. Contrary to expectations, carbonaceous chondrites fit only partially the mass spectra obtained for detected cometary particles. Furthermore, the spectra seem to indicate that the light elements specifically seem to have low mass to volume ratio, with particle densities perhaps as low as

0.01 g cm^{-3}. In general agreement with the results from DIDSY, PIA results indicate that small particles ($m < 10^{-17}$ g) are more abundant than implied by the dust models (Kissel *et al.* 1986).

The neutral mass spectrometer (NMS) determined the nature of Halley's volatile components, measuring the elemental and isotopic composition of the coma. As expected, water was the dominant parent molecule in the coma, with a density of 4.7×10^7 molecules cm^{-3} at 1000 km. The radial gas outflow velocity was determined to be 0.9 km s^{-1}, with a gas production rate of 6.9×10^{29} molecules s^{-1}. (Krankowsky *et al.* 1986). These data are of relevance to the *in-situ* dust data for the purpose of determining the dust:gas ratio.

The optical probe experiment (OPE) was able to measure the optical properties of the cometary dust by comparing the polarized spectral radiance observed at several points along the trajectory. Thus, together with results obtained from the Halley multicolour camera (HMC) (Keller *et al.* 1986) and ground-based observations (Green & Hughes 1987), the amount of light scattered by the dust or emitted by the gas can be directly measured. Initial results from OPE indicate that the emissivity of the dust follows an r^{-2} law for $r > 2000$ km from the nucleus, but increases at a greater rate closer in, probably due to sunward side enhancement (Levasseur-Regourd *et al.* 1986*b*) and with a flatter distribution at $r > 10000$ km. This is comparable in form with the DIDSY particulate cross section distribution and when related together the data will establish the particle albedo.

The cometary dust and gas fluence was also deduced from the deceleration of *Giotto* due to atmospheric drag by measuring the Doppler shift of the signals arriving from *Giotto* (radio science experiment). The total velocity change was approximately 16.7 cm s^{-1} implying that the total impacting mass on *Giotto* lies in the range 0.1–1 g (Edenhofer *et al.* 1986*b*), not conflicting with the DIDSY data. A detailed comparison of these data will follow now that extended telemetry records for the close encounter period are available.

By combining the results from all experiments, composition, distribution, mass and dynamics of the dust halo round the nucleus will be determined. This will allow the emission dynamics of dust and gas from the nucleus to be studied and more accurate assumptions to be made as to the genesis of comets, and the general accretion of matter into protoplanets in the prehistory of the Solar System.

5. SIGNIFICANCE OF ENCOUNTER DATA

Although the data from the various spacecraft encounters are still at a very early stage of analysis, some broad conclusions can already be drawn from the data that refer specifically to the dust. Referring to the nucleus-mapping data, the first of which are presented in figure 5 for masses of around 4×10^{-13} kg, there appears to be a general increase in emitted flux as one proceeds further into the sunlit hemisphere of the nucleus. This may be indicative of a thermal-lag effect past the dawn terminator. From this data, we estimate the bright to dark side integrated flux ratio to be of the order of 20 to 30. As discussed earlier, existing models tend to adopt a smoothly varying source function of sinusoidal form. The function described by (10) predicts that this ratio is of the order of only 2. Additionally, camera images from *Giotto* and *Vega* show quite clearly that the emission is dominated by forward jet emission. Detailed analysis of the dust impact profiles will enable the relative importance of emission from the jets and the general background emission to be determined.

During the very close encounter period, we can use the DIDSY impact rates and derived mass spectrum, combined with a $1/r^2$ model for the radial dependence of the dust distribution to determine the amount of dust in the line of sight from the spacecraft to the nucleus. For example, at a distance of 2000 km, we estimate that in a line-of-sight column to the nucleus, the integrated geometrical cross-sectional area of dust represents some 0.02 of the cross-sectional area of the column. This is consistent with the visibility of the nucleus as determined by HMC images (Keller *et al.* 1986).

Integration of the DIDSY flux rates yields a dust production rate of 3 t s^{-1}. Combined with data from NMS (Krankowsky *et al.* 1986), this gives a total mass rate of some 19 t s^{-1} at the time of *Giotto* encounter. Results from *Vega* 1 and 2 yield slightly different results, possibly indicative of variations in emission rates due to the dominance of jet activity. The *Giotto* results implies an erosion rate of some 0.5 m lost per perihelion passage. If we assume that Comet Halley was injected into the inner Solar System some 20000 years ago (Hughes 1985), this implies that the present mass is approximately half its original value. This mass-loss rate also suggests that the topology and current shape of the nucleus are dominated by ablation processes over the comet's lifetime in the inner Solar System, since the impact erosion rate on lunar rocks at 1 AU heliocentric distance is of the order of 1 mm Ma^{-1} (McDonnell & Ashworth 1972), i.e. corresponding to a negligible loss at each perihelion; evaporation is more than eight magnitudes more significant than impact erosion averaged over the orbit of Halley.

Several results of significance are suggested by preliminary analyses of the dust mass spectrometers (PIA on *Giotto* and PUMA on *Vega*). For example, one type of dust particle seems to be composed entirely of low-Z elements. Another type is only roughly similar to carbonaceous chondrites. Additionally some grains appear to have a low density. Also of considerable significance is that the dust observations of OPE, which yield the product of the dust local spatial density and the scattering cross section and the observations of DIDSY, when combined, will yield the grain albedo along *Giotto*'s trajectory.

From a technical viewpoint, the results from DIDSY have demonstrated the viability of the concept and design of the dual dust protection shield. This is potentially of great significance for future cometary missions. Results imply that approximately 1 % of the shield's surface area was ablated by dust impacts and that the total area penetrated by impacts was of the order of 1.5 cm^2.

Acknowledgements are due to all DIDSY coinvestigators and their collaborators, in particular those who supplied flight equipment or ground support hardware, namely W. M. Burton, E. Bussoletti, R. J. L. Grard, E. Grun, B. A. Lindblad, J.-C. Mandeville, R. F. Turner, J. G. Firth and G. C. Evans. The considerable efforts of the European Space Agency project team, the *Giotto* prime contractor, British Aerospace and cocontractors are also gratefully acknowledged. Financial support from the Science and Engineering Research Council to the Unit for Space Sciences, University of Kent is also acknowledged.

REFERENCES

Combes, M., Moroz, V. I., Crifo, J. F., Lamarre, J. M., Chara, J., Sanko, N. F., Soufflot, A., Bibring, J. P., Cazes, S., Coron, N., Crovisier, J., Emerich, C., Encrenaz, T., Gispert, R., Grigoryev, A. V., Guyot, G., Krasnopolsky, V. A., Nikolsky, Yu. V. & Rocard, F. 1986 *Nature, Lond.* **321**, 266.
Delsemme, A. H. & Swings, P. 1952 *Annls Astrophys.* **15**, 1.
Delsemme, A. H. & Miller, D. C. 1971 *Planet. Space Sci.* **19**, 1229.

Divine, N. 1981 ESA special publication no. SP-174, p. 25.
Divine, N., Fechtig, H., Gombosi, T. I., Hanner, M. S., Keller, H. U., Larson, S. M., Mendis, D. A., Newburn, Ray L. Jr, Reinhard, R., Sekanina, Z. & Yeomans, D. K. 1986 *Space Sci. Rev.* **43**, 1.
Edenhofer, P., Bird, M. K., Brenkle, J. P., Buschert, H., Esposito, P. B., Porsche, H. & Volland, H. 1986*a* ESA special publication no. SP-1077 p. 173.
Edenhofer, P., Bird, M. K., Brenkle, J. P., Buschert, H., Esposito, P. B., Porsche, H. & Volland, H. 1986*b* *Nature, Lond.* **321**, 355.
Fertig, J., Hechler, F. & Schwehm, G. H. 1984 *ESA bull.* **38**, 36.
Green, S. F. & Hughes, D. W. 1987 (In preparation.)
Hanner, M. S. 1980 In *The Solar System* (ed. I. Halliday & B. McIntosh), p. 223.
Hanner, M. S. 1982 In *Comets* (ed. L. Wilkening), p. 341. Tucson: University of Arizona Press.
Hughes, D. W. 1985 *Mon. Not. R. astr. Soc.* **213**, 103.
Keller, H. U., Arpigny, C., Barbieri, C., Bonnet, R. M., Cazes, S., Coradini, M., Cosmovici, C. B., Delamere, W. A., Huebner, W. F., Hughes, D. W., Jamar, C., Malaise, D., Reitsema, H. J., Schmidt, H. U., Schmidt, W. K. H., Seige, P., Whipple, F. L. & Wilhelm, K. 1986 *Nature, Lond.* **321**, 320.
Kissel, J. 1986 ESA special publication no. SP-1077, p. 67.
Kissel, J., Brownlee, D. E., Büchler, K., Clark, B. C., Fechtig, H., Grün, E., Hornung, K., Igenbergs, E. B., Jessberger, E. K., Krueger, F. R., Kuczera, H., McDonnell, J. A. M., Morfill, G. M., Rahe, J., Schwehm, G. H., Sekanina, Z., Utterback, N. G., Völk, H. J. & Zook, H. A. 1986 *Nature, Lond.* **321**, 336.
Krankowsky, D., Lämmerzahl, P., Herrwerth, I., Woweries, J., Eberhardt, P., Dolder, U., Herrmann, U., Schulte, W., Berthelier, J. J., Illiano, J. M., Hodges, R. R. & Hoffman, J. H. 1986 *Nature, Lond.* **321**, 326.
Krasnopolsky, V. A., Gogoshev, M., Moreels, G., Moroz, V. I., Krysko, A. A., Gogosheva, Ts., Palazov, K., Sargoichev, S., Clairemidi, J., Vincent, M., Bertaux, J. L., Blamont, J. E., Troshin, V. S. & Valníček, B. 1986 *Nature, Lond.* **321**, 269.
Krishna Swamy, K. S. & Donn, B. 1979 *Astron. J.* **84**, 697.
Levasseur-Regourd, A.-C., Bertaux, J. L., Dumont, R., Festou, M., Giese, R. H., Giovane, F., Lamy, P., Llebaria, A. & Weinberg, J. L. 1986*a* ESA special publication no. SP-1077, p. 187.
Levasseur-Regourd, A.-C., Bertaux, J. L., Dumont, R., Festou, M., Giese, R. H., Giovane, F., Lamy, P., Le Blanc, J. M., Llebaria, A. & Weinberg, J. L. 1986*b* *Nature, Lond.* **321**, 341.
Massone, L., Fertig, J., Grun, E., Schwehm, G. H. 1985 In *Asteroids, comets and meteors II* (Meeting at Uppsala University) (ed. C.-I. Lagerkvist, B. A. Lindblad, H. Lundstedt & H. Rickman), p. 407.
McDonnell, J. A. M. & Ashworth, D. G. 1972 *Adv. Space Res.* **13**, 333.
McDonnell, J. A. M., Alexander, W. M., Burton, W. M., Bussoletti, E., Clark, D. H., Evans, G. C., Evans, S. T., Firth, J. G., Grard, R. J. L., Grun, E., Hanner, M. S., Hughes, D. W., Igenbergs, E., Kuczera, H., Lindblad, B. A., Mandeville, J.-C., Minafra, A., Reading, D., Ridgeley, A., Schwehm, G. H., Stevenson, T. J., Sekanina, Z., Turner, R. F., Wallis, M. K. & Zarnecki, J. C. 1986*a* ESA special publication no. SP-1077, p. 85.
McDonnell, J. A. M., Alexander, W. M., Burton, W. M., Bussoletti, E., Clark, D. H., Grard, R. J. L., Grün, E., Hanner, M. S., Hughes, D. W., Igenbergs, E., Kuczera, H., Lindblad, B. A., Mandeville, J.-C., Minafra, A., Schwehm, G. H., Sekanina, Z., Wallis, M. K., Zarnecki, J. C., Chakaveh, S. C., Evans, G. C., Evans, S. T., Firth, J. G., Littler, A. N., Massone, L., Olearczyk, R. E., Pankiewicz, G. S., Stevenson, T. J. & Turner, R. F. 1986*b* *Nature, Lond.* **321**, 338.
Mukai, T. 1977 *Astron. Astrophys.* **61**, 69.
Sagdeev, R. Z., Szabó, F., Avanesov, G. A., Cruvellier, P., Szabó, L., Szegö, K., Abergel, A., Balazs, A., Barinov, I. V., Bertaux, J.-L., Blamont, J., Detaille, M., Demarelis, E., Dul'nev, G. N., Endroczy, G., Gardos, M., Kanyo, M., Kostenko, V. I., Krasnikov, V. A., Nguyen-Trong, T., Nyitrai, Z., Reny, I., Rusznyak, P., Shamis, V. A., Smith, B., Sukhanov, K. G., Szabó, F., Szalai, S., Tarnopolsky, V. I., Toth, I., Tsukanova, G., Valníček, B. I., Varhalmi, L., Zaiko, Yu K., Zatsepin, S. I., Ziman, Ya. L., Zsenei, M. & Zhukov, B. S. 1986 *Nature, Lond.* **321**, 262.
Schmidt, W. K. H., Keller, H. U., Wilhelm, K., Arpigny, C., Barbieri, C., Beirmann, L., Bonnet, R. M., Cazes, S., Cosmovici, C. B., Delamere, W. A., Huebner, W. F., Hughes, D. W., Jamar, C., Malaise, D., Reitsma, H., Seige, P. & Whipple, F. L. 1986 ESA special publication no. SP-1077, 149.
Whipple, F. L. 1950 *Astrophys. J.* **111**, 375.
Whipple, F. L. 1951 *Astrophys. J.* **113**, 464.

Discussion

Sir Bernard Lovell, F.R.S. (*Nuffield Radio Astronomy Laboratories, Cheshire, U.K.*). There are two meteor streams ascribed to Comet Halley as a source. How do your data relate to these streams?

J. A. M. McDonnell. From the *Giotto* DIDSY data we have established two input parameters that would assist in this correlation, namely (i) the size spectrum or mass index and (ii) the total mass input potentially available for such streams generated by Comet Halley during its 1986 perihelion passage. The latter is, of course, determined only up to a limiting sensitivity corresponding to some 40 mg, but certainly corresponds to a second-magnitude photographic meteor or brighter. The mass index of the dust distribution may well differ from that observed in the streams because of observational factors and evolutionary changes within the perhaps 20 000 year old particles in the stream.

Sir Fred Hoyle, F.R.S. (*Department of Applied Mathematics and Astronomy, University College, Cardiff, U.K.*). In passing from particle masses to their densities, volumes are needed. How were the volumes obtained? Were they from areas, taken to be $(\text{area})^{\frac{3}{2}}$?

J. A. M. McDonnell. Particle density is deduced by the particle impact analyser (PIA) (principal investigator Dr J. Kissel, Heidelberg) by consideration of the ionization from incident particles compared with the impact ionization which that particle causes from the silver target material. Such a ratio is a measure of the average mass per unit area of the particles, derived by their effective penetration into the PIA silver target. Interpretation is tentative as yet (not published).

S. Durrani (*Department of Physics, University of Birmingham, U.K.*). Is there any information on the material strength or density of the particles impacting the *Giotto* spacecraft from the DIDSY experiment?

J. A. M. McDonnell. At the *Giotto* impact speed of 68 km s^{-1}, the particle strength has a very minor effect on penetration ability and therefore inferring particle density from the DIDSY data is not straightforward, especially for particles penetrating the shield. For one range of masses, however, the high-sensitivity DIDSY IPM-P sensor has two separate parts, identical except for a 2.5 µm foil covering one part. From comparison of these with the subunits, the penetration limit may be obtained but such data are still under analysis.

Note added in proof (30 *May* 1987). In the more recent analysis of the data from DIDSY and PIA the database has been extended and instrumental calibration factors reassessed. It is now seen that the flux of the smaller particles on the IPM-M and P sensors before encounter was suppressed because of the malfunction of a thin protective Mylar cover (22 µm); this was indeed abraded totally at 10 s before encounter but does explain some of the anomalies characteristic of the pre-encounter DIDSY mass distribution (figure 3) where fluxes plotted at masses 10^{-20}, 10^{-18} and 10^{-17} kg are now seen to correspond to residues from larger penetrating particles in the mass range 10^{-15} to 10^{-12} kg. Revised and extended data are published in J. A. M. McDonnell *et al.* ESA Special Publication SP250 (1986).

Phil. Trans. R. Soc. Lond. A **323**, 397–404 (1987)

Printed in Great Britain

Comet Halley: the gas composition derived from space missions

By Thérèse Encrenaz

Observatoire de Paris-Meudon, Section d'Astrophysique, 92190 Meudon, France

Important results have been obtained by the *Vega* and *Giotto* missions concerning Comet Halley's gas composition. Water vapour and carbon dioxide have been identified with respective production rates of about 10^{30} s^{-1} and 10^{28} s^{-1}. In addition, there is evidence for the presence of hydrocarbons and/or carbonaceous material in large amounts in the immediate vicinity of the nucleus.

1. Introduction

Before the 1986 apparition of Comet Halley very little was known about the nature of parent molecules directly outgassed from cometary nuclei. Visible and ultraviolet spectra of various comets had given a large amount of information upon the nature and relative abundance of secondary products – ions, radicals and daughter molecules – which come from the dissociation of the parent molecules by the solar flux. In contrast, very little was known about the parent molecules because their observation is very difficult. First, they have a limited lifetime, so that they can be found only in the immediate surrounding of the nucleus, in a region corresponding to a few seconds of arc in diameter as observed from the Earth. Secondly, the strongest transitions expected for the candidate parent molecules occur in the infrared or millimetric range, where observations are more difficult.

On the basis of indirect arguments, as well as marginal detections, a few molecules had been selected as best candidates. The first one is H_2O because of: (1) the amount of OH and H, the most abundant radicals observed in comets; (2) the presence of the H_2O^+ ion; (3) the tentative detection of H_2O in the radio range, on Comet IRAS–Araki–Alcock (Altenhoff *et al.* 1983). HCN was considered as a possible parent of the CN radical, observed on all comets; it was also tentatively detected (Huebner *et al.* 1974), as well as CH_3CN (Ulich & Conklin 1974), on Comet Kohoutek. S_2 was also detected as a minor parent molecule on the uv spectrum of Comet IRAS Araki-Alcock (A'Hearn *et al.* 1983). The abundance of carbon and the form in which it is incorporated (gas or grains) is a major puzzle for cometary research. Although CO_2^+ was observed in many cases, CO_2 was never observed before 1986; CO, in contrast, was observed on several comets, as well as CO^+, but in very variable amounts, and it was not clear whether CO was a parent or a daughter molecule. More generally, the presence of parent carbonaceous molecules were needed to explain the abundances of CN, CH, C_2 and C_3 regularly observed in the visible spectra of all comets.

Ground-based observations of Comet Halley during the 1986 apparition already provided some new results for this research. First, the H_2O molecule was unambiguously detected, thanks to the progress of infrared airborne astronomy (Mumma *et al.* 1986), with an abundance comparable to the expected value. Water vapour has thus been confirmed as one of the most (if not the most) abundant parent molecules in comets. The second important result came from the detection of HCN, obtained in the millimetric range (Despois *et al.* 1986). The derived mixing ratio HCN/H_2O, in the order of 10^{-3}, is also in agreement with previous expectations. These discoveries have provided a major step in cometary research, with the first non-ambiguous detection of two parent molecules, but, very clearly, there were still other abundant parent molecules which remained to be found.

2. THE SPACE MISSIONS

Among the five space missions devoted to the study of Comet Halley, three had experiments to determine the gas composition of the coma: *Vega* 1, *Vega* 2 and *Giotto*. The two *Vega* probes encountered Comet Halley on 6 March 1986 and 9 March 1986, with respective miss distances of 8890 km and 8030 km. The instruments for gas analysis were two spectrometers, a neutral-mass spectrometer and an ion-mass spectrometer. The *Giotto* probe was somewhat more ambitious, in that it encountered Comet Halley on 14 March 1986 with a miss distance of only 605 km. Instrumentation directly devoted to *in situ* gas measurements consisted of ion- and neutral-mass spectrometers; an optical photopolarimeter was also included in the payload. In addition, several instruments were added on the three probes for the study of dust (dust-impact analysers, dust-mass spectrometers) and plasma (ion and electron analysers, magnetometers).

For the study of gas composition, information of different kinds was obtained. First, data were obtained on the spatial distribution of radicals in the inner coma; these results come from the trichannel spectrometer of *Vega* 2 (TKS) and the *Giotto* optical photopolarimeter (OPE). Second, information on parent molecules was obtained by the infrared spectrometer IKS aboard *Vega* 1; third, direct information on the gaseous constituents was derived from the *Vega* and *Giotto* mass spectrometers (ING aboard *Vega*, NMS and IMS aboard *Giotto*).

3. SPATIAL DISTRIBUTION OF RADICALS

The TKS spectrometer on Vega was designed to record the spectrum of the comet in three channels (UV, visible and near IR) with one spatial dimension; the other spatial dimension is achieved by scanning, so that monochromatic images can be obtained in the central coma. The spatial resolution is about 20 km at the fly-by point. The instrument operated successfully on *Vega* 2, except in the UV channel, during the encounter sequences and during two other sequences, one day before and after encounter.

By integrating the signal in the near infrared channel, during 10 mn, at the time of encounter, the H_2O v_1+v_3 band at 1.38 µm has been detected. The (0–0) band of the CN red system seems to be present at 1.1 µm, as well as some structure due to the OH $\Delta V = 2$ band between 1.5 and 1.8 µm. In the visible range, a large number of spectra were recorded, at a rate of 1 every 5 s, between 2800 and 7000 Å†, with a spectral resolution of ~ 25 Å and a very high signal: noise ratio. The following bands are easily identifiable: OH (3090 Å), NH (3360 Å), CN (3883 Å), C_3 (4040 Å), C_2 ($\Delta V = 0$, -1 and -2 at about 5100 Å, 5600 Å and 6100 Å, respectively) and NH_2 (6000–7500 Å). The S_2 signature is possibly present below 3000 Å. At a few hundred kilometres from the nucleus, the continuum increases very strongly following the solar continuum; this effect is due to the rapid increase of the dust distribution, more peaked than the distribution of radicals, towards the nucleus. The first results of the TKS

---

† $1 \text{ Å} = 10^{-1} \text{ nm} = 10^{-10} \text{ m}$.

experiment can be found in Krasnopolsky *et al.* (1986). We extract from this study the table of production rates for the most abundant species (table 1).

The OPE experiment aboard *Giotto* was built on a very different principle. The instrument was a photopolarimeter which operated in seven different visible filters; three in the dust continuum and four in selected emission bands corresponding to different cometary species: OH, C_2, CN and CO^+ (see figure 1). The instrument was oriented backwards and measured the integrated number density of each species along the line of sight. By difference, it is possible to retrieve the density distribution of each constituent as a function of the distance to the nucleus. The instrument operated successfully until the time of miss distance, and data were recorded every 4 s in each filter. The first results are presented in Levasseur-Regourd *et al.* (1986).

For nucleus distances ranging between 5×10^3 and 10^5 km, the dust density is found to follow

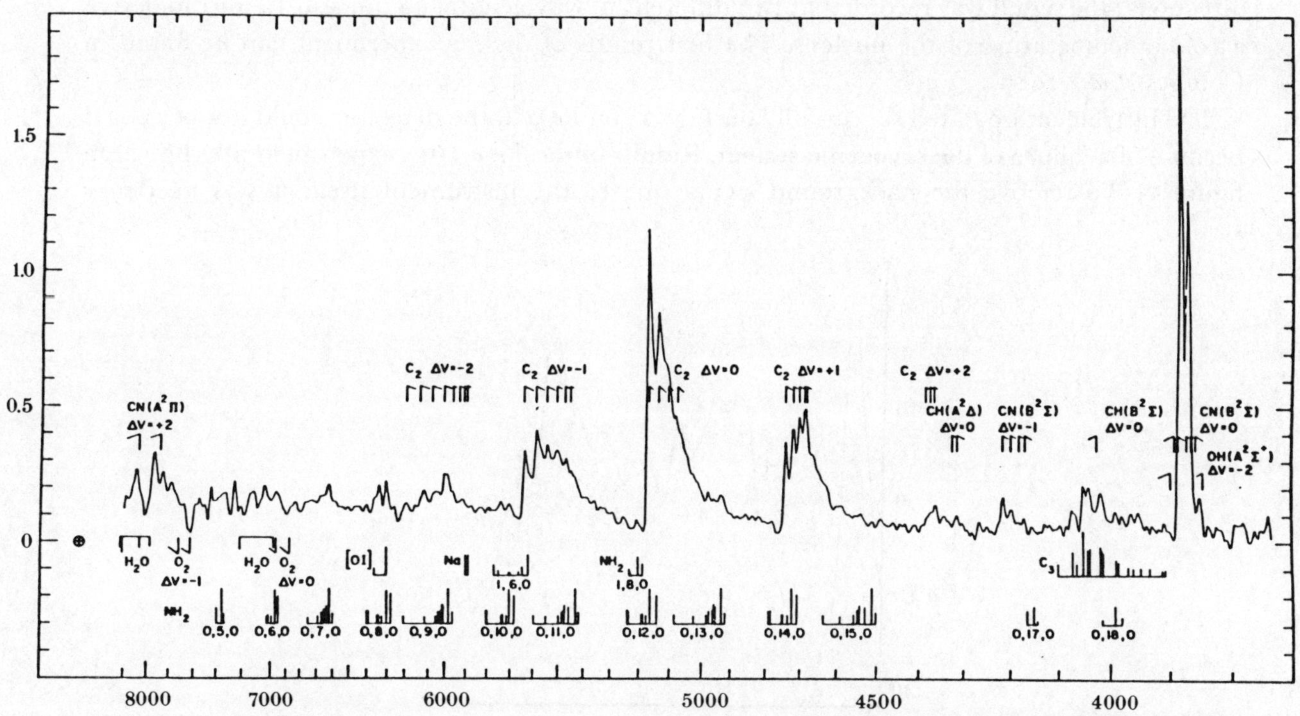

FIGURE 1. The visible spectrum of Comet Kohoutek (1973 XII) (A'Hearn 1983). Above the dust continuum are superimposed emission features due to fluorescence of cometary species: C_2, CN, NH_2 in the visible region, OH in the uv region (3090 Å).

TABLE 1. GASEOUS PRODUCTION RATES DERIVED FROM THE TKS EXPERIMENT (9 MARCH 1986)

(Krasnopolsky *et al.* (1986).)

molecule	production rate/s^{-1}
H_2O	4×10^{29}
OH	2×10^{30}
C_2	6×10^{27}
C_3	3×10^{27}
CN	10^{27}
NH	2×10^{26}

[151]

a r^{-2} distribution (where r is the distance to the nucleus). This result is consistent with the assumption of expansion at constant velocity, currently admitted in the vicinity of the nucleus. Below 5×10^3 km, the dust distribution increases faster than r^{-2} as the nucleus distance decreases; this is in agreement with the TKS result. The density distributions of the radicals follow a slope that is not as steep as in the case of the dust; this is as expected for secondary products.

4. The parent molecules

The IKS experiment aboard *Vega* was especially designed to search for parent molecules in the vicinity of the coma. The cometary flux was recorded in two spectroscopic channels, in the 2.5–5 µm range and 6–12 µm range with a resolving power of 40; the field of view was 1°. In addition, an 'imaging channel' was designed to modulate the infrared signal of the nucleus by two perpendicular grids, to recover the diameter of the nucleus along two perpendicular directions; the signal was recorded at two different IR wavelengths (8 µm and 12 µm) to derive a colour temperature of the nucleus. The first results of the IKS experiment can be found in Combes *et al.* (1986).

The instrument operated successfully on *Vega* 1; on *Vega* 2, the detectors could not be cooled because of a failure of the cryogenic system. Results of the *Vega* 1-IKS experiment are shown on figure 2. To remove the background signal due to the instrument itself, it was necessary

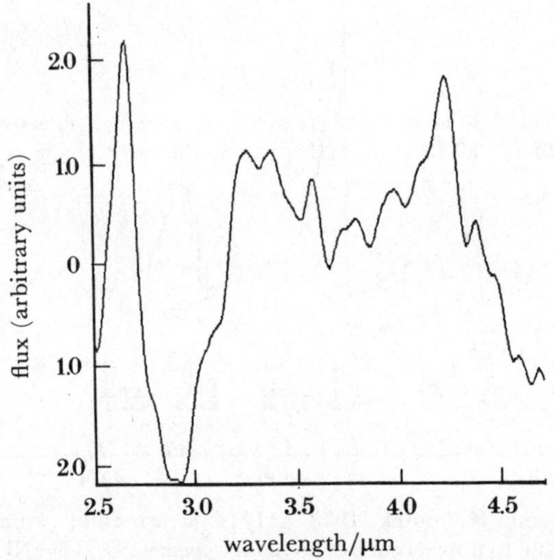

FIGURE 2. The 2.5–5 µm spectrum of P/Halley recorded by the IKS-*Vega* experiment. The spectral signature of H_2O (2.7 µm), CO_2 (4.25 µm) in emission, and H_2O ice (2.9 µm) in absorption, are easily identified. The broad feature centred at 3.3–3.4 µm is attributed to hydrocarbons (from Combes *et al.* 1986).

to take differences of spectra recorded at different distances from the nucleus; by this method, the spectrum obtained is representative of the cometary signal. Additional filtering is used to remove the remaining background. Figure 2 shows the filtered cometary signal between 2.5 and 4.7 µm obtained as the difference of two spectra recorded at 40000 km and 90000 km, respectively, from the nucleus. The spectrum is dominated by the ν_3 band of H_2O at 2.7 µm and the ν_3 band of CO_2 at 4.25 µm excited by fluorescence. These two signatures have the widths expected from theoretical calculations, and correspond to production rates of about

10^{30} s^{-1} and 10^{28} s^{-1} for H$_2$O and CO$_2$, respectively (Crovisier & Encrenaz 1983; Crovisier 1984; Bockelée-Morvan & Crovisier 1986). In addition, there is a broad emission centred at 3.3–3.4 μm, which apparently cannot be interpreted by fluorescence of a single molecule, for example CH$_4$. It is more likely that different species could contribute to give the broad feature. Because the 3.3–3.4 μm signature is typical of C—H bonds, it can be suggested that the feature observed on the cometary spectrum could be due to the presence of hydrocarbons. Finally, we also note that the absorption signature of ice seems to be present at 2.9 μm.

The 13.3 μm emission could be the signature of hydrocarbonaceous grains; these grains could be on the nucleus or in its surrounding; with their very low albedo, they could be at a temperature higher than the surrounding. It is also interesting to note the analogy between the cometary features and the emissions observed in the interstellar medium (figure 3); in the 3 μm regions. A preliminary conclusion is that there seems to be some evidence for carbonaceous material in the immediate vicinity of the nucleus, in the form of hydrocarbons, either in the gaseous or the solid phase.

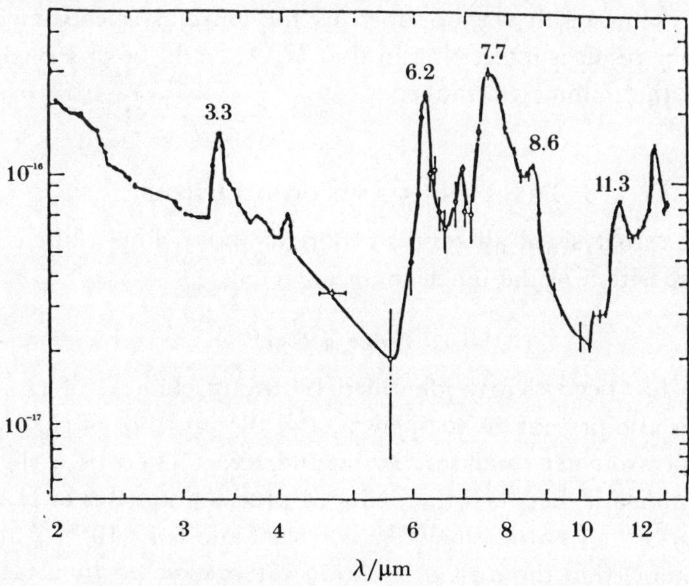

FIGURE 3. The spectrum of M82 between 2 and 14 μm (Willner *et al.* 1977).

5. *In situ* measurements

Additional information upon the composition of cometary gas and dust has been provided by mass spectrometers for both neutrals and ions. Two instruments have to be especially mentioned: the PUMA experiment, aboard the two *Vega* probes and *Giotto*, and the NMS aboard *Giotto*.

The PUMA experiment (Kissel *et al.* 1986) obtained most of its results from the *Vega* 1 encounter, which corresponded to the maximum dust-ejection rate. The PUMA experiment recorded the impacts of about 40000 grains, and recorded 4000 mass spectra. These spectra show evidence for several very distinct classes of particles: (1) the silicate particles either the 'olivine' type (including Mg, Si, Fe) or the 'pyroxene' type (including Mg, Si, Ca, Na); (2) the 'light element' particles dominated by the elements H, C, O and N, these particles being especially abundant (60%); (3) the intermediate particles, with all possible compositions between the two extremes (1) and (2). An additional important result is that the 'light-element' particles have a remarkably low density (0.1 g cm^{-3}) implying some kind of porous material. The unexpectedly large abundance of these 'light-element' particles, rich in carbon and with relative abundances close to the solar abundances, seems to be a good indicator of primordial Solar System matter in Comet Halley.

The NMS experiment aboard *Giotto* recorded mass spectra of neutrals and ionic species at distances ranging from 920 000 to 350 000 km to the nucleus, and mass spectra of neutrals from 350 000 to 750 km to the nucleus (Krankowsky *et al.* 1986). For neutral particles, strong peaks occur around atomic masses of 18 (H_2O group), 28 (CO group), 32 (C_2) and 44 (CO_2); we have to note that the 44 peak may be partly due to the presence of CS. The speed of the species was measured to be 900 ± 200 m s^{-1}. The production rate of H_2O was found to be 6.9×10^{29} s^{-1}, with a R^{-2} density distribution between 10^3 and 10^4 km. The CO_2 production rate was found to be up to 2×10^{28} s^{-1} (the upper limit comes from a possible contribution from CS). The ion mass spectra also show a strong maximum in the H_2O group, with a strong amount of H_3O^+. The same result was obtained by the ion-mass spectrometer (Balsiger *et al.* 1986). This preliminary result is interesting in that H_3O^+ could be observed from the ground by radioastronomy in the millimetric range.

6. DISCUSSION AND CONCLUSION

From a comparative analysis of all results reported above, important conclusions can be drawn about the composition of the inner coma.

6.1. *Water vapour*

The H_2O molecule has been clearly identified by the IR channel of TKS, and by IKS. The H_2O signature of ice is also present on IKS spectra, and the presence of H_2O (ice or gas) is also obvious from mass spectrometer analyses (PUMA and NMS). There is a slight disagreement, which may not be significant, between the deduced production rates of H_2O: 10^{30} s^{-1} at the time of *Vega* 1 (IKS) 4×10^{29} s^{-1} at the time of *Vega* 2 (TKS) and 5.5×10^{29} s^{-1} at the time of *Giotto* (NMS). However, we know that the dust production rate decreased by a factor of about four between *Vega* 1 and *Vega* 2 and by another factor of about 4 between *Vega* 2 and *Giotto*; it is likely that the gas production rate also decreased during the same period; it is worth noting that IUE observations derived a H_2O production rate of 8×10^{29} s^{-1} at the time of *Vega* 2 and 5×10^{29} s^{-1} at the time of *Giotto* (M. Festou personal communication 1986).

6.2. *Carbonaceous material*

The first carbonaceous material unambigously identified is CO_2, with a CO_2/H_2O ratio of about 10^{-2}; this result obtained by the IKS-*Vega* experiment, is also confirmed by the NMS-*Giotto* measurements. This result, expected from theoretical calculations, shows that CO_2 is a possible parent for CO, but other carbonaceous parent molecules still have to be found. It is interesting to note that CO was not observed in large amounts with IKS, although IUE spectra, recorded in a much larger field of view seem to indicate a CO/H_2O ratio of about 0.1. This result might imply that CO is probably not a parent molecule, so that its amount would be very low in a sphere of a few hundred kilometres around the nucleus.

The detection of carbonaceous material in the vicinity of the nucleus is probably the most important result of the space mission concerning the gaseous phase of P/Halley. The result is deduced from, or confirmed by many different experiments: (1) the TV cameras of *Vega* and *Giotto* measured a very low albedo, 4 % or less, implying that the surface of the nucleus is covered by some strongly absorbing material (Keller *et al.* 1986; Sagdeev *et al.* 1986); (2) the imaging channel of IKS measured a high temperature (higher than 300 K) for the surface of the nucleus or its immediate environment (Combes *et al.* 1986); this result is fully compatible with the low albedo of the surface; (3) as mentioned above, the mass spectrometers of *Giotto* and *Vega* detected the presence of carbonaceous particles in large amounts; (4) as mentioned above also, the IKS experiment detected spectroscopic signatures which are likely to be due to hydrocarbons.

It is interesting to compare the infrared signatures of IKS with the spectra obtained at the same wavelengths from ground-based observations. The 3.3–3.4 µm spectral range has been studied in several occasions by a large number of observers. In December 1985 and January 1986, results seem to have been negative (Tokunaga *et al.* 1986), whereas in March and April 1986, an emission feature centred at 3.4 µm was definitely reported by several authors from IRTF, UKIRT, AAT and ESO (Wikramasinghe & Allen 1986). This latter result seems to imply that the hydrocarbons presumably responsible for this emission could have a lifetime of at least 10^3 km so that they could be observable in a field of view of a few seconds of arc from the Earth. The absence of any feature before perihelion is puzzling; the only possible explanation that we see now is a temporal variation.

In conclusion, it can be said that the space missions devoted on Comet Halley have significantly changed our conception of the cometary nucleus. Carbon has been found to be present in large amounts, probably on the surface itself; if the 'dirty snow ball' model of F. Whipple is still valid, the cometary nucleus has to be very dirty: on most of the surface, the ice is covered by (or mixed with) a layer of absorbing material, and water vapour is outgassed from sporadic areas. The exact nature of carbonaceous material has still to be determined. Most likely these preliminary results will open a new field of research in cometary physics.

REFERENCES

A'Hearn, M. F., Feldman, P. D. & Schleicher, D. G. 1983 *Astrophys. J.* **274**, L99.
Altenhoff, W., Huchtmeier, W., Schmidt, J., Stumpff, P. & Walmsley, M. 1983 *Astron. Astrophys.* **125**, L19.
Balsiger, H., Altwegg, K., Bühler, F., Geiss, J., Ghielmetti, A. G., Goldstein, B. E., Goldstein, R., Huntress, W. T., Ip, W.-H., Lazarus, A., Meir, A., Neugebauer, M., Rettenmund, U., Rosenbauer, H., Schwenn, R., Sharp, R. D., Shelley, E. G., Ungstrup, E. & Young, D. T. 1986 *Nature, Lond.* **321**, 330.
Bockelée-Morvan, D. & Crovisier, J. 1986 In *Asteroids, comets, meteors, II* (ed. C. Lagerkvist *et al.*). Uppsala University.
Combes, M., Moroz, V. I., Crifo, J. F., Lamarre, J. M., Charra, J., Sanko, N. F., Soufflot, A., Bibring, J. P., Cazes, S., Coron, N., Crovisier, J., Emerich, C., Encrenaz, T., Gispert, R., Grigoryev, A., Guyot, G., Krasnopolsky, V. A., Nikolsky, Yu. V. & Rocard, F. 1986 *Nature, Lond.* **321**, 266.
Crovisier, J. 1984 *Astron. Astrophys.* **130**, 361.
Crovisier, J. & Encrenaz, T. 1983 *Astron. Astrophys.* **126**, 170.
Despois, D., Crovisier, J., Bockelée-Morvan, D., Schraml, U., Forveille, T. & Gérard, E. 1986 *Astron. Astrophys.* **160**, L11.
Huebner, W. F., Snyder, L. E. & Buhl, D. 1974 *Icarus* **23**, 580.
Kaminski, C. & Griep, D. 1985 IAU circ. no. 4154.
Keller, H. U., Arpigny, C., Barbieri, C., Bonnet, R. M., Cazes S., Coradini, M., Cosmovici, C. B., Delamere, W. A., Huebner, W. F., Hughes, D. W., Jamar, C., Malaise, D., Reitsema, H., Schmidt, H. U., Schmidt, W. K. H., Seige, P., Whipple, F. & Wilhelm, K. 1986 *Nature, Lond.* **321**, 320.
Kissel, J., Sagdeev, R. Z., Bertaux, J. L., Angarov, V. N., Audouze, J., Blamont, J. E., Büchler, K., Evlanov, E. N., Fechtig, H., Fomenkova, M. N., Von Hörner, H., Inogamov, N. A., Khromov, V. N., Knabe, W., Krueger, F. R., Langevin, Y., Leonas, V. B., Levasseur-Regourd, A. C., Managadze, G. G., Podkolzin, S. N., Shapiro, V. D., Tabaldyev, S. R. & Zubkov, B. V. 1986 *Nature, Lond.* **321**, 280.

Krankowsky, D., Lämmerzahl, P., Herrwerth, I., Woweries, J., Eberhardt, P., Dolder, U., Herrmann, U., Schulte, W., Berthellier, J. J., Illiano, J. M., Hodges, R. R. & Hoffman, J. H. 1986 *Nature, Lond.* **321**, 326.
Krasnopolsky, V. A., Gogoshev, M., Moreels, G., Moroz, V. I., Krysko, A. A., Gogosheva, Ts., Palazov, K., Sargoichev, S., Clairemidi, J., Vincent, M., Bertaux, J. L., Blamont, J. E., Troshin, V. S. & Valníček, B. 1986 *Nature, Lond.* **321**, 269.
Levasseur-Regourd, A. C., Bertaux, J. L., Dumont, R., Festou, M., Giese, R. H., Giovane, F., Lamy, P. Le Blanc, J. M., Llebaria, A. & Weinberg, J. L. 1986 *Nature, Lond.* **321**, 341.
Mumma, M. J., Weaver, H. A., Larson, H. P., David, D. S. & Williams, M. 1986 *Science, Wash.* **232**, 1523,
Sagdeev, R., Blamont, J., Galeev, A., Moroz, V. I., Shapiro, V. D., Shevchenko, V. & Szegö, K. 1986 *Nature, Lond.* **321**, 259.
Tokunaga, A. T., Smith, R. G., Nagata, T., Deloy, D. & Sellgren, K. 1986 *Astrophys. Lett.* **310**, L45.
Ulich, B. L. & Conklin, E. J. 1974 *Nature, Lond.* **248**, 121.
Willner, S. P., Soifer, B. T., Russell, R. W., Joyce, R. R. & Gillett, F. C. 1977 *Astrophys. J.* **217**, L121.
Wickramasinghe, D. & Allen, D. 1986 *Nature, Lond*, **323**, 44.

Discussion

W. K. H. SCHMIDT (*MPI für Aeronomie, Lindau, F.R.G.*). I was a bit surprised about the figure of $T = 330$ K that seemed to refer to the dark part of the nucleus. Where did this figure come from?

THÉRÈSE ENCRENAZ. The temperature of the nucleus derived by the IKS-*Vega* instrument is actually a colour temperature, derived from the ratio of the infrared flux in two channels (7–10 μm and 9–15 μm). The result is that the temperature of the central emissive region is higher than 300 K.

D. T. WICKRAMASINGHE (*Australian National University, Canberra, Australia*). Was there any evidence of water ice in the *Vega* infrared spectra between 3 and 4 μm?

THÉRÈSE ENCRENAZ. As mentioned in Combes *et al.* (1986) water ice seems to be present in absorption at 2.9 μm. However, we have to be careful, because the spectrum had to be filtered (to remove the internal background emission) so that the continuum level of the cometary signal is lost. Thus, there is some uncertainty in the definition of the zero level, and in the assignment of emission and absorption features.

Phil. Trans. R. Soc. Lond. A **323**, 405–420 (1987)

Printed in Great Britain

ICE observations of Comet Giacobini-Zinner

By S. W. H. Cowley

Blackett Laboratory, Imperial College, London SW7 2BZ, U.K.

The first spacecraft encounter with a comet took place on 11 September 1985 when the International Cometary Explorer spacecraft passed through the tail of Comet Giacobini-Zinner at a distance of 7800 km from the nucleus. It provided the first definitive *in-situ* information concerning the interaction of a cometary atmosphere with the flowing solar-wind plasma, and the results of initial analyses are reviewed in this paper. Large-scale MHD aspects of the interaction largely conform to prior expectation. The flow surrounding the comet is mass-loaded and slowed by *in situ* ionization and pick-up of heavy cometary neutrals, and the solar-wind magnetic field consequently becomes draped around the obstacle, and forms an induced magneto-tail. Substantial evidence exists for the permanent presence of a weak shock lying in the subsolar mass-loaded region upstream from the comet, through whether the spacecraft itself passed through shocks on the cometary flanks remains controversial. There is no doubt, however, that a sharp boundary was observed both inbound and outbound (centred on *ca.* 09h29 and 12h20 U.T.) whose width is an energetic heavy-ion Larmor radius (*ca.* 10^4 km), where the flow is deflected away from the comet and slowed, and where the magnetic field and plasma become compressed and very turbulent. The location of this boundary is also consistent with that expected for a weak shock based upon the known Giacobini-Zinner water-molecule production rate. An unexpected feature of the interaction was the extreme levels of field and plasma turbulence, and broadband wave activity observed in the region of mass-loaded flow.

1. Introduction

On 11 September 1985 the first ever close encounter took place between a spacecraft and a comet when the International Cometary Explorer (ICE) spacecraft passed through the tail of Comet Giacobini-Zinner at a distance of 7800 km from its central nucleus. The results of initial analyses of the ICE data have recently been published, and in this paper we will present a brief overview. The ICE spacecraft was not specifically designed for cometary studies. In particular it carried no cameras or instruments to investigate the comet's neutral atmosphere, and only rudimentary measurements of the dust were possible. Rather, the spacecraft's prime mission following launch in August 1978 was to study plasma particles and electromagnetic fields in the solar wind upstream from the Earth's magnetosphere as part of the NASA–ESA International Sun–Earth Explorer programme. Consequently, the data returned by ICE from Giacobini-Zinner concern plasmas and fields in its vicinity, and relates principally to the interaction between the comet's atmosphere and the flowing solar-wind plasma.

The nature of this interaction is quite different from that between the solar wind and planetary bodies in the Solar System (see, for example, Mendis & Houpis 1982). When the cometary nucleus approaches the Sun, the ices of which it is composed (mainly water) sublime, and being unrestrained by gravity, the resulting gases expand away at speeds of *ca.* 1 km s^{-1}

producing an extensive neutral atmosphere or coma. Water molecules in the coma are photo-dissociated by sunlight on time scales of *ca.* 10^5 s (at *ca.* 1 AU), so that at distances exceeding a few hundred thousand kilometres the coma becomes dominated by atomic oxygen and hydrogen. Being highly tenuous and nearly collision-free, the solar wind blows continuously through this coma, and there is little interaction between the two populations. However, on timescales of a few million seconds (and hence on spatial scales of a few million kilometres surrounding the nucleus) the atoms in the coma become ionized, either by solar uv radiation or by charge exchange with solar-wind protons. The ions are then 'picked up' by the flow, moving in cycloidal orbits in the crossed electric and magnetic fields of the solar wind. As the 'pick-up' ions gyrate about the magnetic field, their energy in the comet frame (spacecraft frame) varies between essentially zero and a maximum energy,

$$E_{\mathrm{max}} = 4\,A\,\sin^2\alpha E_{\mathrm{sw}}. \tag{1}$$

where A is the ion mass in unified atomic mass units, α is the cone angle between the magnetic field and the velocity vector in the solar wind and E_{sw} is the energy of a proton moving at the solar-wind speed. For typical wind speeds of *ca.* 450 km s^{-1} solar-wind protons have an energy $E_{\mathrm{sw}} \approx 1$ keV, so that 'pick-up' protons have energies extending to *ca.* 4 keV, and 'pick-up' oxygen is energized up to *ca.* 64 keV (for $\sin^2\alpha = 1$). Thus heavy particles, particularly, gain large energies from the pick-up process. This energy is extracted from the solar-wind flow, such that sufficiently close to the comet where the mass density of the pick-up ions starts to become a sensible fraction of the solar-wind mass density, the flow becomes 'mass-loaded' and slowed. Theory (see, for example, Ip & Axford 1982) suggests that a weak shock should lie near the periphery of the mass-loaded region, where the heavy ions first reach *ca.* 1 % of the solar wind by number density. At the time of the ICE encounter the Giacobini-Zinner water production rate was *ca.* (2–3) × 10^{28} molecules s^{-1} (Combi *et al.* 1986; Strauss *et al.* 1986), leading to estimates of the spatial scale of the mass-loaded region of *ca.* 10^5 km. As the impinging flow slows, the frozen-in magnetic field becomes distorted and 'draped' around the central region as first proposed by Alfvén (1957), and as depicted schematically in figure 1. Most of the plasma flow and embedded field slips 'around' the central region, out of the plane of the diagram, but some flux tubes can become 'captured' near to the nucleus by mass-loading, perhaps for intervals of 1 h or more. Because the 'ends' of these field lines remain frozen to the flowing solar-wind plasma away from the comet, these 'captured' flux tubes become stretched out on the antisolar side of the comet to form an induced magnetotail. The form of the tail should be approximately cylindrical, bisected by a current sheet separating fields of opposite polarity (figure 1). Cold cometary plasma flowing along these 'captured' flux tubes then gives rise to the visible 'ion' (type 1) comet tails.

Figure 1 also shows the trajectory of ICE relative to the comet structure. Giacobini-Zinner is a short-period (6.5 years) comet, which moves in a prograde orbit inclined at 31.9° to the ecliptic between the orbits of Jupiter and the Earth (perihelion 1.03 AU). At the time of the encounter the comet was 6 d past perihelion, moving from north to south across the ecliptic plane with little radial component of velocity. Fortuitously, the azimuthal component of the comet velocity nearly matched that of the spacecraft (*ca.* 30 km s^{-1}), such that relative to the comet, ICE passed very nearly directly south to north through the comet structure at a speed of *ca.* 21 km s^{-1}. ICE was targeted to the antisolar side of the comet, and passed centrally through the ion tail with an impact parameter of 7800 km relative to the nucleus

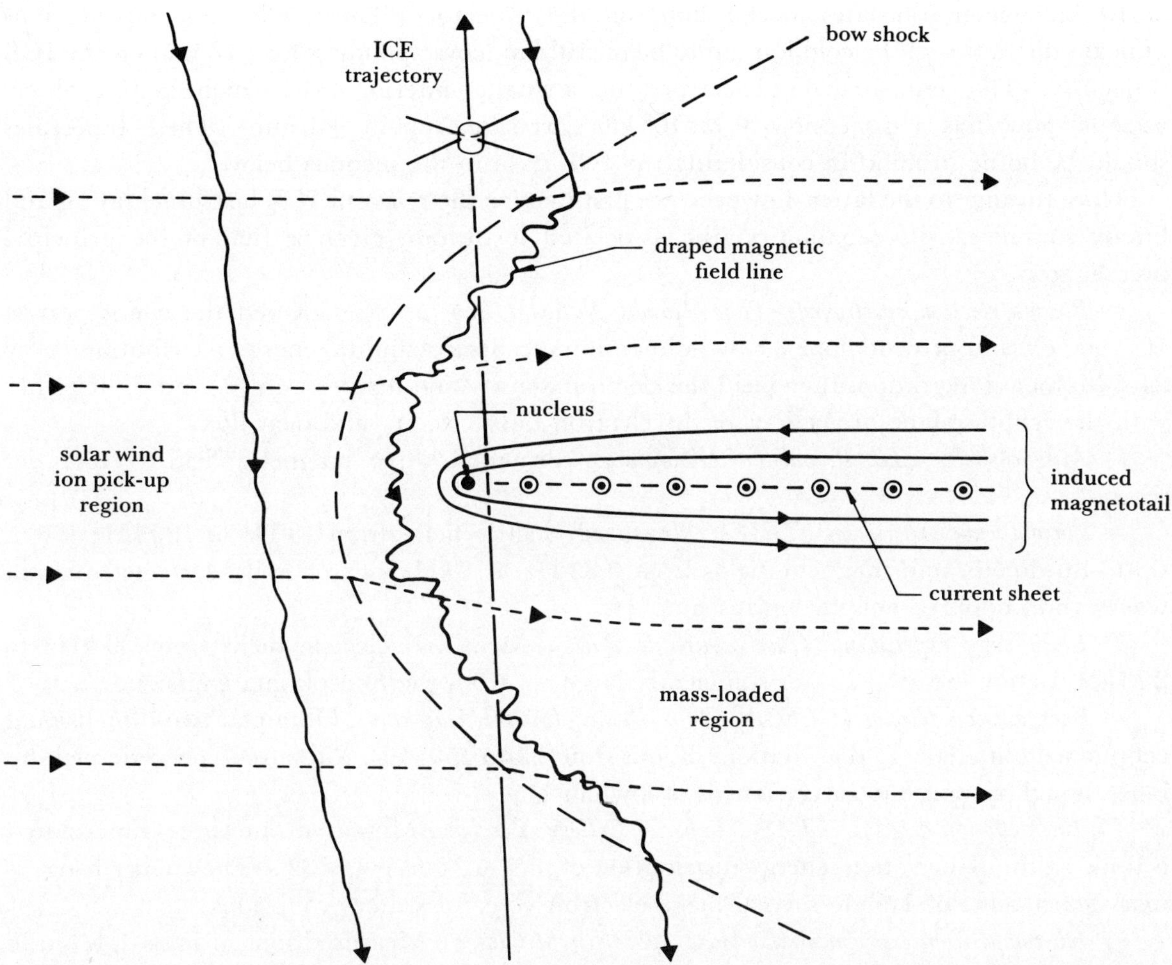

FIGURE 1. Sketch showing the main theoretically expected features of the interaction between the atmosphere of Comet Giacobini-Zinner and the solar-wind plasma, together with the trajectory of the ICE spacecraft relative to those regions. Solid arrowed lines show magnetic field lines, short-dashed arrowed lines are plasma streamlines, and the long-dashed line is the bow shock, lying in the outer part of the region of slowed, mass-loaded flow.

(von Rosenvinge *et al.* 1986). It should be noted for purposes of later discussion that with this relative trajectory the principal signatures of field-draping should depend upon the sign of the north–south component of the solar-wind magnetic field, because the spacecraft would fly approximately along the current sheet for a purely east–west field. For a southerly directed field as depicted in figure 1, the draped field should be deflected antisunward to the south of the comet (inbound) and sunward to the north of the comet (outbound), and vice versa for a northward-directed field.

The spatial scales discussed above can now be translated to timescales on the ICE trajectory, there being three basic regions of interest. First, an 'ion pick-up region' should extend a few million kilometres from the comet, where significant heavy-ion 'pick-up' fluxes should be present in the solar wind, but where densities remain sufficiently small that the flow is not affected. At *ca.* 20 km s^{-1} this translates to plus or minus a few days on the spacecraft trajectory about closest approach. Second, the 'mass-loaded region' with slowed flows has a scale of *ca.* 10^5 km which translates to ± 1 hour on the trajectory. Third, telescopic observations indicate the radius of the cold ion tail to be *ca.* 10^4 km, corresponding to ± 10 min on the ICE trajectory. (The central current sheet presents a smaller interior scale, which, as ICE observations show, has a dimension of *ca.* 10^3 km corresponding to ± 1 min). These timescales should be borne in mind in considering the ICE data in the sections below.

Before turning to the latter, however, we first itemize the relevant ICE instrumentation and briefly summarize its capabilities (the associated institution given is that of the principal investigator).

(*a*) *Plasma electron spectrometer* (*Los Alamos National Laboratory*). Covered the energy range 10–1000 eV and provided one 3 s two-dimensional ecliptic azimuth–energy distribution every 24 s. Moment integrations then yield the electron density and temperature, n_e and T_e, together with the ecliptic plane projections of the electron bulk velocity and heat flux.

(*b*) *Magnetometer* (*Jet Propulsion Laboratory*). Provided three magnetic field vectors per second.

(*c*) *Plasma-wave experiment* (*TRW*). Measured electric fields from 18 Hz to 100 kHz (90 m tip-to-tip dipole) and magnetic fields from 0.32 Hz to 1 kHz (search coil). Detected plasma waves and impulsive noise from dust impacts.

(*d*) *Radio-wave experiment* (*Observatoire de Paris*). Measured electric fields from 30 kHz to 2 MHz. Determined n_e, T_e (for sufficiently large n_e) from thermal-plasma emission.

(*e*) *Energetic-ion detector* (*ULECA*) (*Max-Planck-Institut, Garching*). Obtained two-dimensional ecliptic azimuth–energy distributions of ions from 35 to 150 keV. Measured energetic pick-up ions, though response to heavy ions is of low efficiency.

(*f*) *Energetic-ion detector* (*EPAS*) (*Imperial College, London*). Measured one three-dimensional ecliptic azimuth–elevation–energy distribution of pick-up ions every 32 s. The energy range is mass-dependent; 65 keV to several mega electron volts for water-group ions.

(*g*) *Ion-composition experiment* (*Goddard Space Flight Center*). Measured ions of mass 1.4–3 u in the speed range 237–463 km s^{-1} (appropriate to solar-wind alphas) and ions of mass 14–33 u in the range 80–223 km s^{-1} (appropriate to heavy cometary ions in the slowed mass-loaded region). Long instrument cycle time of 21.1 min.

The major gap in the above list (as regards plasma physics) is the lack of thermal-ion data of high temporal resolution.

The next sections will now describe in turn the main results obtained by this instrumentation concerning the thermal plasma, magnetic field, plasma waves and energetic heavy ions.

2. Thermal-plasma observations

Figure 2 gives an overview of the encounter data obtained by the electron spectrometer (Bame *et al.* 1986). Data for 10 h are shown, with closest approach occurring at 11h03 min 38 s U.T., nearly at the centre of the diagram. The panels display the electron density n_e, temperature T_e and ecliptic plane projection of the bulk speed V_e, and clearly show the presence of three basic regions, though these have been further subdivided by Bame *et al.* (1986) in the panel above the data. At the beginning and end of the interval the spacecraft is located in the solar wind (sw), though effects of heavy-ion pick up are clearly present as will be outlined below. The mass-loaded region where the plasma speed is substantially reduced by pick up is then traversed between *ca.* 09h20 and 12h20 U.T. (marked TR and S), corresponding to

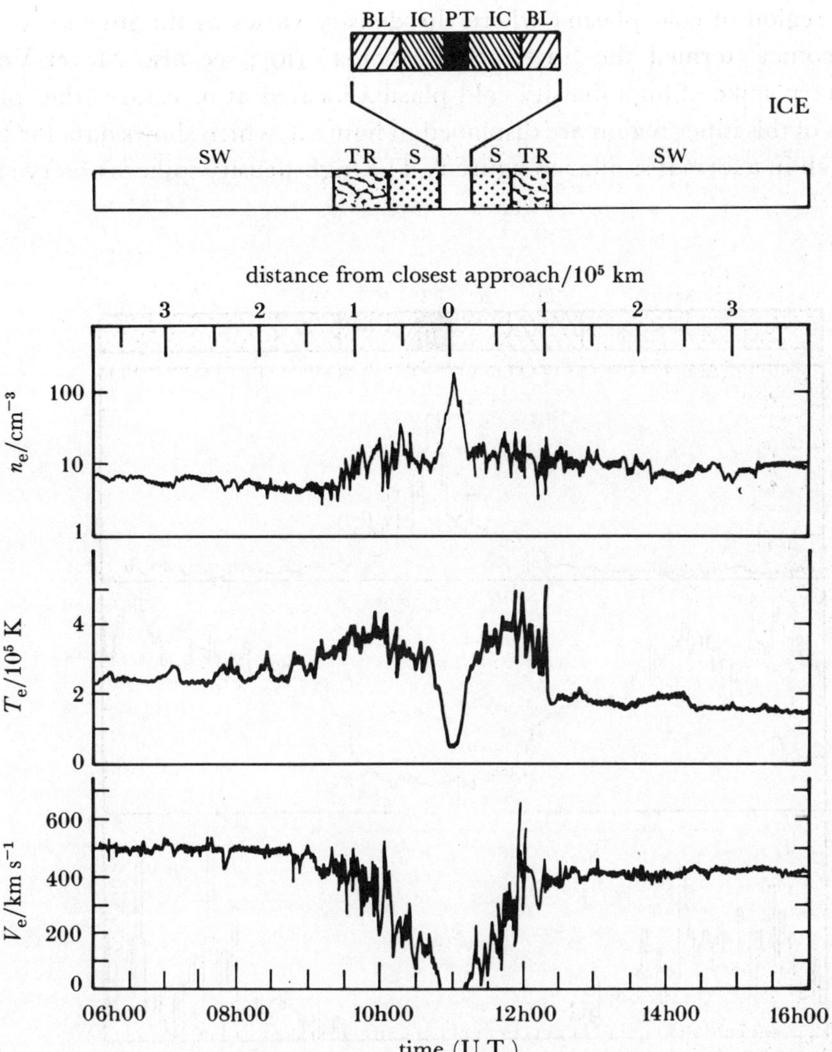

FIGURE 2. Overview of the electron density, temperature and ecliptic plane bulk speed observed by the electron spectrometer during the Giacobini-Zinner encounter. Three-point running averages are shown. The key at the top of the plot identifies the plasma regions encountered, as follows: sw, solar wind; TR, transition region; s, sheath; BL, boundary layer; IC, intermediate coma; PT, plasma tail. (From Bame *et al.* 1986.)

distances of $\pm 10^5$ km about closest approach as indicated by the scale at the top of the diagram, and in the discussion of the preceding section. (It may be noted, however, that small depressions in flow speed are clearly evident for at least 30 min outside the times indicated.) This region is further subdivided by Bame *et al.* (1986) into an outer transition region (TR) and inner sheath (s) for reasons described further below. All of these regions are characterized by flow that is predominantly antisolar (see, for example, Baker *et al.* 1986), and plasma that has at least a significant solar-wind component. In the central 25 min of the encounter, however, ICE entered a region believed to be dominated by dense, cold cometary plasma where the flow speed is smaller than can be detected by this instrument (not more than about 30 km s⁻¹). This region, of width *ca.* 3×10^4 km, is separated from the sheath by a boundary layer (BL) of intermediate characteristics believed to have the nature of a diffuse contact surface, interior to

410   S. W. H. COWLEY

which there is a region of cold plasma where the density varies as the inverse square of the
distance to the comet (termed the 'intermediate coma' (IC); see also Meyer-Vernet *et al.*
1986 *b*) with a sharp spike of high density cold plasma located at its centre (the 'plasma tail'
(PT)). The details of this inner region are displayed in figure 3, which shows data for 1 h centred
on closest approach in a format similar to figure 2. The high-density spike is observed for 3 min

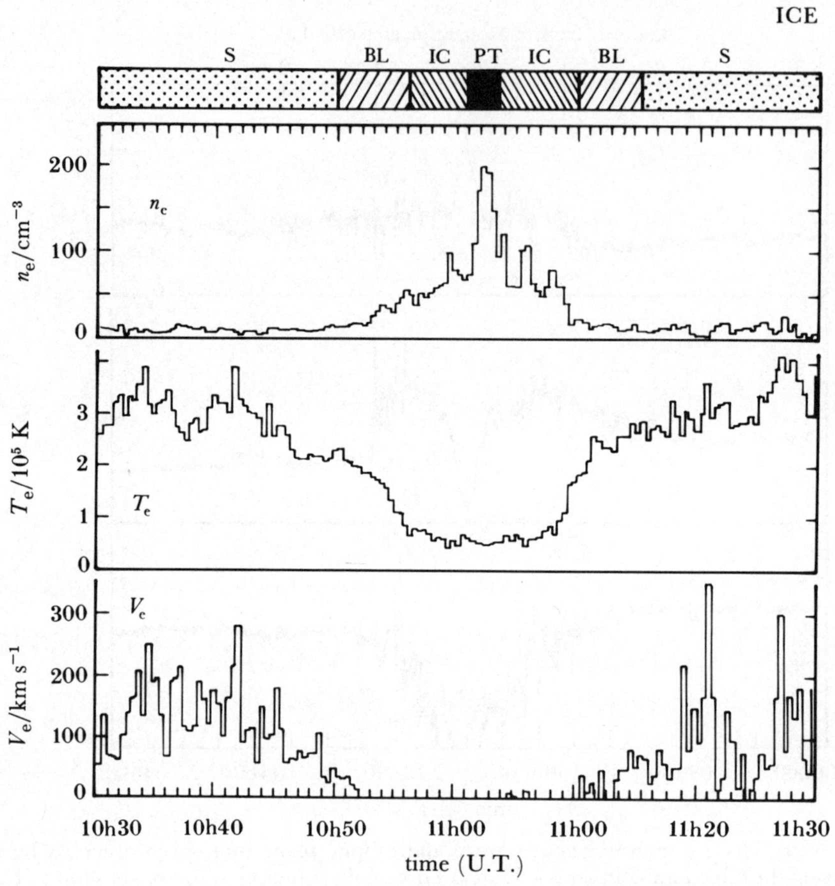

FIGURE 3. Expanded plot of the region nearest closest approach in the same format as figure 2, except that the
data are here unsmoothed. (From Bame *et al.* 1986.)

at the spacecraft, corresponding to a width of 3000 km, though the highest density and coldest
plasma (*ca.* 650 cm^{-3} and *ca.* 1 eV (Meyer-Vernet *et al.* (1986 *a*, *b*)) is observed for only
ca. 1 min, or *ca.* 1200 km along the spacecraft track. In the next section it will be shown that
an induced magnetotail is contained within the 'intermediate coma' region, and that the
'spike' in fact corresponds to a crossing of the central current sheet.

We now discuss further details of the plasma observations, taking each of the three main
regions in turn. The major feature of the solar wind surrounding the comet is the presence of
fluctuations in the electron density and temperature (generally anticorrelated) with characteristic timescales of *ca.* 2 min (Bame *et al.* 1986; Gosling *et al.* 1986). Convected at a solar-
wind speed of *ca.* 500 km s^{-1}, this time corresponds to a spatial scale of *ca.* 6 × 10^4 km. These
fluctuations are undoubtedly due to an electromagnetic instability set up by the pick-up process

[162]

(see next section). They were first observed at distances of *ca.* 10^6 km from the comet, the amplitude increasing as the comet was approached and peaking with $\delta n/n \approx 1$ in the 'transition region', before falling slightly in amplitude in the 'sheath'. Thus, in agreement with the previous section, these results indicate that ion pick up takes place in a region at least *ca.* 10^6 km in extent. The largest density spikes in the 'transition region' have $\delta n/n \approx 3$–5, and are accompanied by large perturbations in the flow speed and direction, though no systematic pattern has been found (Baker *et al.* 1986). Evidence also exists that these large density fluctuations occur not only in the mass-loaded region along the north–south trajectory of the spacecraft, but also along similar path lengths (*ca.* 10^5 km) in the transverse spacecraft–Earth direction. This has been deduced by Steinberg *et al.* (1986) from analysis of the angular broadening of the Earth radio source (kilometric waves from the auroral zone) observed during encounter by the ICE radio-wave experiment.

Turning now to the mass-loaded region, it can be seen that although the plasma is compressed, heated and slowed as it passes from the solar wind into the transition region it is not possible in the presence of the large fluctuations to convincingly identify a single weak shock structure. Nevertheless, sharp features do occur in the density and temperature data, particularly in T_e at *ca.* 12h 20 U.T. Within the 'sheath' region the electron spectra also consistently show evidence that the particles have passed through a weak shock (Thomsen *et al.* 1986). In the solar wind the spectra are of the usual two-component form with a thermal 'core' of 20 eV and a high-energy tail or 'halo' above *ca.* 100 eV. In the 'sheath' the latter population remains relatively constant in form (at least above *ca.* 300 eV), while the core becomes heated and 'flattened', indicative of the effect of the shock potential. Thomsen *et al.* infer that the sheath streamlines pass through a weak shock that is a permanent feature of the subsolar mass-loading region. In the transition region, however, they report variable spectra that are sometimes sheath-like and sometimes solar-wind-like (though heated), and infer that a shock is present only intermittently on the flanks, possibly as a result of the strong fluctuations present in the inflow plasma, as described above. In particular, they infer that no shock was actually crossed by the spacecraft on its north–south trajectory, although this topic will be taken up again later.

The electrons in the high-energy tail of the distribution of the heated plasma in the mass-loaded region escape continuously into the upstream region by flow along the magnetic field lines. When ICE was in the solar wind and magnetically connected to the mass-loaded region it detected this as an approximately sunward-directed field-aligned heat flux of electrons with energies of more than 100 eV. These heat-flux events were observed at distances of up to *ca.* 5×10^5 km from the comet (giving rise to the sporadic enhancements of T_e seen in the solar wind in figure 2, particularly during the inbound pass), and are important in allowing one to sense remotely the shape of the hot-particle source, because the magnetic field should be tangential to the boundary both on entry and exit to the event. An analysis presented by Fuselier *et al.* (1986) in which the boundary was fit to a paraboloid of revolution gives a subsolar stand-off distance of *ca.* 4×10^4 km (approximately the expected subsolar shock location for the Giacobini-Zinner production rate), and a distance along the spacecraft trajectory that is approximately consistent with the *in situ* observations. Thus consistent evidence does exist for the presence of a shock in the flow surrounding Giacobini-Zinner, at least in the subsolar régime.

It should also be mentioned with regard to the mass-loaded region that effects produced by the falling bulk speed were also observed in the alpha-particle observations made by the ion composition experiment (Ogilvie *et al.* 1986). Heavy ions observed in the inner part of the sheath region by this instrument indicate a preponderance of water-group ions (H_2O^+ with some H_3O^+), though some ions of mass *ca.* 30 (probably CO^+ or HCO^+) and *ca.* 23 (possibly Na^+ or C_2^+) were also detected.

Turning now to the central cold-plasma region, the analysis presented by Zwickl *et al.* (1986) shows a three component electron population to be present (see also Bame *et al.* 1986). The two higher-energy components have relatively constant density in this region ($T_e \approx 12$ eV with $n_e \approx 15$ cm^{-3}, and $T_e \approx 100$ eV with $n_e \approx 0.25$ cm^{-3}) and are thought to correspond to locally produced electrons resulting from photoionization of water molecules, and the solar-wind 'halo' population, respectively. If so, the presence of the 'halo' in this region indicates its accessibility to solar-wind particles, presumably by motion along 'open' tail magnetic-field lines. The main density variations in the region are then confined to the coldest population having temperatures below 3–4 eV. Because the lower energy limit of the electron spectrometer is 10 eV, determinations of the parameters of this population are of limited accuracy, leading principally to overestimates of the temperature and underestimates of the density. However, accurate values were obtained in this region from analysis of the thermal noise produced by the electrons, as detected by the radio-wave experiment (Meyer-Vernet *et al.* 1986 *a, b*). These data show in particular that in the density spike (PT), the temperature dropped to *ca.* 1 eV while the density peaked at *ca.* 670 cm^{-3}. Zwickl *et al.* (1986) suggest that these electrons are produced by photoionization close to the nucleus, where they can become collisionally cooled by the cold neutrals before flowing outwards to the vicinity of the spacecraft.

3. MAGNETIC FIELD OBSERVATIONS

Figure 4 shows an overview of the magnetic field observations, complementary to the plasma overview shown in figure 2 (Smith *et al.* 1986). In this case, however, data for 6 h are shown, again nearly centred on closest approach. The four panels show the three field components and the field magnitude. B_x is positive towards the Sun, B_y to the east of the Sun and B_z northward. Comparison with figure 2 shows that there exists direct correspondence between magnetic and plasma features. In particular, in the mass-loaded region the mean field was compressed and draped, as expected, and extreme levels of compressional turbulence were also observed corresponding to the fluctuations in plasma properties, while within the 'intermediate coma' an induced bipolar magnetotail was observed as can be seen from the two large, oppositely directed spikes in B_x in the upper panel. The latter feature thus confirms the theoretical proposals originally made by Alfvén (1957).

At largest distances from the comet, up to *ca.* 10^6 km or more, the field data is characterized by the appearance of waves of period *ca.* 100 s (75–135 s), which are undoubtedly the counterpart of the *ca.* 2 min plasma fluctuations discussed by Gosling *et al.* (1986). Tsurutani & Smith (1986 *b*) suggest that the waves are due to a resonant instability associated with heavy-ion pick up, in which the wave period in the spacecraft frame (approximately the parent neutral frame) is approximately the heavy ion gyroperiod. The predominant heavy ion at large distances is expected to be oxygen, giving a gyroperiod of *ca.* 130 s in a field of 8 nT, in approximate agreement with this suggestion. The waves are linearly polarized, generally have a compressional component, and exhibit non-sinusoidal waveforms. Some published examples

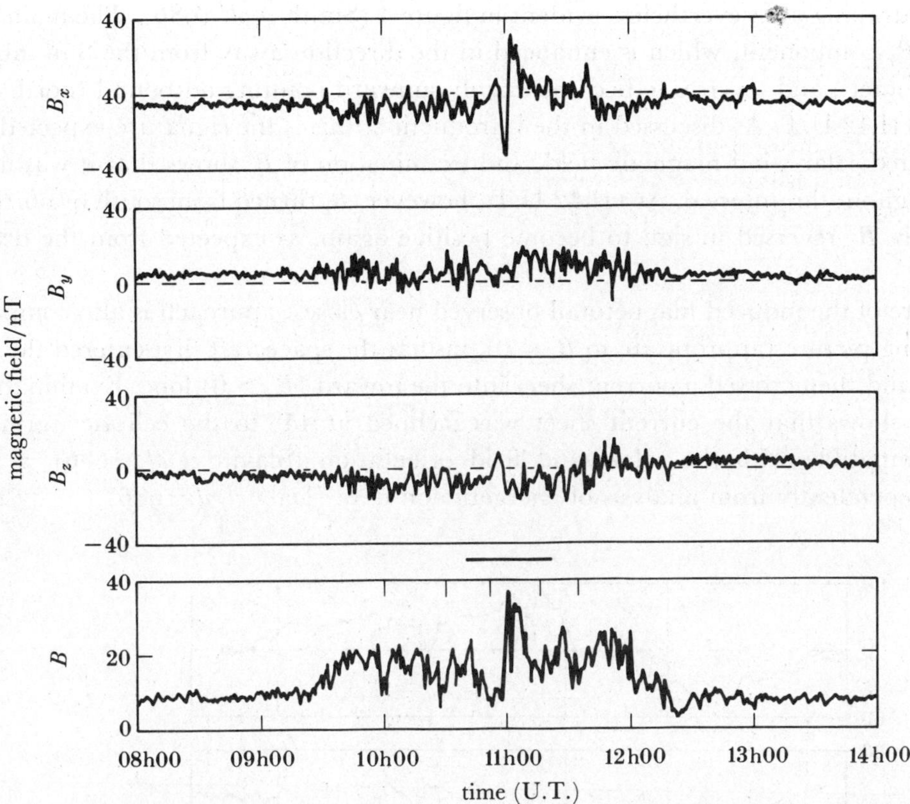

FIGURE 4. Overview of the magnetic field observations (1 min averages) at Giacobini-Zinner on 11 September 1985. Positive B_x points towards the Sun, B_y to the east of the Sun and B_z northward. B is the total field strength. (From Smith *et al.* 1986.)

(e.g. Smith *et al.* 1986; Tsurutani & Smith 1986*b*) appear to exhibit relatively slow approximately linear variations of the field components associated with increases in field strength, followed by more rapid 'returns' of the field and strength reductions. During the latter intervals discrete damped wave packets of waves of period *ca.* 3 s (1–7 s) occur, which at least visually give the impression of 'ringing' of the system resulting from the rapid change in the long-period wave. The wave packets last for *ca.* 15 s, have amplitudes 10–15 nT, and the waves are left-hand circular polarized in the spacecraft frame and propagate nearly parallel to the ambient field (Tsurutani & Smith 1986*b*).

Well away from the comet the power in the long-period magnetic waves is mainly confined to the field components transverse to the magnetic field, and this power steadily grows as the comet is approached (Tsurutani & Smith, 1986*a*). However, the power in the compressional component is strongly enhanced within the mass-loaded region, with sharp onset and termination at 09h30 and 12h20 U.T., respectively, approximately coinciding with the boundaries of the region of compressed average field (Jones *et al.* 1986). These authors have also pointed out that the locations of these boundaries are compatible with those expected for a Mach 2 shock at the Giacobini-Zinner gas production rate (similar to the conclusions of Fuselier *et al.* (1986)), and that the observed field compressions and inferred boundary orientations are not incompatible with this interpretation.

Despite the extreme levels of magnetic turbulence observed in the mass-loaded region, the effects of field-draping are nevertheless evident in figure 4 (Smith *et al.* 1986). The main effect is seen in the B_x component, which is enhanced in the direction away from the Sun inbound (south of the comet), and reverses to become mainly sunward pointing outbound (north of the comet) up to 11h42 U.T. As discussed in the introduction, this is the signature expected for a southerly pointed solar-wind magnetic field, and examination of B_z shows that it was indeed negative throughout this interval. At 11h42 U.T., however, B_z turned from south or north and correspondingly B_x reversed in sign to become positive again, as expected from the draping picture.

The structure of the induced magnetotail observed near closest approach is also compatible with the draping picture (appropriate to $B_z < 0$), in that the spacecraft first entered the away ($B_x < 0$) lobe and then crossed a current sheet into the toward ($B_x > 0$) lobe. Examination of the field data shows that the current sheet was inclined at 45° to the ecliptic during the encounter, compatible with the solar-wind field orientation (Slavin *et al.* 1986), a result confirmed independently from analysis of energetic ion data (Daly *et al.* 1986).

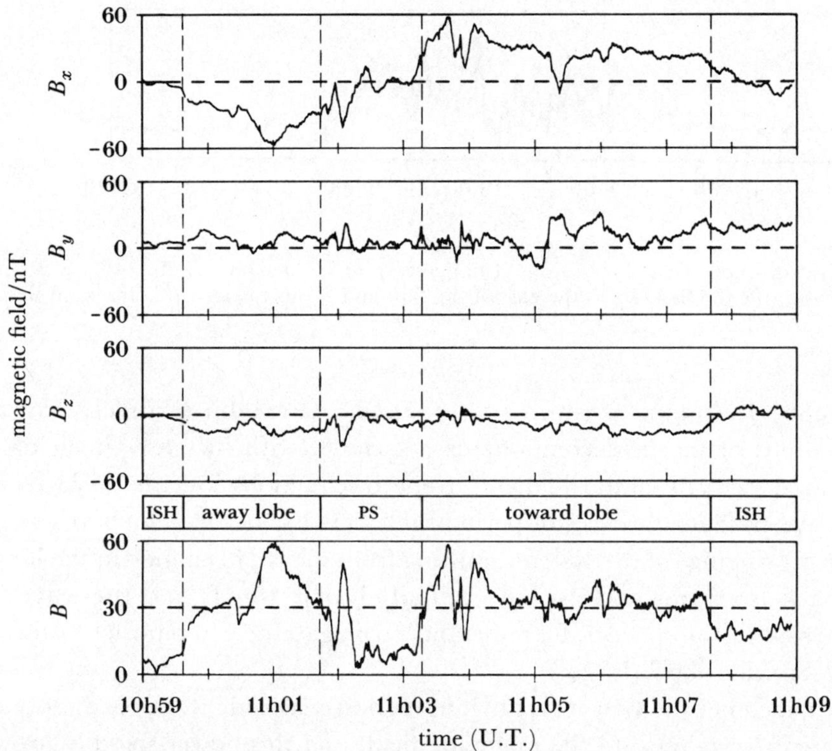

FIGURE 5. High-resolution (three vectors s⁻¹) magnetic-field data obtained during the crossing of the magnetotail. The outer vertical dashed lines mark the boundaries between the tail lobes and the 'ionosheath' (ISH), and the inner dashed lines indicate the location of the current sheet (plasma sheet, PS) region. (From Slavin *et al.* 1986.)

Details of the magnetotail data are shown in figure 5, taken from Slavin *et al.* (1986). The lobes have sharp outer boundaries of width *ca.* 200 km (outer dashed lines), which are flared in direction away from the comet–Sun axis by *ca.* 20° to 40°. Inside the boundary the field increases to *ca.* 60 nT near the edge of the plasma sheet, and then falls to *ca.* 5 nT within the latter. Small-scale field fluctuations are also present, which may be due to tail motion, though other processes may also be present (Slavin *et al.* 1986).

Comparing the magnetotail data in figure 5 with the corresponding plasma data in figure 3 it can be seen that the outbound 'away lobe' corresponds closely to the outbound 'intermediate coma' region (where the plasma density varies as r^{-2}), and that the central density spike corresponds to the plasma sheet (PS in figure 5). However, despite the approximate inbound–outbound symmetry in the 'intermediate coma' plasma data (see also Meyer-Vernet *et al.* 1986*b*), it can be seen that the inbound 'away lobe' occupies only the innermost part of the inbound 'intermediate coma' region. This leads one to suspect that the weak field region observed in the outer part of the latter is physically associated with the tail structure. The relation remains as yet unclear, though several ideas (in terms, for example, of a fragmented nucleus, gas jets, reconnected field lines) have been outlined by Smith *et al.* (1986).

Finally it should be mentioned that despite the large densities observed in the 'spike', the cold (*ca.* 1 eV) electron pressure within it is insufficient to balance the maximum lobe field pressure by approximately an order of magnitude (Smith *et al.* 1986). Because no significant pressure is associated with warm or hot electrons in this region (Meyer-Vernet *et al.* 1986*b*), the lobe pressure must be balanced by ions of temperature *ca.* 10 eV, ten times hotter than the thermal-electron component.

4. PLASMA-WAVE OBSERVATIONS

An overview of the plasma wave data obtained during the Giacobini-Zinner encounter is shown in figure 6, taken from Scarf *et al.* (1986). This shows the wave power against time for a set of frequency channels spanning 31–310 Hz for the magnetic field data in the lower panel, and 310 Hz to 100 kHz for the electric field data in the upper panel. Nine orders of magnitude in wave power is contained between each pair of horizontal lines, so that the vertical scales are

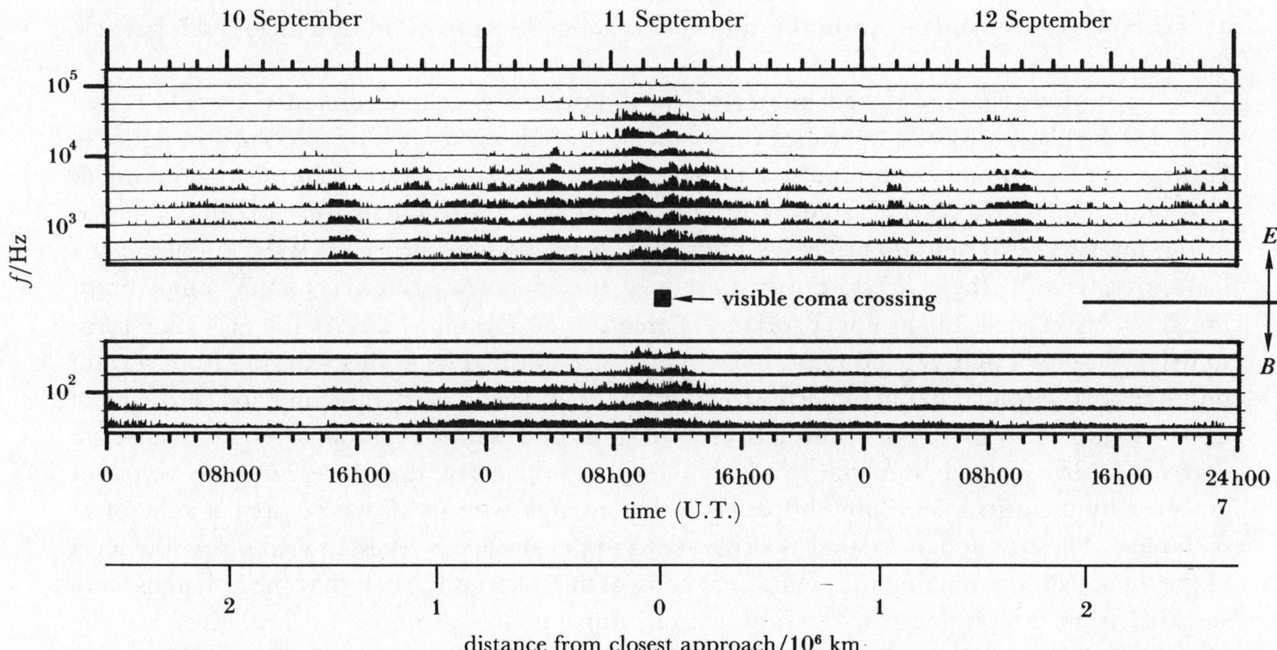

FIGURE 6. Overview of plasma wave observations (128 s peak amplitudes in eleven electric field channels 310 Hz–100 kHz, and five magnetic channels 31–310 Hz) for 10–12 September 1985. The amplitude scale is logarithmic with nine orders of magnitude of signal power between each pair of horizontal lines. (From Scarf *et al.* 1986.)

very compressed. Data for three days are shown, centred on closest approach, such that this scale is much larger than those used for the plasma and magnetic field overviews (figures 2 and 4). This plot therefore emphasizes the large (several million kilometres) scale of the ion pick-up region in which comet-associated waves were nearly continuously recorded.

Three principal wave types can be seen in the figure. Throughout the pick-up region, whistler-mode electromagnetic waves with frequencies less than 100 Hz, and ion-acoustic electrostatic waves with frequencies from a few 100 Hz to 10 kHz were nearly continuously present, though the latter have a more 'bursty' appearance. Like the waves of period of *ca.* 100 s observed in the plasma and magnetometer data over similar spatial scales, these waves are presumably set up by instabilities associated with the pick-up process, though theoretical details have yet to be investigated. Bursts of electron plasma oscillations at frequencies of a few tens of kilohertz can also be seen in the pick-up region nearer the comet, associated with, and presumably generated by, the electron 'heat-flux' events discussed in §2 (Fuselier *et al.* 1986).

As the comet was approached, the amplitudes of the characteristic whistler and ion-acoustic waves generally increased, but close to the comet, in the vicinity of the mass-loaded region, the character of the wave data underwent abrupt changes across thin structures that have the wave characteristics of shock traversals (Scarf *et al.* 1986). Downstream from these structures the waves are impulsive, broadband (extending to a few hundred hertz (approximately the electron gyrofrequency) in the magnetic data and to a few tens of kilohertz (approximately the electron plasma frequency) in the electric field data) and are as intense as any waves previously seen by the ICE wave instrument. Several such wave boundaries were observed inbound to the comet, between 08h27 and 09h11 U.T., which thus appear to relate to the electron 'heat flux' events rather than to the *ca.* 09h20–09h30 U.T. inbound 'bow wave' observed by other instruments (§2, 3 and 5). Outbound, however, the boundary of broadband waves occurred at 12h20 U.T., coincident with simultaneous changes recorded in the field and particle data.

In the central region of the encounter where the plasma became dominated by cold comet particles the plasma wave amplitudes were much reduced, as can just be discerned in figure 6. However, a large number of impulsive broadband signals were observed in this region due to dust grain impact (Scarf *et al.* 1986; Gurnett *et al.* 1986). These bursts were mainly observed in the interval ± 20 min about closest approach, corresponding to $\pm 2.5 \times 10^4$ km along the trajectory. Due to the south-to-north motion of the spacecraft relative to the comet, most impacts were expected on its upper surface, of area 2.4 m². Ground-based IR imagery then leads to expected fluxes that depend upon the assumed size of the dust grains, varying from several hundred per second for micrometre-sized grains to a few tens per second for particles of *ca.* 10–20 μm (Campins *et al.* 1986). By comparison Gurnett *et al.* (1986) report dead-time corrected fluxes peaking at *ca.* 0.5 s⁻¹. They infer that impacts on the spacecraft body were not detected by the instrument, only impacts on the antenna wire itself, whose area is a factor of *ca.* 50 less. The reason may be that the spacecraft upper surface is recessed well below the level of the solar cells surrounding the cylindrical body of the spacecraft, such that the antennas were shielded from the plasma cloud resulting from impacts on the upper surface. Even so, the observed rates suggest unexpectedly large dust-grain sizes in the range 10–20 μm. This result probably relates to the observation that meteors associated with Giacobini-Zinner (the Giacobinid or Draconid showers) are of unusually low density.

5. Energetic ion observations

In the previous sections we reviewed the effects observed by ICE in the plasma and electromagnetic field that result from mass-loading of the solar wind flow by heavy cometary ions. It will be recalled from the introduction that these ions gain considerably energy on being picked up by the solar-wind flow. The peak energy depends on the solar-wind speed and the cone angle α between the magnetic field and solar-wind velocity vectors, (1), but is typically *ca.* 4 keV u^{-1}, such that for heavy ions of the water group (O^+, OH^+, H_2O^+ or H_3O^+), energies of several tens of kiloelectronvolts are expected. ICE carried two detectors, which measured ions of these energies, and which provided complementary information on their properties. The EPAS instrument provided measurements above 65 keV (for water-group ions) in three dimensions in velocity space with a resolution of 32 s (Hynds *et al.* 1986), while the ULECA sensor was much less sensitive to heavy ions and provided two-dimensional (ecliptic-plane) measurements with reduced time resolution (Ipavich *et al.* 1986). On the other hand the ULECA energy range extends down to 35 keV, such that heavy ion bulk parameters are more easily obtained than with EPAS, and the detector also provides some information about ion mass. Ipavich *et al.* (1986) showed that the ULECA responses were consistent with the detection of singly charged heavy (more than 12 u) ions, and by comparing the ion angular anisotropies with expectations based upon the thermal electron bulk speed, deduced that the observed ion mass was 18 ± 6 u, consistent with water-group ions.

Figure 7 gives an overview of the EPAS observations, taken from Hynds *et al.* (1986). This shows fluxes in the sunward-looking sector of a telescope viewing 30° above the ecliptic plane for four energy channels: 65–95 (E_1), 95–140 (E_2), 140–205 (E_3), and 205–310 (E_4) keV. The interval 10–13 September 1985 is shown, spanning distances of several million kilometres about the comet, with closest approach being indicated by the vertical line. It can be seen that antisunward-streaming pick-up ions were indeed observed throughout this region with fluxes that generally increase as the comet is approached, as also found by Ipavich *et al.* (1986). The fluxes show considerable variability, however, and are clearly asymmetrical about closest approach with higher fluxes being observed outbound than inbound. Sanderson *et al.* (1986) have shown that these effects result from variations in the solar-wind speed and magnetic-field direction which modulate the energy gained by the pick-up ions according to (1). It is inferred that the pick-up ions are always present, but not always with sufficient energy to be detected above the 65 keV threshold of the lowest energy channel. Broadly speaking, the inbound–outbound asymmetry was due to the solar-wind speed being much lower inbound (*ca.* 350 km s^{-1} until *ca.* 05 h U.T. on 11 September) than outbound (*ca.* 500 km s^{-1}), whereas the shorter timescale variations are related to changes in the direction of the magnetic field. The latter result shows that ions in the pick-up region are not strongly pitch-angle scattered in the solar-wind rest frame, because otherwise only velocity, but not magnetic field, modulation would be present.

Figure 8 gives details of the EPAS flux measurements closer to the comet, together with ion bulk parameters deduced from angular anisotropies and ion spectra (Richardson *et al.* 1986). The vertical dashed lines at 09h27 U.T. and 12h24 U.T. mark the locations where sharp changes in ion properties take place, thus approximately coinciding with the locations of rapid changes in plasma and magnetic field properties. The upper panel shows the direction-averaged ion flux in the lowest energy channel (65–95 keV). Inside the dashed lines, fluxes vary much

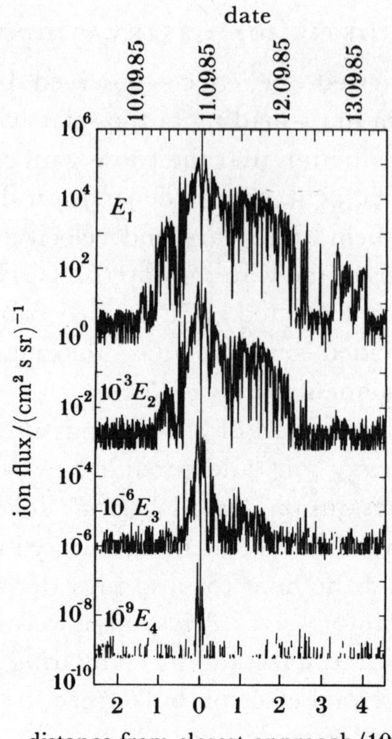

FIGURE 7. Overview of energetic ion observations by the EPAS instrument, showing the fluxes in telescope 2, inclined 30° north of the ecliptic, in sunward-pointing azimuth sector 1 against time. The four energy channels (top to bottom) correspond to water-group ions of energy 65–95, 95–140, 140–205 and 205–310 keV. The fluxes have been scaled to avoid overlap. The vertical line gives the time of closest approach. (From Hynds *et al.* 1986.)

more smoothly than outside (though with continued variability on timescales of a few minutes), and do not show strong magnetic-field modulation. This indicates that the ions have become isotropized in the plasma bulk frame in this region. Close to the comet, in the central part of the mass-loaded region, fluxes decline in strength, with a central 'slot' of *ca.* 10 min of reduced flux corresponding to the induced magnetotail structure.

The next three panels show the ion bulk speed and direction (deduced assuming that the observed ions are O^+); the dashed line shows a smoothed version of the electron bulk speed (figure 2) for comparison. The elevation angle of the flow with respect to north is represented by θ (0° to 90° indicates flow to the north, 90° to 180° flow to the south), and ϕ is the flow azimuth (180° is flow away from the Sun, 90° is flow from west to east of the Sun). These show that at the boundaries marked by the vertical lines the flow is deflected in angle away from the comet-tail axis, to the south inbound and to the north outbound, as expected from MHD models of the flow (see, for example Ip & Axford 1982 and references therein), and that the flow is also slowed by *ca.* 100 km s^{-1}. Tranquille *et al.* (1986) have examined these regions in more detail and have shown that as the plasma approaches the comet it is first deflected in elevation angle (09h24–09h28 U.T. inbound, 12h20–12h25 U.T. outbound) and then slowed (09h28–09h34 U.T. inbound, 12h14–12h20 U.T. outbound). It is the latter interval that coincides with the onset of compressional turbulence and enhanced magnitudes in the magnetic field. The full width of the transitions is *ca.* 10^4 km, corresponding to an energetic O^+ gyroradius.

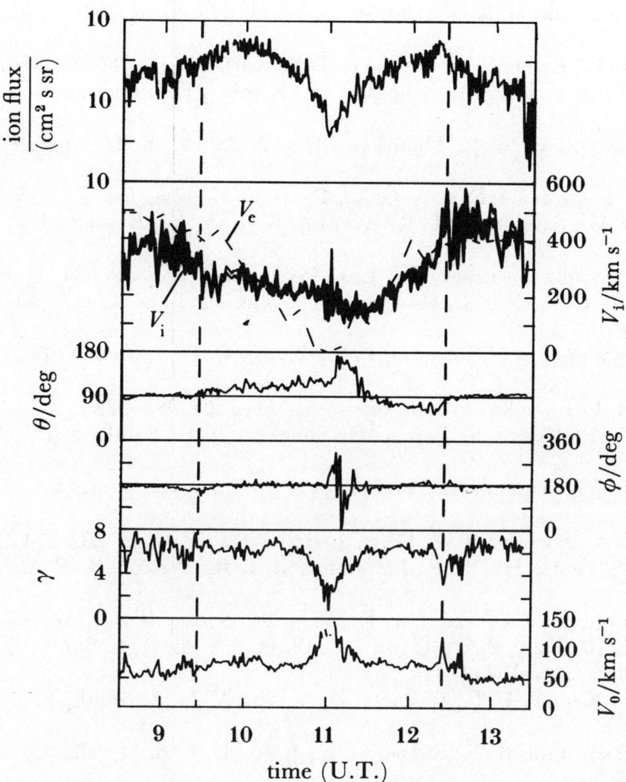

FIGURE 8. EPAS energetic ion observations and deduced ion bulk parameters for the interval 08h30 to 13h30 U.T. on 11 September 1985 spanning the mass-loaded region and induced magnetotail. The top panel shows the spin-averaged 65–95 keV ion flux, and the second to fourth panels show the deduced ion bulk speed V_i and direction (θ, ecliptic elevation, and ϕ, ecliptic azimuth), together with (dashed line) the smoothed thermal electron bulk speed V_e. The fifth and sixth panels show the power law (γ) of a fit to the spectrum of the form $\mathrm{d}J/\mathrm{d}E \sim E^{-\gamma}$ and the characteristic speed V_0 of a fit of the form $f \sim \exp(-V/V_0)$ respectively. (From Richardson *et al.* 1986.)

Nearer to the comet the ions flow increasingly from north to south, and the bulk speed does not fall to very low values as indicated by the electron data. Ions are observed in this region mainly because of finite Larmor radius-penetration from the outside, such that the angular anisotropies in this region (± 30 min of closest approach) are related to density gradients rather than to plasma flows (Daly *et al.* 1986).

Finally, heavy-ion densities have been determined for the interval 06h00–16h00 U.T. from the ULECA data by Gloeckler *et al.* (1986). Their results are consistent with a water-production rate of 2.6×10^{28} mol s^{-1} and a photoionization time of *ca.* 10^6 s.

REFERENCES

Alfvén, H. 1957 *Tellus* **9**, 92–96.
Baker, D. N., Feldman, W. C., Gary, S. P., McComas, D. J. & Middleditch, J. 1986 *Geophys. Res. Lett.* **13**, 271–274.
Bame, S. J., Anderson, R. C., Asbridge, J. R., Baker, D. N., Feldman, W. C., Fuselier, S. A., Gosling, J. T., McComas, D. J., Thomsen, M. F., Young, D. T. & Zwickl, R. D. 1986 *Science, Wash.* **232**, 356–361.
Campins, H., Telesco, C. M., Decher, R., Mozurkewich, D. and Thronson, H. A. Jr 1986 *Geophys. Res. Lett.* **13**, 295–298.
Combi, M. R., Stewart, A. I. F. & Smyth, W. H. 1986 *Geophys. Res. Lett.* **13**, 385–388.

Daly, P. W., Sanderson, T. R., Wenzel, K.-P., Cowley, S. W. H., Hynds, R. J. & Smith, E. J. 1986 *Geophys. Res. Lett.* **13**, 419–422.

Fuselier, S. A., Feldman, W. C., Bame, S. J., Smith, E. J. & Scarf, F. L. 1986 *Geophys. Res. Lett.* **13**, 247–250.

Gloeckler, G., Hovestadt, D., Ipavich, F. M., Scholder, M., Klecker, B. & Galvin, A. 1986 *Geophys. Res. Lett.* **13**, 251–254.

Gosling, J. T., Asbridge, J. R., Bame, S. J., Thomsen, M. F. & Zwickl, R. D. 1986 *Geophys. Res. Lett.* **13**, 267–270.

Gurnett, D. A., Averkamp, T. F., Scarf, F. L. & Grün, E. 1986 *Geophys. Res. Lett.* **13**, 291–294.

Hynds, R. J., Cowley, S. W. H., Sanderson, T. R., Wenzel, K.-P. & van Rooijen, J. J. 1986 *Science, Wash.* **232**, 361–365.

Ip, W.-H. and Axford, W. I. 1982 In *Comets* (ed. L. L. Wilkening), pp. 588–634. University of Arizona Press.

Ipavich, F. M., Galvin, A. B., Gloeckler, G., Hovestadt, D., Klecker, B. & Scholer, M. 1986 *Science, Wash.* **232**, 366–369.

Jones, D. E., Smith, E. J., Slavin, J. A., Tsurutani, B. T., Siscoe, G. L. & Mendis, D. A. 1986 *Geophys. Res. Lett.* **13**, 243–246.

Mendis, D. A. & Houpis, H. L. F. 1982 *Rev. Geophys. Space Phys.* **20**, 885–928.

Meyer-Vernet, N., Couturier, P., Hoang, S., Perche, C., Steinberg, J. L., Fainberg, J. & Meetre, C. 1986*a Science, Wash.* **232**, 370–374.

Meyer-Vernet, N., Couturier, P., Hoang, S., Perche, C. & Steinberg, J. L. 1986*b Geophys. Res. Lett.* **13**, 279–**272.**

Ogilvie, K. W., Coplan, M. A., Bochsler, P. & Geiss, J. 1986 *Science, Wash.* **232** 374–377.

Richardson, I. G., Cowley, S. W. H., Hynds, R. J., Sanderson, T. R., Wenzel, K.-P. & Daly, P. W. 1986 *Geophys. Res. Lett.* **13**, 415–418.

von Rosenvinge, T. T., Brandt, J. C. & Farquhar, R. W. 1986 *Science, Wash.* **232**, 353–356.

Sanderson, T. R., Wenzel, K.-P., Daly, P. W., Cowley, S. W. H., Hynds, R. J., Smith, E. J., Bame, S. J. & Zwickl, R. D. 1986 *Geophys. Res. Lett.* **13**, 411–414.

Scarf, F. L., Coroniti, F. V., Kennel, C. F., Gurnett, D. A., Ip, W.-H. & Smith, E. J. 1986 *Science, Wash.* **232**, 377–381.

Slavin, J. A., Smith, E. J., Tsurutani, B. T., Siscoe, G. L., Jones, D. E. & Mendis, D. A. 1986 *Geophys. Res. Lett.* **13**, 283–286.

Smith, E. J., Tsurutani, B. T., Slavin, J. A., Jones, D. E., Siscoe, G. L. & Mendis, D. A. 1986 *Science, Wash.* **232**, 382–385.

Steinberg, J. L., Fainberg, J., Meyer-Vernet, N. & Hoang, S. 1986 *Geophys. Res. Lett.* **13**, 407–410.

Strauss, M. A., McCarthy, P. J. & Spinrad, H. 1986 *Geophys. Res. Lett.* **13**, 389–392.

Thomsen, M. F., Bame, S. J., Feldman, W. C., Gosling, J. T., McComas, D. J. & Young, D. T. 1986 *Geophys. Res. Lett.* **13**, 393–396.

Tranquille, C., Richardson, I. G., Cowley, S. W. H., Sanderson, T. R., Wenzel, K.-P. & Hynds, R. J. 1986 *Geophys. Res. Lett.* **13**, 853–856.

Tsurutani, B. T. & Smith, E. J. 1986*a Geophys. Res. Lett.* **13**, 259–262.

Tsurutani, B. T. & Smith, E. J. 1986*b Geophys. Res. Lett.* **13**, 263–266.

Zwickl, R. D., Baker, D. N., Bame, S. J., Feldman, W. C., Fuselier, S. A., Huebner, W. F., McComas, D. J. & Young, D. T. 1986 *Geophys. Res. Lett.* **13**, 401–404.

Phil. Trans. R. Soc. Lond. A **323**, 421–436 (1987)

Printed in Great Britain

The origin of dust in the Solar System

By S. V. M. Clube

Department of Astrophysics, South Parks Road, Oxford OX1 3RQ, U.K.

The assumption that the Zodiacal Cloud is a predominantly meteoritic rather than a meteoroidal complex is questioned. On the basis of (i) the observed exposure ages of interplanetary dust particles collected from the stratosphere, (ii) the compressive strength of the commonest fireballs, (iii) the existence of a broad ecliptic stream centred on the Taurids and (iv) the observation of substantial short-lived meteoroid swarms therein, a suitably consistent replenishment model is constructed in which the Zodiacal Cloud appears to derive from a now defunct large comet that arrived in an Earth-crossing orbit *ca.* 10–100 ka ago. A corollary of this model is that the latter's remnant, a surviving large meteoroid, may be reactivated as a comet at intervals of *ca.* 1 ka giving rise to a variety of observable effects such as Zodiacal Cloud enhancements and rare multiple bombardments of the Earth by many bodies with masses at least 10^{11} g, which typify a general process throughout Earth history responsible for climatic excursions and extinction events. It is recommended that a search be conducted for the large meteoroid or minor planet responsible for the dust now in the Solar System, to place our understanding of the latter's evolution on a secure quantitative basis. If verified, this model would have profound implications so far as our understanding of the origin of comets is concerned because most of the cometary mass would apparently be contained in large differentiated bodies.

1. Modelling the interplanetary dust complex

The Zodiacal Cloud is now generally taken to be a *meteoritic* complex in which typical fireball progenitors of mass up to 10^3 g (density *ca.* 2.5 g cm^{-3}, albedo *ca.* 0.1 and compressive strength similar to that of basalt) are degraded by catastrophic collisions and lost from the Solar System in the form of micrometeorite particles through the combined effects of Poynting–Robertson drag (chiefly in the mass range $10^{-9} \leqslant m \leqslant 10^{-3}$ g) and solar radiation pressure (in the range $10^{-18} \leqslant m \leqslant 10^{-12}$ g) (Grun *et al.* 1986; cf. Dohnanyi 1978; Le Sergeant & Lamy 1980; Leinert *et al.* 1983). But for this picture to be acceptable, it is also necessary that it be consistent with recent *in situ* measurements of the interplanetary dust concentration (Zook & Berg 1975; Hoffmann *et al.* 1975; McDonnell 1978; Grun *et al.* 1980) and the long-term microcratering flux recorded by the Moon (Fechtig *et al.* 1975; Morrison & Clanton 1979). According to the recent calculations by Grun *et al.* (1986), this combination of constraints allied to brightness measurements of the Zodiacal Cloud cannot be met with a simple steady-state model. They therefore suggest that the concentration of low-mass particles in the range $10^{-9} \leqslant m \leqslant 10^{-3}$ g may be growing on a timescale of *ca.* 100 ka and that the excess microcratering on the Moon by particles of less than 10^{-12} g might be understood in terms of the effects of secondary cratering due to oblique-angle hypervelocity impacts. They also point out, however, that the material involved may not be wholly similar to basalt; thus it is possible that a substantial fraction of the net flux through the meteoritic complex comprises porous *meteoroidal* material of lower density and much lower compressive strength, and that this material is intermittently replenished on timescales very much shorter than 100 ka. Current *in situ* measurements of the interplanetary dust concentration (β-meteoroids, radio meteors and photographic meteors) would then be close to the long-term background flux whereas the integrated cratering and microcratering flux on the Moon would reflect more the occasional rapid throughputs of much larger amounts of material from the mostly very weak bodies whose mass distribution corresponds to that observed for the larger fireballs ($10^3 < m < 10^8$ g (McCrosky 1968)):

$$\mathrm{d}n\,(m + \mathrm{d}m, m) \propto m^{-\alpha}\,\mathrm{d}m; \quad \alpha = 1.67. \tag{1}$$

If such bodies are a factor of *ca.* 10^2–10^3 weaker than basalt (*cf.* Wetherill & ReVelle 1982), the current growth of the Zodiacal Cloud then implies a substantial fragmentation *ca.* 1 ka ago.

This picture is broadly consistent with measurements of the iridium and osmium content of deep-sea sediments (Barker & Anders 1968) and of trace elements on the lunar surface (Anders *et al.* 1973). Thus, these indicate that the accretion rate of meteoritic and meteoroidal matter on Earth is spasmodic and may reach *ca.* 100 kt a^{-1} for periods of *ca.* 100 ka, including the most recent, some two orders of magnitude greater than the underlying background flux of *ca.* 1 kt a^{-1}. Such values bracket the current annual terrestrial infall rate of *ca.* 10 kt a^{-1} (Grun *et al.* 1986; *cf.* Naumann 1966; Zook *et al.* 1970; Hughes 1978) and it seems likely that the enhancements in the terrestrial influx may therefore be due to the intermittent break-up of single large bodies whose fragmentation products are effectively suspended in the Zodiacal Cloud for periods of up to 100 ka (Barker & Anders 1968). To calculate the approximate size of the bodies responsible for such episodic infall, we assume a prefragmentation density $\rho \sim 0.1$–1 g cm^{-3} and a radius R. Taking the Zodiacal Cloud volume to be $V \sim 10^{41}$ cm^3, the rate $\dot{Q}$ at which material is collected by the Earth (radius R_E) during periods of increased infall (incident velocity $u \sim 5$ km s^{-1}, say), is given by $\dot{Q} \sim \pi R_\mathrm{E}^2 u \, \tfrac{4}{3}\pi R^3 \rho V^{-1}$ whence it follows that $R \sim 100$ km. With meteoroidal material *ca.* 10^2–10^3 weaker than basalt, each episode of *ca.* 100 ka then corresponds to a sequence of (tinted) Zodiacal Cloud enhancements as separate fragments of the parent body undergo degradation over periods of *ca.* 0.1–1 ka. According to this picture, the Zodiacal Cloud is sustained by successive very large *meteoroids* at intervals of *ca.* 100 ka to 10 Ma interspersed with less frequent but very large *meteorites* at intervals of *ca.* 10 Ma to 1 Ga consistent with the observation (Anders 1971) that most meteorite finds on the Earth due to falls during the past gigayear are derived from a limited number (*ca.* 10) of large parent bodies. It is implicit moreover that the Zodiacal Cloud is now fed by a very large meteoroidal body.

To be specific, it is assumed that this large meteoroid belongs to the family of very low albedo, red-black bodies (D-class asteroids), which includes 'the Trojan asteroids, Hidalgo, Chiron, comets and several small, dark outer satellites' and which contains 'kerogen-like, low-temperature carbonaceous condensates' (Hartmann 1986). The proposed body is thus looked upon as one of the most primitive objects in the Solar System, while its presumed evolution through a cometary phase into very friable, dust-inducing meteoroids is consistent with a primordial state comprising a very porous refractory matrix whose interstices are filled with a volatile carbonaceous medium. The systematic elimination of this medium from meteoroidal particles in the solar vicinity, leaving fluffy chondritic aggregates, implies that Zodiacal-Light particles are likely to be physically similar to C-type asteroidal regoliths (cf. Lumme & Bowell 1985).

2. MODELLING THE SOURCE OF THE DUST COMPLEX

Because the rate at which long-period comets are deflected into short-period orbits is *ca.*
1 ma^{-1} (Everhart 1972) and because $\int \phi(m \geqslant 10^{21}$ g$)$ d$m/\int \phi(m)$ d$m \sim 10^{-3}$ based on the observed mass function $\phi(m)$ of comets (see, for example, Hughes & Daniels 1982), it follows that
the current arrival rate of 'giant comets' with $m \geqslant 10^{21}$ g is *ca.* 1 Ma^{-1}. Giant comets may
therefore plausibly be thought of as identical to large meteoroids ($R \sim 100$ km), the most recent
having arrived in the inner Solar System up to 100 ka ago. A significant corollary of this
hypothesis is that the otherwise enigmatic overabundance of short-period comets (see, for
example, Fernandez & Ip 1983) based on the above deflection rate and lifetime of short-period
comets (i.e. *ca.* 1–10 ka), may be due to the early break-up of this most recent giant comet and
the dispersal of many of its cometary fragments during or after its arrival in a sub-jovian orbit.
If so, the giant comet and many other cometary fragments would be expected to have evolved
by now into a variety of undetected meteoroids in Earth-crossing subjovian orbits (Apollo
asteroids and potential fireballs). The hypothesis thus implies that a now defunct giant comet
in an Encke-type orbit is responsible for the present Zodiacal Cloud, a suggestion that has
already been made by several authors e.g. Whipple (1967), Delsemme (1976) and Kresak
(1980). For overall consistency, however, the spasmodic input of meteoroidal material must
correspond to a long-term giant-comet arrival rate of 10^{-1} Ma^{-1} implying that comets probably
reach us in episodes or showers, the latest of which may be active now (cf. Clube & Napier
1984; 1986a).

According to this picture therefore, the interplanetary dust complex may be in one of several
different states (see table 1) depending on the epoch and duration of observation. It is implicit

TABLE 1. CHARACTERISTIC RATES OF ACCRETION OF METEOROIDAL MATERIAL BY THE EARTH
DURING VARIOUS STATES OF THE INTERPLANETARY DUST COMPLEX

state of interplanetary dust complex	average over Solar System life time	comet shower	disintegrating giant meteoroid	inactive period during disintegration
duration/a	$\sim 3 \times 10^9$	$\sim 3 \times 10^6$	$\sim 10^5$	$\lesssim 10^3$
flux/(kt a^{-1})	~ 1	$\sim 10^2$	$\sim 10^2$	$\geqslant 1$

that the Zodiacal Cloud will frequently be enhanced above a certain base level but, in general,
we may expect the mass spectrum of the interplanetary dust complex to divide, as at present,
into three principal régimes corresponding to β-meteoroids, meteors and fireballs. Each régime
is characterized by an approximately constant mass distribution index ($\alpha_r, \alpha_{cc}, \alpha_e$) determined
by the dominant process of mass depletion (radiation pressure, catastrophic collisions, erosive
collisions). The current near-Earth state of the complex has α_r ($m \leqslant 10^{-9}$ g) ~ 1.5, α_{cc}
(10^{-9} g $\leqslant m \leqslant m'$) ~ 2.2, α_e ($m' \leqslant m \leqslant m''$) ~ 1.67 where $\{m', n(m')\} \sim \{10^2$ g, 10^{-28} cm$^{-3}\}$. The
observed $n(m)$ is therefore not in equilibrium (Dohnanyi 1978), values in the fireball régime
in particular being greater than the time-averaged lunar cratering flux. This could be due to
the recent fragmentation of a large meteoroid of unknown mass, but because large meteoroids
are rare, and we are unable to observe them directly, it is not possible to specify the upper limit
m''. Nevertheless, the observed fireball spectrum $n(m \geqslant 10^2$ g$)$ defines one section of the present
'base level' of the Zodiacal Cloud, the rest being normalized to m' which fluctuates in response

to the above depletion processes and the random disintegration of large meteoroids. It follows that the lowest base level (i.e. $m' \sim 10$ g) corresponds to a total terrestrial influx of *ca.* 1 kt a^{-1}, whereas the current level (i.e. $m' = 10^2$ g) corresponds to *ca.* 10 kt a^{-1}.

Because a very large body of mass $m'' \sim 10^{21}$ g (i.e. $R \sim 100$ km), comprising weak meteoroidal material of the postulated kind, may be catastrophically destroyed by an incident total mass Σm where $\Sigma m \geqslant (10^4\,\Gamma)^{-1}\,m''$ and $\Gamma \sim 10^6$ (cf. table IV in Grun *et al.* 1986), a single encounter with a missile of *ca.* 10^{11} g may serve to reduce it substantially to dust. Such an encounter with a typical Tunguska missile might take place within *ca.* 100 ka but partial erosion can also be expected on this timescale through encounters with bodies greater than *ca.* 10^3 g, the limit for observed fireballs above which the effect of erosive collisions becomes dominant over catastrophic collisions. Thus, for complete destruction within 100 ka by erosion, encounters by bodies of at least 10^3 g would have to occur at the rate $\nu \sim 10 \times (10^{11}/10^3)^{\alpha-1}$ Ma^{-1}, giving an equivalent lunar encounter rate of *ca.* 500 a^{-1}, in tolerably close agreement with the observed rate during 1971–6 (Dorman *et al.* 1978). It follows that the background flux of weak meteoroidal material entering and leaving the Zodiacal Cloud annually owing to encounters with a hypothetical large body would be *ca.* $f_1 f_2\,10^3\,(10^4\Gamma)\,\nu$ g, where the factors f_i (a few) allow for the mass spectrum of the target material ($10^3 \leqslant m \leqslant 10^{21}$ g) and that of the eroding material ($m \leqslant 10^3$ g) respectively. A total influx of *ca.* 10^{14} g a^{-1} is thus implied, consistent with the observationally constrained Zodiacal Cloud outflow given by Grun *et al.* (1986). Superimposed on this fairly regular background flux, however, is the more intermittent but larger flux due to erosive encounters by bodies greater than 10^3 g. Over intervals of *ca.* 10^3–10^5 years, these enhancements produce a significantly greater average flux, *ca.* 10^{16} g a^{-1}.

In principle therefore, the scheme put forward by Barker & Anders (1968) is capable of providing a simple, self-consistent explanation of the interplanetary dust complex, but for it to work, a substantial portion of the micrometeoroid complex must be at least a factor of 10^{-4} weaker than basalt, with an age of 10 a–100 ka. There is also an expectation of discrete replenishment events of *ca.* 10^{15}–10^{17} g (10 a)$^{-1}$ sustaining the Zodiacal Cloud. The question arises, therefore, whether there is any independent evidence relating to the strength and age of the meteoroidal complex, and whether replenishment events of the predicted kind take place. These separate issues will be considered in §§3 and 4, respectively.

3. Dispersed meteoroids and interplanetary dust

From the systematic observation of fireballs, it is known that most of their progenitors ($10^2 \leqslant m \leqslant 10^6$ g) are in a variety of subjovian Earth-crossing orbits similar to those of many Apollo–Amor asteroids. Although some proportion of the Apollo–Amor asteroids may originate from the asteroid belt (Wetherill 1974), many of them are probably cometary disintegration products (see, for example, Wasson & Wetherill 1979). Most fireballs thus derive from comets through an asteroidal phase, a view that has now gained further support from new asteroid discoveries (see, for example, Rickman 1985), which indicate, with due allowance for the overall discovery rate of Earth-crossing asteroids (see, for example, Shoemaker *et al.* 1979), that the short-period comets may be accompanied by as many asteroids with similar orbital parameters.

A few fireballs have been associated with known meteorite falls, however (Ceplecha 1961; McCrosky *et al.* 1971; Halliday *et al.* 1981), and a study of intermediate strength fireballs from transjovian orbits (Wetherill & ReVelle 1982; *cf.* Ceplecha & McCrosky 1976) has shown that their peak dynamic pressures take up a continuum of values in the range *ca.* 10^4–10^6 Pa. Laboratory measurements of the compressive strength of meteorites (*ca.* 10^6–10^8 Pa; Buddhue 1942) give average values that are evidently an order of magnitude above the strongest fireballs in this continuum. The commonest fireballs on the other hand, which are often too brief to be well observed, have compressive strengths that are probably an order of magnitude below the weakest in the continuum. It follows that fireballs reveal a broad range of compressive strengths lying between extremes for stones and meteoroids, which are on the order of and 10^{-4}–10^{-3} weaker than basalt respectively. Ablation studies indicate that for a given mass, the former decelerate more slowly and penetrate further into the atmosphere as a single body by reason of their robust composition, whereas the latter are mostly friable, low-density objects whose visible flight through the atmosphere terminates abruptly at high altitude as the meteoroid fragments into many small pieces. The overall picture that emerges is thus of a fireball population fed by two separate but merging sources, cometary asteroids and meteorites, the former more copious than the latter, which probably arrive in Earth-crossing orbits by similar routes, namely through successive jovian deflections and the subsequent action of non-gravitational forces (Kresak 1980).

The existence of meteoroids in the same mass range $10^2 \leqslant m \leqslant 10^6$ g is also inferred from observations outside the upper atmosphere, through near coincidence between otherwise randomly distributed small particles ($10^{-16} \leqslant m \leqslant 10^{-8}$ g) registered by *in situ* detectors (Fechtig 1982). The parent meteoroids are evidently very weakly bound because they undergo electrostatic fragmentation within the auroral plasma region. They also represent some 30% of the total flux of material flowing through the region and evidently correspond to the low mass end of the eroded meteoroid population. The physical similarities between the co-incident and random groups indicate moreover that these parent meteoroids are the likely intermediaries feeding the interplanetary dust. The individual dust particles in the range $10^{-16} \leqslant m \leqslant 10^{-8}$ g differ among themselves, however, as judged by their cratering properties and hence their compressive strength, some 25% being of very low density up to 1 g cm^{-3} and probably very porous.

A rather similar distinction emerges among particles in the intermediate mass range $10^{-6} \leqslant m \leqslant 10^2$ g, especially between meteor streams which self-evidently derive from different cometary or asteroidal sources. Thus brighter meteors at appropriate pre-entry velocities are due to relatively compact single particles, whereas fainter meteors are thought to be due to very porous, fragile structures, which disintegrate into separately ablating smaller grains (Jacchia 1955). More specifically, given their common spectroscopic characteristics, the latter may be pictured as open assemblies of chondritic grains ($m \geqslant 10^{-6}$ g) held together by a lower melting point 'glue' (Hawkes & Jones 1975; Beech 1984). Indeed, it is clear that the ablation of both meteor dust particles and their larger meteoroidal counterparts responsible for fireballs, probably involves the removal of this less refractory 'glue' by heating and melting, thereby causing the parent body in each case to completely fragment.

Microparticle, meteor and fireball observations thus indicate a common meteoroidal material that comprises a skeleton of finely grained refractory substances (melting points *ca.* 500–2000 K), whose differing bulk densities imply varying degrees of porosity. These variations in structure can tell us something about the cometary material from which meteoroidal material derives. Thus, if the primordial structure has been relatively unaltered, differing concentrations of the volatile component may be implied; this might suggest various degrees of differentiation in cometary material, possibly consistent with an origin from a much larger body. On the other hand, the structure may reflect subsequent processing in the vicinity of the Sun: various modes of crust formation may be implied. The porosity thus remains an ambiguous factor so far as the formation and evolution of comets are concerned; however, the skeletal structure is no longer in doubt since it has also been observed in small samples of interplanetary dust collected from the stratosphere.

The masses of stratospheric particles are commonly in the range $10^{-12} \leqslant m \leqslant 10^{-9}$ g and apart from the *ca.* 5% of irregular and spherical chondritic grains that may have a predominantly meteoritic origin, most of these particles seem to be small, relatively unaltered pieces of skeletal structure (Fraundorf *et al.* 1982). Thus, over half the particles are fluffy aggregates of exceedingly small grains held together in a porous and very fragile black matrix whose cumulative composition is opaque and similar to primitive CI/CM meteorites. The remainder in the same mass range seem to be larger, non-porous examples of some of the sulphur-bearing and non-sulphur-bearing grains that make up the common aggregates. Some of the very smallest grains are found to be unmetamorphosed enstatite whiskers and platelets indicative of direct gas-to-solid condensation in a highly reducing environment. But a variety of oxidation states evidently existed during the formation process and the presence of amorphous crystals is suggestive of processing at lower temperatures as well. Indeed, the presence of lower melting point volatiles in the pores would require that there were several temperature régimes during formation and it may be that the presence of amorphous crystals accounts for the intrinsically weak structures (Smoluchowski 1980). Either a gathering of particles from a wide range of initial environments is indicated therefore, or condensation takes place in a medium whose average physical and chemical state undergoes significant evolution during the formation process. Thus, the connected skeletal structure may be due to separate interstellar grains, with volatile mantles, that merged during cold accretion or it may have been built up from the separate refractory cores that developed in a dense cooling medium. The choice between such alternatives is not yet settled but there seems to be little doubt now that the stratospheric particles studied in the laboratory do indeed enter the atmosphere from the Zodiacal Cloud and that they typify the material of which meteors and fireballs are largely composed.

High concentrations of implanted helium, neon and argon are commonly detected in stratospheric particles consistent with exposure to the solar wind for up to 50 years (Fraundorf *et al.* 1982), though a primordial bombardment of individual particles before their possible accretion cannot also be excluded. On either of the above modes of formation however, it is clear that the lifetimes of typical dust particles in the interplanetary medium are considerably shorter than the collision times associated with ordinary basaltic material. Searches have also been conducted for nuclear tracks in interplanetary dust particles due to solar flares but with varying success (Fraundorf *et al.* 1982). Thus, for most of the porous aggregates of chondritic particles examined, where something like 30 tracks μm^{-2} would be expected for a residence time of 10 ka, none have been identified. Nevertheless, in two cases, *ca.* 10% of those examined, high track densities have been found, indicating exposure ages of *ca.* 1–10 ka (Bradley *et al.* 1984). The absence of identifiable tracks in the other examined bodies could be due to track annealing on atmospheric entry but this is considered to be unlikely given the generally rather low level of physical alteration observed; alternatively the average space exposure of these particles at 1 AU is orders of magnitude less than 10^4 ka, as might be expected for bodies with their demonstrably fragile structure. These investigations are of course still very much in their infancy but the evidence from laboratory examinations does not apparently exclude a mixed interplanetary dust complex comprising a longer lived meteoritic component together with a rapidly evolving meteoroidal component.

4. Replenishing the dust complex

According to the recent calculations by Grun $et\ al.$ (1986), there is an unreplenished loss of $ca.\ 10^9$ t a^{-1} from particles in the interplanetary dust complex with $10^{-5} \leqslant m \leqslant 10^2$ g, assuming $\alpha \geqslant 2$ (see (1)). Evidently such a loss may be compensated by the occasional erosion of much larger bodies of low compressive strength provided that $1 \leqslant \alpha \leqslant 2$, as observed. Thus, if we consider target bodies of mass of at least m_1 undergoing catastrophic collisions with masses of at least m_2, where

$$m_1 = \Gamma m_2 \tag{2}$$

and we tentatively assume $\Gamma \approx 10^{10}$ for material whose compressive strength is $ca.\ 10^{-4}$ that of basalt (cf. table IV in Grun $et\ al.$ 1986), then the number of masses of at least m_1 involved in destructive collisions per unit time is given approximately by:

$$N(\geqslant m_1) \approx Vn(\geqslant m_1)\, n(\geqslant m_2)\, r_{12}^2 u_{12}, \tag{3}$$

where $n(\geqslant m_i)$ is the space density of bodies of mass of at least m_i and radius of at least r_i, V is the volume of the Zodiacal Cloud and u_{12} is an appropriate encounter velocity. Normalizing to the space density of objects of 10^4 g striking the Moon during the period 1970–6 (Dorman $et\ al.$ 1978),

$$n(\geqslant 10^4 g) \approx 10^{-29}\ \mathrm{cm}^{-3} \tag{4}$$

and assuming a material density of $ca.\ 0.1$–1 g cm^{-3}, $V \approx 10^{41}$ cm^3 and $u_{12} \approx 20$ km s^{-1}, we find that $N\ (\geqslant 10^{13}$ g) and $N\ (\geqslant 10^{17}$ g) are of the order of one per year for values of α equal to 1.75 and 1.5 respectively. These values bracket the observed mass distribution index of fireballs and indicate that if we make due allowance for the radial dependence of $n(\geqslant m_i)$ in the Zodiacal Cloud, we might expect masses of at least 10^{16} g, which are destroyed at the rate of approximately one per year, would be more than adequate to sustain the current loss from the existing interplanetary dust complex.

For unrestrained dispersal, however, the collision energy has to be distributed more or less uniformly throughout the disrupted body and exceed the gravitational binding energy. Masses of $ca.\ 10^{16}$ g are in fact close to the upper limit of $m_1 \approx r_1^2 u_{12}\, \Gamma^{-1} G^{-1}$ capable of total disruption, the velocity of dispersal being at least 10^{-4} km s^{-1}. It follows that bodies of at least 10^{17} g undergoing erosive collisions somewhat less frequently than one per year and dispersing material at at least 10^{-3} km s^{-1} are the most probable source replenishing the Zodiacal Cloud at $ca.\ 10^{15}$ g a^{-1}. It is not to be expected of course that the collision energy will be uniformly distributed among the eroded fragments and escape velocities considerably greater than 10^{-3} km s^{-1} may be anticipated, even as large as 1 km s^{-1}. Such velocities result in the dispersal of material into a broad tube around a typical short-period orbit within the course of only a very few circuits giving a typical cross-section radius at the Earth of at least 0.5 AU (see, for example, table I of Plavec 1954). The question naturally arises therefore whether there is any evidence for these postulated events, involving the production and rapid dispersal of $ca.\ 10^{15}$ g meteoroid swarms every few years.

Comet Biela, first detected in 1772, was observed disrupting violently in 1845 and since then has never been seen (see, for example, Lovell 1954). However, in 1872 and 1885, when the Earth crossed the orbit of the vanished comet, there were tremendous displays of meteors, apparently comparable to the famous Leonid showers in 1799 and 1833. The disappearance of Comet Biela, together with its strange disruption, gives the Andromedid (i.e. Bielid) shower a unique position in meteor astronomy and is widely regarded as evidence of an event that involved the more or less total break-up of a body whose mass would have been, if it were an average short-period comet, $ca.$ 10^{16} g. On the face of it, therefore, the demise of Comet Biela could have been a particularly conspicuous example of the proposed replenishment events. To set this example in perspective, however, we also list all the passages of the Earth through dense meteor swarms since 1800 and the current passages through major tube-shaped meteor swarms: see table 2, adapted from Kresak (1980). The quantities D and E refer respectively to the width of the tube and the maximum relative enhancement above the sporadic background. Some of the streams are so narrow and the measurements so uncertain that the values of E must be regarded as indicative rather than true; nevertheless, even here, the Biela shower is clearly a notable event.

At the same time though, it is also apparent that if the Taurid stream were confined to a

TABLE 2. DENSEST METEOR SWARMS SINCE 1800 AND MAJOR ANNUAL METEOR STREAMS[†]

swarm/stream		parent comet	P/years	q/AU	D/AU[‡]	E[§]
Andromedids	1885	Biela	6.6	0.87	0.0070	230
Draconids	1933	Giaccobini–Zinner	6.6	1.00	0.0025	180
Andromedids	1872	Biela	6.7	0.87	0.0070	120
Draconids	1946	Giaccobini–Zinner	6.6	1.00	0.0015	60
Leonids	1966	Tempel–Tuttle	32.9	0.98	0.0008	40
Leonids	1833	Tempel-Tuttle	33.1	0.98	0.0010	14
Andromedids	1892	Biela	6.6	0.86	0.0010	6
τ Herculids	1930	Schwassman–W.3	5.4	1.01	0.0015	2.3
Andromedids	1847	Biela	6.6	0.86	0.0007	2.3
Bootids	1916	Pons–Winnecke	5.9	0.97	0.0003	2.1
Draconids	1952	Giaccobini–Zinner	6.4	0.99	0.0002	1.8
Andromedids	1838	Biela	6.6	0.88	0.0004	1.6
Andromedids	1899	Biela	6.7	0.86	0.0004	1.6
Lyrids	1803	Thatcher	415.5	0.92	0.0005	1.4
Leonids	1867	Tempel–Tuttle	33.5	0.98	0.0002	1.4
Leonids	1965	Tempel–Tuttle	32.9	0.98	0.0030	1.4
Geminids		?	1.6	0.13	0.14	0.19
Quadrantids		?	5.4	0.98	0.03	0.17
ξ Perseids		?	2.0	0.34	0.12	0.11
β Taurids		Encke	3.3	0.34	0.20	0.07
Arietids		?	2.0	0.09	0.18	0.06
Taurids		?	3.3	0.34	0.85	0.05
δ Aquarids		?	4.6	0.08	0.30	0.05
Perseids		Swift-Tuttle	120.0	0.96	0.30	0.035
Ursids		Tuttle	13.8	1.02	0.12	0.023
Lyrids		Thatcher	415.5	0.92	0.04	0.014
μ Aquarids		Halley	76.1	0.59	0.20	0.010
Orionids		Halley	76.1	0.59	0.25	0.007
Leonids		Tempel–Tuttle	32.9	0.98	0.02	0.003

† Based on Kresak (1980).
‡ Typical half-width of stream.
§ Relative enhancement above sporadic background.

volume similar to that of the Andromedids, its *annual* relative enhancement would be the equal of the Andromedids in 1885. Thus, to the extent that visible phenomena associated with typical disruption events may survive for a decade or so (cf. Comet Biela), it is evidently possible that the Taurid stream is sustained by lesser events of the same kind which happen at average intervals of a few years. There is of course no question of *comet* disruption in the case of the Taurid stream since such bodies are not observed (except Comet Encke), but that does not exclude the possibility that unsighted dead comets with freshly black surfaces may be involved. If this is indeed the case, such events would correspond to those predicted above and would lead one to concur with Kresak's comment (Kresak 1980) that the Taurid stream, by reason of its width, is quite abnormal and potentially a uniquely rich source for the interplanetary meteoroidal complex. More recently, Stohl (1983) has shown that a substantial portion of the sporadic meteor flux, commonly associated with the Zodiacal Cloud, may be concentrated in an extremely broad ecliptical stream of meteoroids centred on the Taurids. In fact, he demonstrates that the broad stream is double, so it may have arisen from the asteroid belt fragmentation experienced five thousand years ago by the progenitor of Comet Encke (Whipple & Hamid 1952), already believed by several authors to have been a very large body (Whipple 1967; Delsemme 1976; Kresak 1980; Clube & Napier 1984; Olsson-Steele 1986). There seems to be little doubt that the evidence favours strongly the Taurid stream as the likely principal source of the postulated events. But such data do not prove that the postulated events actually take place.

Recently, however, there appear to have been at least three well-documented observations of conspicuous short-lived meteoroid swarms within the broad ecliptical stream; see table 3. In each case, the observational technique employed has been listed together with estimates of

TABLE 3. ENCOUNTERS WITH SHORT-LIVED METEOROIDAL SWARMS

date	detection technique	$\dfrac{\sigma}{\text{cm}^{-3}}$	$\dfrac{m}{\text{g}}$	$\dfrac{V_{\text{s}}}{\text{cm}^3}$	$\dfrac{M_{\text{s}}}{\text{g}}$	Taurids ?
April 1964	meteor radar[1,2]	10^{-23}	10^{-6}	10^{41}	$10^{15.5}$	no
November 1974	explorer 46[3] (impact detector)	10^{-13}	10^{-15}	10^{37}	$10^{15.5}$	yes
June 1975	lunar seismic detector[4]	10^{-29}	$10^4 - 10^6$	10^{37}	10^{16}	yes

References: 1. Ellyett & Keay 1964; 2. McIntosh & Millman 1964; 3. Singer & Stanley 1980; 4. Dorman *et al.* 1978.

the space density of particles observed (σ), their mass (m) and the size of the swarm (V_{s}). In those instances where $m \leqslant 10^2$ g, the equivalent space density of 10^2 g objects has been deduced taking the relative particle fluxes from table 4 as standard. If these inferred objects or observed 'particles' with $m \geqslant 10^2$ g represent the low mass end of a meteoroid distribution $(\alpha = 1.67)$ dominated by a single large body, the cumulative mass M_{s} of the swarm may be also readily deduced. In each independent case, it may be noted that the mass of the swarm M_{s} is comparable with that predicted for a typical erosion event every few years. In practice, of course, it seems very likely that the second and third events in table 3 may not be physically independent, though this fact has not previously been remarked upon, and that the differing values of M_{s} give some indication of the likely precision of the present calculations, which can

TABLE 4. RELATIVE PARTICLE FLUX†

m/g	$n(m)/cm^{-3}$	α
10^{-17}	10^{-11}	
		1.5
10^{-9}	10^{-15}	
		2.2
10^{2}	10^{-28}	
		1.67

† Based on Dohnanyi (1978).

indeed be no more than suggestive given the approximations in the modelling. Nevertheless, on the basis of quite simple assumptions, consistent with what is known about interplanetary dust, these unexpectedly observed swarms during the most recent decades, seem to be impressively like the ones anticipated from general arguments regarding the origin of the Zodiacal Cloud. The speculation that the interplanetary dust complex in the Solar System is maintained by the regular fragmentation of substantial meteoroids in the Taurid stream seems to be not unjustified.

5. PAST AND FUTURE IMPLICATIONS

Arguments based on (i) the exposure ages of interplanetary dust particles collected from the stratosphere, (ii) the compressive strength of the commonest fireballs, (iii) the existence of a broad ecliptic stream centred on the Taurids and (iv) the observation of substantial short-lived meteoroid swarms therein, have led to the suggestion in this paper that the Zodiacal Cloud is a predominantly meteoroidal complex and not a meteoritic complex as commonly supposed. A replenishment model based on this assumption has thus been constructed which is consistent with (i) *in situ* measurements of the interplanetary dust, (ii) Zodiacal Cloud brightness, (iii) lunar microcratering, (iv) siderophile deposits in deep-sea sediments, and (v) current physical and dynamical relations between comet, asteroid and fireball populations. It is a feature of the model that the Taurid meteor stream is a prominent source of large eroding meteoroids and that fireballs associated with the core of the stream are likely to be weaker on average than those that have survived collisionally induced dispersal into the broad ecliptic stream. This is, in fact, an observed property of the Taurid fireball distribution (Wetherill & ReVelle 1982; section III; cf. Hindley, 1972) though their strength in general somewhat exceeds that of random fireballs on account of their possibly greater carbonaceous content.

If, in accord with the strength of these fireballs, we assume that a typical swarm decays exponentially on a characteristic timescale of *ca.* 10 years (cf. the collisional timescale of *ca.* 100 ka derived by Grun *et al.* (1986), assuming material like basalt), it is readily shown that enhancements of the Taurid fireball flux by factors of a few over periods of 10^{2}–10^{3} years would arise at corresponding average intervals of 1–10 ka following the disruption of individual meteoroid masses of *ca.* 10^{3}–10^{4} times that of the commonly observed swarms that occur at *ca.* 10 year intervals and are responsible for maintaining the background stream. A conspicuous centuries-long enhancement of the Taurid fireball flux, typical of such a disruption within the last 1–2 millennia, has been recorded in the eleventh century A.D. (Astapovic & Terenteva 1968), implying the existence over an indeterminate period at this time of a particularly dense meteoroid swarm. This finding clearly places in a new light possible coincidences with the daytime Taurids around this epoch, particularly a terrestrial encounter in A.D. 793 when it was noted by a chronicler that '...excessive whirlwinds, lightning storms and fiery dragons were seen flying in the sky' (see, for example, Brondsted 1965) and a lunar encounter in 1178 when a significant crater-forming event may have been observed (see, for example, Brecher 1984; Clube & Napier 1984, note added in revision). It also raises questions concerning the attention given to fireball activity during much the same period (*ca.* 600–1000) by Chinese astronomers (Schafer 1977; cf. Biot 1848). In addition, it is significant that the late classical period is known to have coincided with a similar period of considerable meteoric activity (Cornford 1952), following an earlier time when the Zodiacal Cloud was probably enhanced. Thus, the discussion by Aristotle concerning the properties of the 'Milky Way' up to 400 B.C. is clearly anomalous (Jaki 1973) and if its proposed replenishment by comets and location in the apparent path of the Sun are taken *inter alia* at face value, there is good evidence that an earlier, conspicuous zodiacal cloud was being described (Bailey *et al.* 1986; Clube & Napier 1984). Despite the unavoidably circumstantial nature of much of this evidence, the three most recent millennia are not inconsistent with a model in which the interplanetary dust complex is replenished by a stochastic time-sequence of disintegrating meteoroid swarms in the Taurid stream.

If the Moon encounters *ca.* 40 meteoroids of mass at least 10^5 g (Dorman *et al.* 1978) during the passage of a disintegrating swarm and *ca.* 10 swarms of this or greater magnitude are produced during the course of a century, the total number of such bodies produced during the course of a century, the total number of such bodies encountered by the Earth is approximately $20\,000\ (100\ \mathrm{a})^{-1}$. Extrapolating the observed mass distribution of fireballs, this corresponds to one or two bodies of mass at least 10^{11} g striking the Earth per 100 years, in satisfactory accord with the single landfall at Tunguska during the twentieth century (Krinov 1966). The lunar surface on which craters are counted is at least 4 Ga old and the inferred flux of meteoroids of at least 10^{11} g is usually time averaged over this period. But if the episodic theory referred to previously (Clube & Napier 1986*a*) is correct, most of this cratering will be bunched in impact episodes during which the average flux will be at least an order of magnitude higher. Because the overall 'Tunguska' flux is approximately 1 per 500 years (Shoemaker 1983), the implied current rate is again 1–2 per century and in good agreement with the land fall this century. Thus, both the 'recent' historical past and the long-term terrestrial–lunar record are compatible with regularly disintegrating meteoroid swarms sustained by an eposidic sequence of giant comets.

It needs to be emphasized, however, that although the model arrived at here is constrained by available observational evidence, it is also predicated upon an *assumed* compressive strength for meteoroidal material for which there is no laboratory measurement. Nevertheless, if the theory is accurate at the order-of-magnitude level, there are several significant predictions which, failing these independent measurements, would now serve to test its authenticity:

(1) Relatively high cosmic dust deposition rates at least $10^5\ \mathrm{t\ a^{-1}}$ will have occurred within the most recent $\sim$ 10–100 ka corresponding to the dispersal of the latest giant comet. These may correlate with the most recent ice-age *ca.* 10–20 ka ago (*cf.* La Violette 1983; Clube & Napier 1984; 1986*a*).

(2) At least one multiple Tunguska bombardment of the Earth (by 10^2–10^3 meteoroids of at least 10^{11} g) will have occurred during the last 5 ka, giving rise to widespread incineration and a probable climatic recession (Clube & Napier 1986*b*). Such a process bears comparison with a similar event at the Cretaceous–Tertiary boundary (Wolbach *et al.* 1985) and indicates that

rare swarm encounters are probably typical low-level extinction events. Although the involvement of a *ca.* 10 km missile with the K–T boundary (see, for example, Alvarez *et al.* 1980) is not excluded by this hypothesis, the failure so far to discover any large crater corresponding to this epoch may not be surprising.

(3) *Future* multiple Tunguska bombardments are still possible on the present model. Their prediction would be facilitated by a search for the meteoroidal remains of the latest giant comet. These may include a large carbonaceous asteroid ($R \lesssim 100$ km, albedo *ca.* 0.02) in the Taurid meteor stream, which could be reactivated as a weak comet from time to time by particularly violent impacts. It is not known now whether any reactivations have been observed during the last few thousand years but a search in the Taurid stream for the putative minor planet (for which the name Chronos is proposed since its regular motion will have been observed in the past) could bring a return on the investment of telescope time.

(4) If the stochastic erosion model for the most recent giant comet, otherwise proto-Encke, is correct, Comet Encke may itself be an asteroid that was reactivated two centuries ago. Also, because a node of the latter's orbit passed through the Earth's orbit at the time of Christ (cf. Whipple & Hamid 1952), this body may have been unexpectedly visible at this time.

6. Conclusion

A wide range of evidence now indicates that a large cometary or D-class asteroid in the Taurid meteor stream may be the primary source of the interplanetary meteoroidal dust complex. It is recommended that a search be conducted for this particular minor planet to place models of the dust in the immediate solar environment on a secure quantitative basis. If Chronos is found, the probably dominant role of large differentiated comets in Earth history would be reinforced and the question of the origin of such bodies, in which most of the cometary mass resides, would overtake that of their more common but less significant counterparts, ordinary comets. Thus, it has been suggested in this paper that the fragile, skeletal structure associated with the stratospheric particles from the Zodiacal Cloud may be understood in terms of condensation as much as accretion. It follows that the search would not only be of interest for its own sake but for its potential contribution to the understanding of a fundamental astrophysical process.

References

Alvarez, L. W., Alvarez, W., Asaro, F. & Michel, H. V. 1980 *Science, Wash.* **208**, 1095–1105.

Anders, E. 1971 NASA special publication no. 267, p. 429.

Anders, E., Ganapathy, R., Krahenbuhl, U. & Morgan, J. W. 1973 *The Moon* **8**, 3–24.

Astapovic, I. S. & Terenteva, A. K. 1986 In *Physics and dynamics of meteors* (ed. L. Kresak & P. M. Millman), pp. 308–319. Dordrecht–Holland: Reidel.

Bailey, M. E., Clube, S. V. M. & Napier, W. M. 1986 *Vistas Astronaut.* **29**, 53–112.

Barker, J. L. & Anders, E. 1968 *Geochim. cosmochim. Acta* **32**, 627–635.

Beech, M. 1984 *Mon. Not. R. astr. Soc.* **211**, 617–620.

Biot, E. 1848 *Mém. Acad. Sci. Inst.* **10**, 129–352.

Bradley, J. P., Brownlee, D. E. & Fraundorf, P. 1984 *Science, Wash.* **226**, 1432–1434.

Brecher, K. 1984 *Bull. Am. astr. Soc.* **16**, 476.

Bronsted, J. 1965 *The Vikings*, p. 32. London: Penguin Books.

Buddhue, J. D. 1942 *Contr. Soc. Res. Meteorites* **3**, 39–40.

Ceplecha, Z. 1961 *Bull. astr. Insts Csl.* **12**, 21–47.

Clube, S. V. M. & Napier, W. M. 1984 *Mon. Not. R. astr. Soc.* **211**, 953–968.

Clube, S. V. M. & Napier, W. M. 1986*a* In *The galaxy and the solar system* (ed. M. S. Matthews, R. Smoluchowski & J. Bahcall). Tuscon: University of Arizona Press.

Clube, S. V. M. & Napier, W. M. 1986*b* *Interdisciplinary science reviews* (ed. A. R. Michaelis), vol. 11, pp. 3, 236–247.

Cornford, F. M. 1952 *Principium sapientiae*. Cambridge University Press.

Delsemme, A. H. 1976 *Interplanetary dust and Zodiacal Light* (ed. H. Elsassen & H. Fechtig), pp. 481–484. Berlin: Springer-Verlag.

Dohnanyi, J. S. 1978 In *Cosmic dust* (ed. J. A. M. McDonnell), pp. 527–605. Chichester: Wiley.

Dorman, H. J., Evans, S., Nakamura, Y. & Latham, G. V. 1978 In *Proc. Lunar Planet. Sci. Conf.* 9, 3615–3626. New York: Pergamon.

Duennebier, F. K., Nakamura, Y., Latham, G. V. & Dorman, H. J. 1976 *Science, Wash.* **192**, 1000–1002.

Ellyett, C. D. & Keay, C. S. 1964 *Science, Wash.* **146**, 1458.

Everhart, E. 1972 *Astrophys. Lett.* **10**, 131–135.

Fechtig, H. 1982 In *Comets* (ed. L. L. Wilkening), pp. 370–382. Tuscon: University of Arizona Press.

Fechtig, H., Gentner, W., Hartung, J. B., Nagel, K., Neukum, G., Schneider, E. & Storzer, D. 1975 NASA special publication no. 370, pp. 585–603.

Fernandez, J. A. & Ip, W.-H. 1983 In *Asteroids, comets, meteors* (ed. C.-I. Lagerkist & H. Rickman), pp. 387–390. Uppsala University Press.

Fraundorf, P., Brownlee, D. E. & Walker, R. M. 1982 In *Comets* (ed. L. L. Wilkening), pp. 383–409. Tuscon: University of Arizona Press.

Grun, E., Pailer, N., Fechtig, H. & Kissel, J. 1980 *Planet. Space Sci.* **28**, 333–349.

Grun, E., Zook, H. A., Fechtig, H. & Giese, R. H. 1986 *Icarus* **62**, 244–272.

Halliday, I., Griffin, A. A. & Blackwell, A. T. 1981 *Meteoritics* **16**, 153–170.

Hartmann, W. K. 1986 In *Asteroids, comets, meteors II* (ed. C.-I. Lagerkvist, B. A. Lindblad, H. Lundstedt & H. Rickman), pp. 191–193. Uppsala University Press

Hawkes, R. L. & Jones, J. 1975 *Mon. Not. R. astr. Soc.* **173**, 339–356.

Hindley, K. B. 1972 *J. Br. astr. Assoc.* **82**, 287–298.

Hoffmann, H.-J., Fechtig, H., Grun, E. & Kissel, J. 1975 *Planet. Space Sci.* **23**, 985–991.

Hughes, D. W. 1978 In *Cosmic dust* (ed. J. A. M. McDonnell), pp. 123–185. Chichester: Wiley.

Hughes, D. W. & Daniels, P. A. 1982 *Mon. Not. R. astr. Soc.* **198**, 573–582.

Jacchia, L. G. 1955 *Astrophys. J.* **121**, 521–527.

Jaki, S. L. 1973 *The Milky Way*. Newton Abbot: David and Charles.

Kresak, L. 1980 *Solid particles in the Solar System* (ed. I. Halliday & B. A. McIntosh), pp. 211–222. Dordrecht: Reidel.

Krinov, E. L. 1966 *Giant meteorites* (tr. M. M. Beynon). Oxford: Pergamon Press.

LaViolette, P. 1983 Ph.D. Thesis. Portland State University.

Leinert, C., Roser, S. & Buitrago, J. 1983 *Astron. Astrophys.* **118**, 345–357.

Le Sergeant, L. B. & Lamy, P. L. 1980 *Icarus* **43**, 350–372.

Lovell, A. C. B. 1954 *Meteor astronomy*. Oxford: Clarendon Press.

Lumme, K. & Bowell, E. 1985 *Icarus* **62**, 54–71.

McCrosky, R. E. 1968 Smithson Astrophys. Obs. special report no. 280.

McCrosky, R. E., Posen, A., Schwartz, G. & Shao, Y. 1971 *J. Geophys. Res.* **76**, 4090–4108.

McDonnell, J. A. M. 1978 In *Cosmic dust* (ed. J. A. M. McDonnell), pp. 337–426. Chichester: Wiley.

McIntosh, B. A. & Millman, P. M. 1964 *Science, Wash.* **146**, 1457.

Morrison, D. A. & Clanton, U. S. 1979 In *Proc. Lunar planet. sci. conf.*, vol. 10, pp. 1649–1663. New York: Pergamon Press.

Naumann, R. J. 1966 NASA technical note no. 3717.

Olsson-Steele, D. 1986 *Mon. Not. R. astr. Soc.* **219**, 47–74.

Plavec, M. 1954 *Bull. astr. Insts Csl.* **5**, 15–21.

Rickman, H. 1985 *Dynamics of comets: their origin and evolution* (ed. A. Carusi & G. B. Valsecchi), pp. 149–172. Dordrecht: Reidel.

Schafer, E. H. 1977 *Pacing the void*. Berkeley: University of California Press.

Shoemaker, E. M. 1983 *A. Rev. Earth Planet. Sci.* **11**, 461–494.

Shoemaker, E. M., Williams, J. G., Helin, E. F. & Wolfe R. F. 1979 In *Asteroids* (ed. T. Gehrels), pp. 253–282. Tuscon: University of Arizona Press.

Singer, S. F. & Stanley, J. E. 1980 *Solid particles in the Solar System* (ed. I. Halliday & B. A. McIntosh), pp. 329–332. Dordrecht: Reidel.

Smoluchowski, R. 1980 *Solid particles in the Solar System* (ed. I. Halliday & B. A. McIntosh), pp. 381–384. Dordrecht: Reidel.

Stohl, J. 1983 In *Asteroids, comets, meteors* (ed. C.-I. Lagerkvist & H. Rickman), pp. 419–424. Uppsala University Press.

Wasson, J. T. & Wetherill, G. W. 1979 In *Asteroids* (ed. T. Gehrels), pp. 926–974. Tucson: University of Arizona Press.

Wetherill, G. W. 1974 *A. Rev. Earth Planet. Sci.* **2**, 303–331.

Wetherill, G. W. & ReVelle, D. O. 1982 In *Comets* (ed. L. L. Wilkening), pp. 297–319. Tucson: University of Arizona Press.
Whipple, F. L. 1967 NASA special publication no. 150, pp. 409–426.
Whipple, F. L. & Hamid, S. E. 1952 *Helwan Obs. Bull.* **41**, 1–28.
Wolbach, W. S., Lewis, R. S. & Anders, E. 1985 *Science, Wash.* **230**, 167–170.
Zook, H. A. & Berg, O. E. 1975 *Planet. Space Sci.* **23**, 183–203.
Zook, H. A., Flaherty, R. E. & Kessler, D. J. 1970 *Planet. Space Sci.* **18**, 953–964.

Discussion

M. A. SAUNDERS. In the April 1986 issue of *Geophys. Res. Lett.*, L. A. Frank and co-workers at Iowa suggest that small cometesimals of mass *ca.* 100 t are impacting the Earth's upper atmosphere at a global rate of *ca.* 20 min^{-1}. Is such a global mass accretion rate of *ca.* 10^{12} kg a^{-1} feasible in view of our current knowledge of matter distribution in space?

S. V. M. CLUBE. An accretion rate of *ca.* 10^{12} kg a^{-1} seems now to be irreconcilable with either the observed chondritic deposition rate or the exospheric diffusion rate of hydrogen. The cometesimal masses may therefore have been overestimated by a factor of *ca.* 10^6. The actual observations are of large decreases in atmospheric UV dayglow due to the transient appearance of *ca.* 50 km absorbing molecular clouds at *ca.* 300 km altitude. It may be noted that the discoverers have not excluded the possibility that substantially smaller masses of incoming material may be involved, needing only to catalyse recombination of the dominant OI species in the atmosphere. However, no reasonable catalyst has yet been identified.

J. DARIUS (*Science Museum, London, U.K.*). Surely Dr Clube's interpretation of the aristotelian Milky Way, for which he would have us read Zodiacal Cloud, is inconsistent with injection of large comets on a timescale of *ca.* 100 ka, which he considers to be linked to the lifetime of the Zodiacal Cloud.

S. V. M. CLUBE. The phenomena referred to by Aristotle in his *Meteorologica* and apparently by previous natural philosophers as well (see Bailey *et al.* 1986) are thought to be due to a temporary enhancement of the Zodiacal Cloud during the preceding millennium. If so, it may be understood in terms of the *continuing* fragmentation of the most recent giant comet, whose remnant is now believed to be still circulating in the Taurid stream.

J. DARIUS. I was mildly perturbed at the prospect that a large comet should have been coincidentally deposited in the 4th century B.C. just in time for Aristotle to compose his *Meteorologica*!

SIR BERNARD LOVELL, F.R.S. (*Jodrell Bank, Macclesfield, Cheshire, U.K.*). The daytime meteor stream of the β-Taurids of late June and the autumn meteor stream of the Taurids have orbits similar to that of Encke's Comet. Thus any of the special phenomena mentioned by Dr Clube, which he suggests are related to the comet, should be identifiable both in June and October.

S. V. M. CLUBE. Weakly bound material emanating from the progenitor of Comet Encke into the broad tube surrounding the Taurid stream may well be observed at both intersections with the Earth's orbit.

F. L. Whipple (*Smithsonian Institution, Washington, D.C., U.S.A.*). Encke's Comet still seems to be important to the Zodiacal Cloud. A more direct measure might be the discovery in the historical records of intense meteor streams when the Earth crossed the comet's orbit within about 200 years of the birth of Christ. These magnificent showers should have been separated 10 years for a few repetitions.

E. Anders (*Enrico Fermi Institute, University of Chicago, U.S.A.*). There exist some recent data bearing on the contribution of large bodies to the total meteoritic influx. Kyte & Wasson (1986) have measured iridium in a deep-sea core covering the period 33–67 Ma ago, and found only a single peak, at 65 Ma, representing the terminal Cretaceous (K–T) impact. Everywhere else the influx rate of extraterrestrial Ir remained essentially constant, corresponding to a global influx of $(7.7 \pm 2.5) \times 10^{10}$ g a^{-1} of chondritic matter. This agrees closely with an earlier value of $(9.6 \pm 4.8) \times 10^{10}$ g a^{-1} (Barker & Anders 1968), based on five cores representing the past 1–2 Ma, which has been recalculated for revised sedimentation rates (Ku *et al.* 1968). Apparently there has been no secular change in the influx rate.

It is curious that no other peaks show up in the record. The K–T body, of mass 1×10^{18} g (Alvarez *et al.* 1980) gave an Ir peak 30 times above background. A body of one tenth this mass would have given a peak 3 times above background, which would be readily detectable. From the relation of Wetherill & Shoemaker (1982), four bodies of at least 10^{17} g should have fallen during the 34 Ma interval represented by this core, yet only one (the K–T body) shows up in the data. Perhaps this is merely a statistical fluke; on the other hand, it is conceivable that bodies smaller than 10 km do not distribute their debris globally, but leave much of it near the impact site. This seems to be true of a 0.1–0.5 km body that fell 2.3 Ma ago (Kyte & Brownlee 1985), but without further data, one cannot tell at which size global distribution of ejecta commences.

These data also speak against the hypothesis that a shower of comets was responsible for the K–T 'event'. Davies *et al.* (1964) estimate a cloud of some 2×10^{9} comets, yielding some 25 Earth impacts in 1–3 Ma. However, such widely spaced impacts – or even the dust from the remaining comets – would give a broad peak rather than the sharp spike observed.

Additional references

Davies, M., Hut, P. & Muller, R. A. 1984 *Nature, Lond.* **308**, 715.
Kyte, F. T. & Brownlee, D. E. 1985 *Geochim. cosmochim. Acta* **49**, 1095–1108.
Kyte, F. T. & Wasson, J. T. 1986 *Science, Wash.* **232**, 1225.
Ku, T. L., Broecker, W. S. & Opdyke, N. 1968 *Earth planet. Sci. Lett.* **4**, 1.
Wetherill, G. W. & Shoemaker, E. M. 1982 *Spec. Pap. geol. Soc. Am.* **190**, 1.

S. V. M. Clube. The standard deviation of the global influx is apparently not instrumental in origin (Barker & Anders 1968), thus the available observations are consistent with a stochastic time sequence of at least 10^{16} g global inputs of chondritic material over periods of *ca.* 100 ka upon a background level up to 10^{10} g a^{-1}, each input being due to the degradation of a separate large meteoroid of at least 10^{21} g. On this hypothesis, the K–T event may be due to a particularly destructive fragmentation producing an exceptionally dense meteor stream and thus only coincidentally correlated with a large crater-producing impactor, if at all. More importantly, dense atmospheric veils of global extent are, on the present theory, comparatively frequent (*ca.* 10 per large meteoroid), the effects probably saturating at inputs of at least 10^{16} g.

Even on the present theory therefore the terminal Cretaceous Ir has to be attributed to a singular event but it is possible that the specially drastic effects associated with the K–T event are not specifically due to an atmospheric veil, as suggested by Alvarez *et al.*

Whether or not the data speak against 'comet showers' depends on whether long-term galactic modulations (e.g. 30, 250 Ma) are detectable in the otherwise stochastic time sequence degraded by bioturbation. Any such finding would not conflict with the 'narrowness' of the K–T event since the latter may take place within a concentration of lower Ir peaks extended over several million years, consistent with an evolutionary decline before the event itself. Although the relation between such predictions and the Ir cores is unavoidably speculative at this time, 'comet showers' of the kind proposed by Davis *et al.* can almost certainly be ruled out because they involve hypotheses superfluous to the astronomical requirements, namely a companion star and an 'inner cloud' of comets, while also overlooking the dominant role played by asteroids in the cratering record (cf. Clube & Napier 1984).

Additional reference

Clube, S. V. M. & Napier, W. M. 1984 *Nature, Lond.* **311**, 635–636.

Note added in proof (*June* 1987). Since this paper was completed, comet 'trails', as distinct from meteor streams, have been detected with IRAS (Sykes *et al.* 1986). It has been suggested that these trails may be sustained by material released during perihelion passages, as for meteor streams, though it is possible that they are more closely associated with swarm production of the kind envisaged in this paper.

Additional reference

Sykes, M. V., Lebofsky, L. A., Hunten, D. M. & Low, F. 1986 *Science, Wash.* **232**, 1115–1117.

Phil. Trans. R. Soc. Lond. A **323**, 437–446 (1987)

Printed in Great Britain

The nature of comets

By J. C. Brandt†

*Laboratory for Astronomy and Solar Physics, NASA-Goddard Space Flight Center,
Greenbelt, Maryland 20771, U.S.A.*

[Plates 1–4]

The vast scientific campaign associated with the 1986 return of Halley's Comet has greatly improved and expanded our knowledge of comets. An overview of the first results is presented here with emphasis on the large-scale structure, the chemistry, and the nucleus.

Biermann and Alfvén's basic large-scale picture involving the interaction with the solar wind was confirmed. The interaction extends over very large distances and involves the draping of magnetic field lines from the solar wind around the head region. The near-nuclear region is essentially free of magnetic field. The cometary environment is a rich plasma physics laboratory as well as the site of spectacular disconnection events.

As Whipple proposed, the chemical composition of the nucleus is largely water, and the breakup of the water molecule produces the large hydrogen-cloud surrounding the comet. Minor constituents with high molecular mass have been observed in the comet. The composition of the dust generally resembles carbonaceous chondrites enriched in the elements H, C, N and O. The interest in the cometary chemistry stems from the belief that cometary material is probably the best remnant of the solar nebula's original composition.

The nucleus is monolithic, as predicted by Whipple's icy-conglomerate model. Far from spherical, the nucleus is irregular and peanut- or potato-shaped. The surface is very dark, and the emission of gas and dust occurs in jets on the sunward side. Irregular erosion of the surface, which is covered by a dust crust, could lead to many interesting possibilities for outbursts or splitting.

Even with our current enhancement of knowledge, comets will continue to excite scientific curiosity. Future research on comets should be very fruitful.

Introduction

The sight of a bright comet in the sky, such as the view shown in figure 1, plate 1, is fascinating to scientists and non-scientists alike. A tail tens of millions of kilometres long stretching many degrees requires explanation. Such explanation is being supplied through a massive group enterprise.

This enterprise, which culminated during 1985 and 1986, has resulted in a fundamental change in the field of cometary research. The standard bearer of the change was, of course, the direct exploration of comets by spacecraft. These are indicated in table 1. Six spacecraft were involved; five were probes for Halley and the sixth probed Comets Halley and Giacobini–Zinner. Roughly 50 experiments aboard these spacecraft gathered information

† Present address: Laboratory for Atmospheric and Space Physics, Box 392, University of Colorado at Boulder, Boulder Colorado 80309-0392, U.S.A.

438   J. C. BRANDT

TABLE 1. SPACE MISSIONS TO COMETS

spacecraft	comet	approximate distance at closest approach/km	date of closest approach	imaging of nucleus?
ICE (NASA)	Giacobini–Zinner	8000, tailward	11 September 1985	no
Vega 1 (U.S.S.R.)	Halley	9000, sunward	6 March 1986	yes
Suisei (Japan)	Halley	150000, sunward	8 March 1986	no
Vega 2 (U.S.S.R.)	Halley	8000, sunward	9 March 1986	yes
Sakigake (Japan)	Halley	7000000, sunward	11 March 1986	no
Giotto (ESA)	Halley	600, sunward	14 March 1986	yes
ICE (NASA)	Halley	28000000, sunward	25 March 1986	no

ranging from images of the nucleus to plasma waves far from the nucleus. We should note our extreme good fortune for the successful launches of all the comet missions and the proper functioning of nearly all the scientific instruments.

As impressive as the direct exploration has been, our progress in understanding comets does not rest solely on these missions. Very important data have been obtained from spacecraft in orbit around the earth (the *International Ultraviolet Explorer*, the *Solar Maximum Mission*, and the *Dynamics Explorer* 1) and around Venus (*Pioneer Venus Orbiter*), from rocket flights, from the *Kuiper Airborne Observatory*, and from the vast ground-based networks of the *International Halley Watch*.

Current preliminary reports involve a considerable amount of 'cream skimming'. Our definitive model will ultimately involve compatibility with all data gathered regardless of method. The goal is a complete observational effort for a highly variable object that leads to a detailed synthesis. Successful completion will surely take several more years.

This paper is organized into three general sections: (1) large-scale structure–plasma physics; (2) chemistry and (3) the nucleus.

LARGE-SCALE STRUCTURE–PLASMA PHYSICS

The view that the solar-wind interaction is important in the large-scale structure of comets stems from the work of L. Biermann in the early 1950s and from H. Alfvén's elaboration founded on the importance of the magnetic field. The physical picture is based on the idea that a strong interaction occurs when sublimated neutral molecules stream away from the nucleus and are ionized and trapped onto the solar-wind magnetic-field lines. The field lines are decelerated by the additional mass from the 'pickup ions' and wrap around the comet. This process produces the plasma tails of comets, which are seen when the trapped ionized molecules fluoresce when illuminated by the sun. The tail structure is normally attached to the comet's head region and has a bi-lobed magnetic configuration. The regions of opposite polarity should be separated by a current sheet. In addition, the comet is an obstacle in the solar-wind flow (which is supersonic and superalfvénic) and, hence, a bow shock was expected.

For Comet Giacobini–Zinner, the *International Cometary Explorer* (ICE) established that our ideas were basically correct, as summarized in figure 2. Magnetic field draping was confirmed. Dust impacts were recorded and the mass spectrometer showed that the principal ions were from the water group (HO^+, H_2O^+, H_3O^+). The plasma tail was dense and cold.

Some results were not expected. The reversal of the magnetic polarity was detected as the

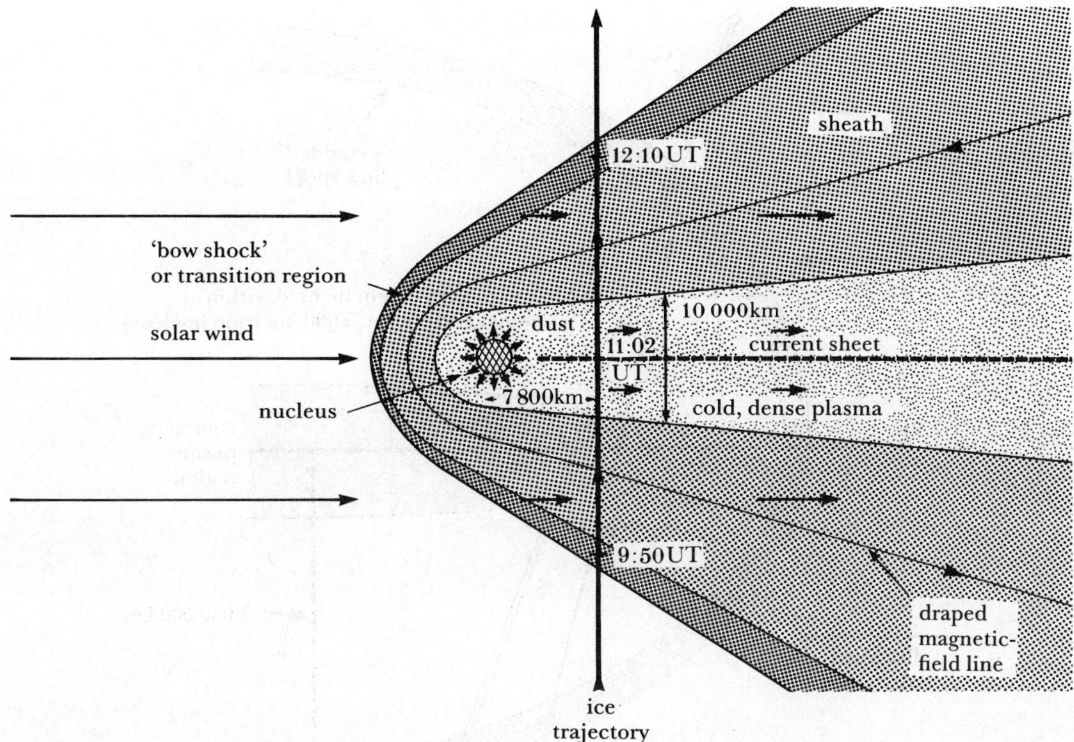

FIGURE 2. Summary schematic for the *International Cometary Explorer* encounter with Comet Giacobini–Zinner, not to scale. The vertical line represents the spacecraft trajectory marked with times on 11 September 1985. See text for discussion.

spacecraft crossed the current sheet. The surprise here exists because of expected unfavourable geometry. Both the normal orientation of the current sheet and the ICE trajectory were approximately perpendicular to the ecliptic. Solar-wind conditions rotated the current sheet and facilitated detection. The size of the interaction region was immense as evidenced by magnetic-wave activity and the measurement of pickup ions. A major (but not universal) surprise was the nature of the bow wave. The classical, abrupt changes associated with a bow shock were not observed. Rather, the deceleration is gradual and takes place over a considerable distance. The function of the expected bow shock or its surrogates is the same in any case, namely to slow the solar wind so that it can flow around the obstacle. This function is achieved for Comet Giacobini–Zinner and Comet Halley in a manner different from other Solar System bodies.

The large interaction region for Comet Giacobini–Zinner gave reason to expect a very large one for Comet Halley. In the range of total gas production rates applicable to these comets, large-scale plasma structures have dimensions that scale linearly with the production rate (*not* the square root). This feature may arise from the collective nature of the solar-wind interaction. Thus, both *Sakigake* at 7 M km sunward and ICE at 28 M km sunward were expected to directly detect Comet Halley, and they did. The results for Comet Halley are illustrated in figure 3. As a rough approximation, the Halley results are similar to the results for Comet Giacobini–Zinner scaled up by a factor of 7. The shock is diffuse, and the total interaction region extends to about 35 M km from the nucleus. The flow speed of the plasma measured by the

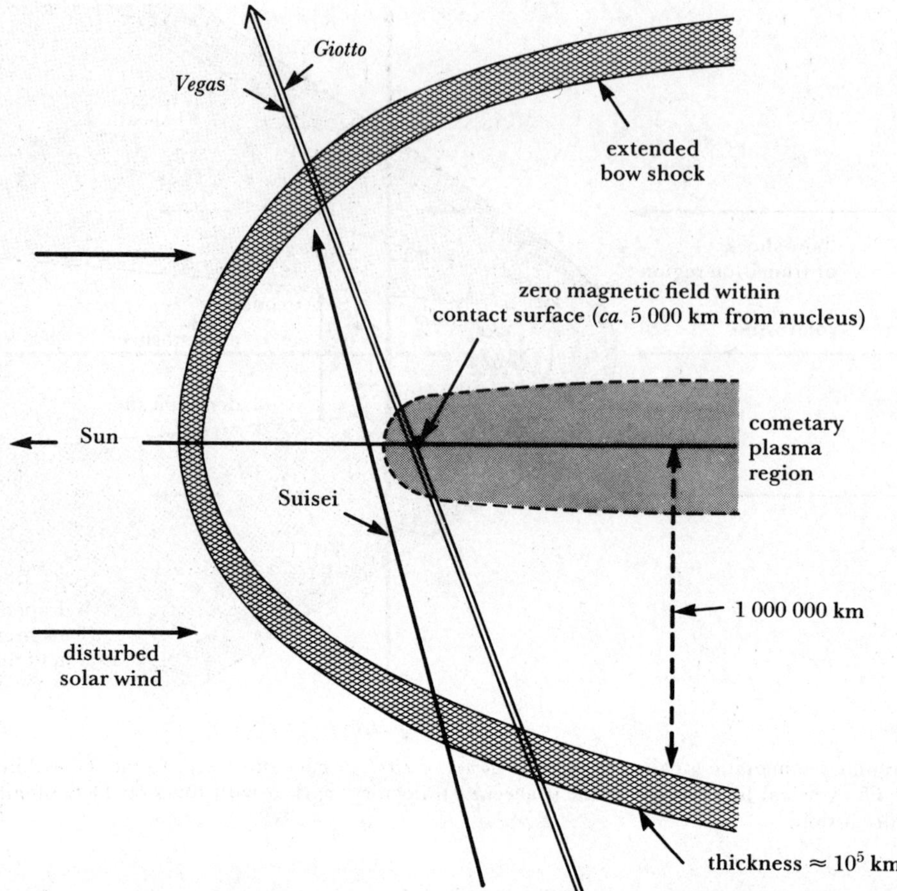

FIGURE 3. Summary schematic for plasma structures in Halley's Comet drawn approximately to scale. The trajectories of *Giotto* and the *Vega*s are displaced for clarity. Cometary ions dominate in the cometary plasma region and *Giotto* found an apparently magnetic-field-free cavity around the nucleus.

various spacecraft varies from the solar-wind speed well away from the comet to much lower speeds of not more than about 10 km s^{-1} near the nucleus. Near the nucleus of Halley's Comet, the measurement showed H_3O^+ to be the most abundant ion.

In addition, *Giotto* passed sufficiently close to the nucleus to verify the existence of the nearly field-free region (where the solar-wind magnetic field is excluded). Measurements at both comets showed the existence of ions at energies of *ca.* 500 keV. At these energies, acceleration processes other than the pickup of cometary ions by the solar wind flow are required.

Ground-based data have recorded some dramatic examples of large-scale phenomena, and correlations of imaging with the *in situ* data are in progress. The images in figure 4, plate 1, show an example of a disconnection event (DE) occurring in Comet Halley. In this event, the entire plasma tail disconnects from the comet, moves in the antisolar direction, and the comet begins to form a new tail. The leading model by Niedner and Brandt predicts that DES should occur at the sector boundaries in the solar wind where the magnetic polarity reverses. Evidence available until now indicates that this model is holding up well.

Much of this discussion has focused on the plasma-physics aspects of comets. This subject is important to understanding comets and, conversely, comets are beautiful plasma-physics

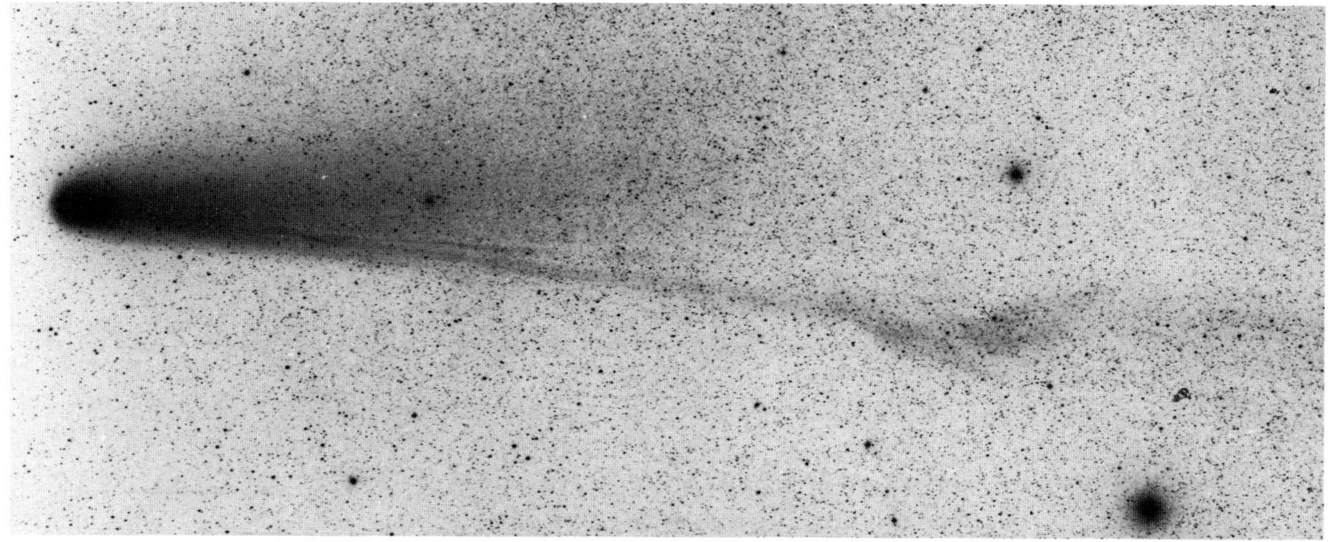

FIGURE 1. Comet Halley as photographed on 22 March 1986 showing the dust tail (above) and the plasma tail (below) with a bend caused by a disturbance in the solar wind. The tail stretches approximately 8° across the sky, or some 20×10^6 km. (Photograph taken by E. P. Moore at the Joint Observatory for Cometary Research, operated by the Laboratory for Astronomy and Solar Physics, NASA/Goddard Space Flight Center and the New Mexico Institute of Mining and Technology.)

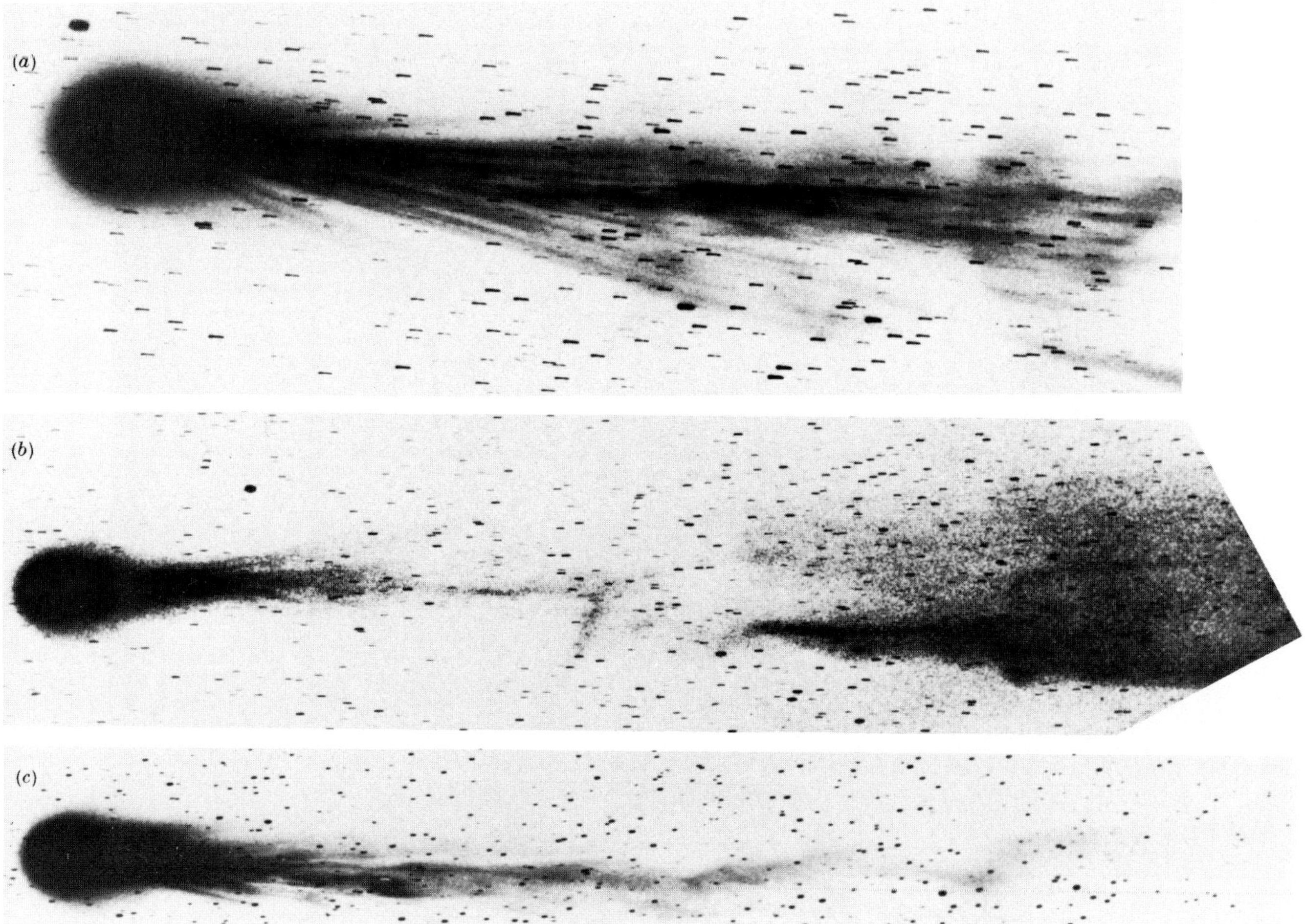

FIGURE 4. Disconnection event in Comet Halley. The individual photographs are: (*a*) 9 January 1986, Calar Alto Observatory, Spain (Max-Planck-Institut für Astronomie, Heidelberg); (*b*) 10 January 1986, Calar Alto Observatory, Spain (Max-Planck-Institut für Astronomie, Heidelberg and (*c*) 11 January 1986, Haute-Provence Observatory (C.N.R.S. – University of Liège). The disconnected tail is clearly shown on 10 January (*b*) along with the usual tail the day before and the day after. Full tail length shown is approximately 15×10^6 km.

FIGURE 5. The hydrogen cloud of Comet Halley in early February 1986 as observed from the *Pioneer Venus Orbiter*. The false-colour image is based on brightness contours in hydrogen Ly-α at 1216 Å (121.6 nm). The image covers an area 20×10^6 km by 23×10^6 km, and the white disc in the lower left corner is the size of the Sun. (I. A. F. Stewart, University of Colorado.)

FIGURE 7. Pseudocolour or false-colour (but not contour) image of the nucleus of Halley's Comet obtained by *Giotto* at a distance of 18270 km. The frame is 30 km by 30 km. The nucleus is the dark object at upper left seen in silhouette against the bright background. The bright jets point toward the Sun (as indicated by the sun pointer). The bright feature in the centre of the nucleus could be due to an elevated feature on the night side of the terminator that is illuminated by sunlight. On images taken closer to the nucleus, a crater-like circular feature is clearly seen. Compare with figures 6 and 8. (Halley Multicolor Camera, Giotto Project, Max-Planck Institut für Aeronomie.)

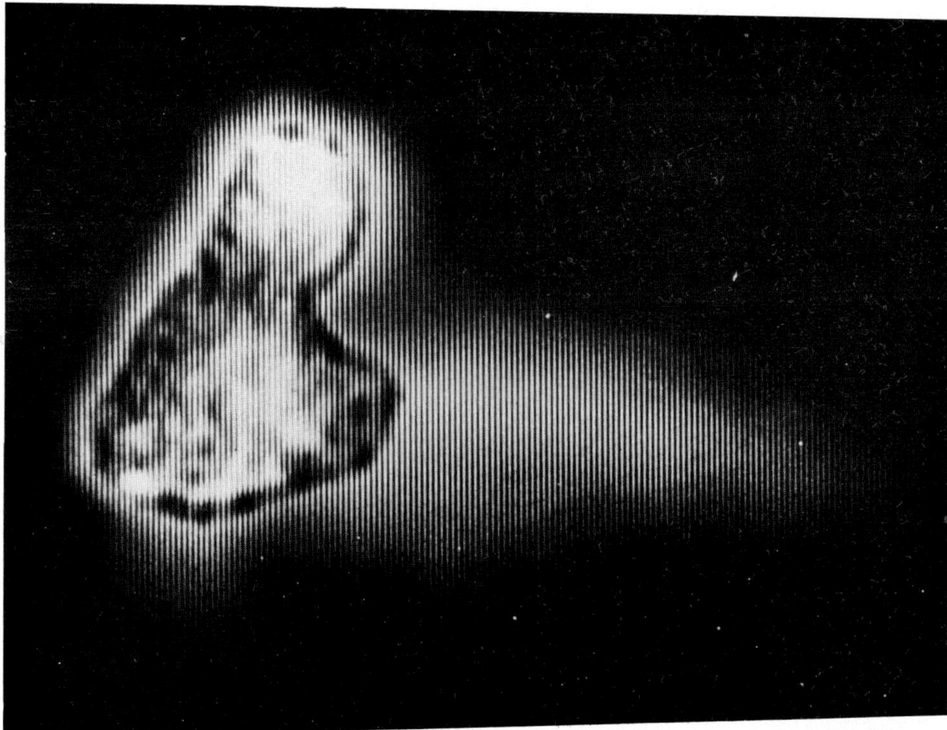

FIGURE 6. Enhanced, grey-scale image of the nucleus of Halley's Comet obtained by *Vega 2* at a distance of 8030 km showing the solid body and a prominent jet. Compare with figures 7 and 8. Note that the jet shown is sunward as are nearly all the near-nuclear structures (figures 6 and 7). The appearance contrasts with views of the entire comet (figures 1, 4 and 9) where the structures (e.g. the tail) are antisunward. (*Vega* Project.)

FIGURE 9. Halley's Comet on 16 March 1986 from a site near Washington, D.C. Approximately 5° of tail is shown. This bittersweet photograph was obtained by E. Grayzeck with the flight spare Wide Field Camera for the *Astro 1* Mission (which was scheduled for launch on 6 March 1986). The dark spot at the centre is a blemish on the photocathode of the image intensifier.

laboratories. Approximately half the experiments sent to comets in 1985–1986 have related to plasma physics. This aspect of cometary physics provides unique opportunities to obtain knowledge of cosmic plasmas, which, after all, constitute the state of the overwhelming majority of matter in the Universe.

Chemistry

The large-scale structure discussion is briefly continued to introduce a topic in chemistry. The icy-conglomerate model of the nucleus is based on water ices as the major volatile constituent. All relevant evidence, whether remote or *in situ*, confirms this view; the ices are at least 80% water. A spectacular manifestation of this composition is the hydrogen cloud consisting of H atoms from H_2O molecules that have been torn apart. The dimensions of the cloud are large, as illustrated in figure 5, plate 2. Monitoring of Comet Halley's hydrogen cloud has been carried out from *Suisei*, *Pioneer Venus Orbiter*, and from *Dynamics Explorer* 1, and the data are extensive. The record contains evidence for major changes in size and scale height. These variations have been described as 'breathing'. Observations from *Suisei* found brightness fluctuations with a period of 2.2 d. This value has been interpreted as the rotation period of the nucleus and, indeed, the value is consistent with the aspect of the nucleus as recorded by the *Vega*s and *Giotto*.

The water vapour from the nucleus has been unambiguously detected by infrared observations at 2.65 μm from the *Kuiper Airborne Observatory*, by Mumma and his associates. These observations are important for at least two reasons. (1) They provide a straightforward way to routinely detect gaseous water in comets and to determine the production rate. (2) They provide a measurement of the temperature of the comet's interior because the ratio *ortho*:*para* water – an observable – is temperature-dependent. Although the water is observed in vapour form, the ratio *ortho*:*para* reflects the temperature of the ice in the interior because the time scale for changing the spin states is very long. For Halley's Comet, the temperature is *ca.* 35 K. The origin of the temperature is unclear. It could be approximately the black-body equilibrium temperature around aphelion or it could have some other explanation. We need a larger base of data for more comets that the relative ease of these observations makes possible.

The second most abundant species in the atmosphere is carbon monoxide (CO). The abundance is in the range 10–15%, a fact established by Feldman and his associates from ultraviolet spectra obtained on rocket flights.

Clearly, there are minor constituents in the cometary composition. Carbon dioxide (CO_2) is present at the 3.5% level and HCN among others has been reported. Detailed results and models are required to proceed. The reason for this is simple. *In situ* composition measurements with mass spectrometers give densities at a particular mass unit. For example, at atomic mass 16, the measurements could refer to O, CH_4, NH_2, etc. The interpretation requires detailed models and synthetic spectra.

In addition, ions in the range up to 200 u have been reported. Here our imagination may be strongly stimulated. The possibility that the interiors of comets are the ultimate storehouse of unprocessed material from the formation of the Solar System has intrigued scientists for years. Included in this interest is the hope that molecules sufficiently complex to encourage evolutionary processes leading to life might be found. The high mass observations might have several causes: (1) metals, these would be relatively uninteresting in this context; (2) cluster ions, these would be of the form $I^+(H_2O)_n$ and should indicate conditions in the region of formation; (3) complex organics, these would be of interest for their evolutionary possibilities and again a detailed analysis is required to proceed.

The albedo of the nucleus (see next section) may support the existence of organic molecules in the nucleus, but they need not have high molecular masses. If the low albedo values *ca.* 0.02 are correct, they may not be explainable by a surface created by vacuum welding of chondritic powers, even though the albedos of the powders of some carbonaceous chondrites are quite dark. Lower albedos could be obtained by polymerization, a process well known to investigators working in the vacuum ultraviolet region. Many simple molecules polymerize easily, including formaldehyde (H_2CO) which is an important minor constituent in some models of the nucleus (e.g. those of Delsemme). Of course, other chemistries such as the ones studied by Greenberg could be responsible for the low albedo. A rough surface could also help to lower the albedo because 'reflection' would involve multiple scattering in a (say) porous surface structure.

The composition of the dust has been reported to resemble carbonaceous chondrites. This result was not unexpected, but the situation is actually somewhat complex. The dust particle compositions have been divided into three groups: (1) particles composed predominantly of H, C, N and O; (2) particles with a silicate composition and (3) particles with compositions that are a mixture of groups (1) and (2). Cosmic ray irradiation may play a role in the chemistry of these particles. Group (3) is the largest. Their compositions could be considered to be like carbonaceous chondrites enriched in light elements, and they would resemble the Brownlee particles collected in the earth's atmosphere. A possible clue to the dust particle composition and origin may be the measured distribution of masses. Contrary to expectations, there was no peak in the number density with decreasing mass. Rather, the number density with decreasing mass increased down to the measurement limit of 10^{-17} g.

Note, however, that there is no conflict between the low albedo of the surface of the nucleus and the high scattering efficiency of the dust after it leaves the surface. The diffraction part of the scattering coefficient does not depend on the composition.

Nucleus

The confirmation of the existence of a monolithic nucleus, even though assumed by almost all contemporary cometary scientists, is a major triumph of the missions to Halley's Comet. Images (figure 6, plate 4, and figure 7, plate 3) from the *Vega*s and *Giotto* clearly show an irregular, very dark central body with very bright jets largely confirmed to the sunward side. We can look forward to construction of a three-dimensional model of Halley's nucleus from all available aspect data. A schematic summary is shown in figure 8. Note that the surface temperature was measured to be *ca.* 330 K. This value is compatible with the equilibrium temperature of a slowly rotating blackbody at the distance of the *Vega 1* encounter (0.8 AU). The standard formula $T = 289\ \mathrm{K}/r^{\frac{1}{2}}$, r in AU, gives $T = 325$ K.

Some of the false-colour contour displays used to present the initial imaging results were confusing and lead to some erroneous reports (e.g. that the nucleus was double). Although such displays can be useful and attractive in many applications (see figure 5), their use in displaying the first images of a very dark object of unknown shape and orientation can be misleading. The brightness contours draw attention to the jets, not the nucleus. Surprisingly, we make relatively minor use of brightness information in our mental processing of scenes. Rather, the

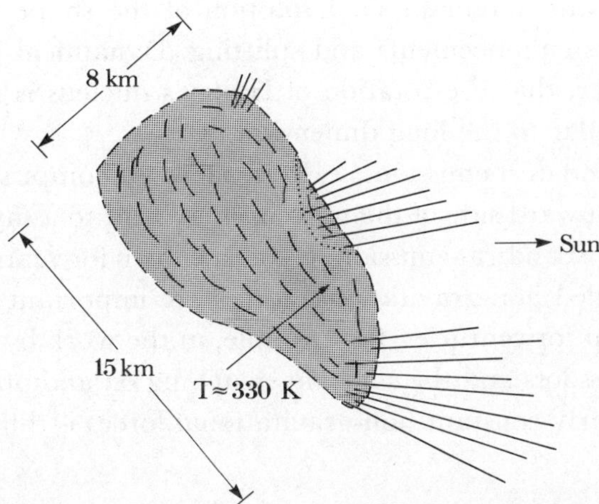

Figure 8. Schematic drawing of the nucleus with dimensions and some jets indicated. The dotted line records the fact that in some views the *Vega* interpretation has a narrower waist than the *Giotto* interpretation. Compare with figures 6 and 7.

brain makes maximum use of edges and their orientation to construct a visual image containing sizes, shapes, and spatial orientations. Grey-scale renditions present the data in readily processable form.

The dimensions of the nucleus are roughly 15 km × 8 km, and the shape has been compared to a peanut or potato. The smaller dimension of 8 km could be larger, say 10 km, because of uncertainties on the night side of the nucleus. The surface has been described as darker than coal or black velvet. The geometric albedo has been quoted in the range 0.02–0.05. These values make the nucleus of Halley's Comet one of the darkest objects in the Solar System. Even though some chondritic powders are very dark and the surface is probably porous, some other process such as polymerization may be needed. Clearly, a critical evaluation of the albedo determination needs to be undertaken before further progress can be made.

The low albedo and the high temperature for the surface lead immediately to the conclusion that most of the comet is covered by a dust crust. Such a crust is a feature of many previously published models of comets. The time scale for heat conduction inward must be much less than the rotation period (to explain the relative lack of nightside sublimation seen in the *Giotto* and *Vega* imagery). Initial estimates for the crust thickness are *ca.* 1 cm, but larger values are also possible. Some previously published models featured an extensive dust cloud that heated all sides of the nucleus by infrared radiation. The concentration of the gas and dust emission on the sunward side calls such models into question.

The overall picture of sublimation in Halley's Comet involves sunlight heating the dark surface to temperatures of *ca.* 330 K, conduction through the crust to the ices, sublimation beneath the surface, and escape of the gases and dust usually in jets covering *ca.* 10% of the surface. The collimation of the jets as seen in sunlight scattered by dust particles (the gas does not remain collimated) requires that the width be approximately equal to the depth.

The evolution of the pits as sublimation continues could easily produce the Brownlee-type particles as the crust at the pit edges breaks away. On a larger scale, it is easy to see how the irregular shape of the nucleus could have been produced. Indeed, after a little reflection, we realize that the spherical nucleus seen so often in our models cannot exist. Even if the shape is initially spherical, it cannot remain so. Evolution of the shape of the nucleus may offer possibilities for brightness enhancements and splitting. Dynamical instabilities could also be important. Note, however, that the rotation of Halley's nucleus is now stable, with rotation about an axis perpendicular to the long dimension.

The nature of the gas and dust emission, i.e. from a few jets comprising *ca.* 10 % of the surface and active only on the sunward side of the nucleus, may lead to a mystery. The reaction force on the nucleus from the dust and gas emission has been known for years, and has been accurately determined. These so-called non-gravitational forces are important in orbit calculations and have been nearly constant for centuries, for example, in the work by Yeomans. Because of the discrete nature of the mass loss and the constant switching on and off of the jets as the nucleus rotates, the history of nearly constant non-gravitational forces is difficult to understand.

Discussion

Because of their diverse physical processes, wide ranges in energy and dimensions, and important impacts on other fields of study, comets continue to excite many investigators' scientific curiosity. Some of these processes, dimensions and impacts are summarized in table 2; the lists are not intended to be complete.

Table 2. The nature of comets

energy

	T/K	T/eV
nucleus (interior)	35	3×10^{-3}
nucleus (surface)	330	3×10^{-2}
ions	6×10^{9}	5×10^{5}

processes	dimensions/km	impacts
condensation	nucleus, 10	cosmic chemistry
sublimation	coma, 10^{5}	origin of solar system
complex chemistry–	plasma tail,	origin of life
gas phase reactions	interaction	meteoritics
fluid flows	region 3×10^{7}	meteors
solar-wind interaction		plasma physics–solar wind

Many of our fundamental ideas for understanding comets date from the 1950s. These are: the Biermann–Alfvén view of the solar-wind interaction and the large-scale plasma structure; the icy-conglomerate model of the nucleus proposed by Whipple; and the storage of comets in the Oort cloud. Only the latter has not been directly and severely tested by the investigations in 1985 and 1986.

Despite the intrinsic interest in the study of comets, the general euphoria over the recent activities, and the anticipation of analysis and intercomparison over the next few years, there may be a tendency toward post-encounter melancholy. Currently, there is no approved comet mission anywhere on Earth, although a retargeted Halley spacecraft may encounter another comet or an asteroid. Yet this depression will probably not last long. The next good comet will cure us. As long as scientists and non-scientists are motivated to obtain images like the one shown in figure 9, plate 4, all is well.

Dr William Butler centuries ago said of strawberries:

'Doubtless God could have made a better berry, but doubtless God never did.'

Comet scientists now might summarize their feelings by paraphrasing Dr Butler:

'Doubtless God could have made a better celestial object, but doubtless God never did.'

I am indebted to Dr A. Delsemme, Dr M. B. Niedner, Professor D. Brownlee and Professor F. Whipple for conversations and clarifications during the preparation of this paper.

FURTHER READING

Our understanding of comets is evolving from the picture presented here. The literature and recent reviews should be consulted. In particular, a controversy has developed over the true rotation period of the nucleus of Halley's Comet. Both a 2.2 d and a 7.4 d period are now (December 1986) being advocated. A comprehensive pre-1985–1986 view of the subject is contained in J. C. Brandt & R. D. Chapman 1981 *Introduction to Comets*. Cambridge University Press (1981). The first spacecraft results for Comet Giacobini–Zinner are in the 18 April 1986 issue of *Science* (vol. 232, pp. 353–385), and for Comet Halley in the 15 May 1986 issue of *Nature* (vol. 321, pp. 259–366). Also see E. Grün (ed.) Comets Halley and Giacobini–Zinner. *Adv. sp. Res.* **5** (1986).

The major presentation of results was in *Exploration of Halley's Comet* (20th ESLAB Symposium). A preliminary proceedings has been published by ESA as *Proc. 20th ESLAB Symposium on the Exploration of Halley's Comet* (in three volumes) ESA SP-250 (1986) and a final proceedings will be published in *Astron. Astrophys.* (scheduled for late summer 1987).

Discussion

E. ANDERS (*University of Chicago, The Enrico Fermi Institute, Chicago, U.S.A.*). Has anyone tried, by computer modelling, to reproduce the elongated shape of Halley's nucleus? I think this shape – with an axial ratio of 2 – is a potentially significant clue to the origin of Halley. Of the three possible mechanisms – cratering, coalescence and sublimation – the first seems unlikely, as comets probably always existed in places where impact rates, or at least impact velocities, were low. Coalescence, by low-velocity collisions, seems more likely, and has actually been proposed for the Trojan asteroid 624 Hektor, with an axial ratio of 3.4 (Dunlap & Gehrels 1969). Sublimation at locally higher rates can also do it, either by a self-accelerating process (such as the preferential evaporation from hollows mentioned by Whipple this symposium), or by some initial compositional heterogeneity (which itself would need to be explained).

Reference

Dunlap, J. L. & Gehrels, T. 1969 *Astron. J.* **74**, 796–803.

J. C. BRANDT. I know of no detailed computer models attempting to reproduce the shape of Halley's nucleus, but I am sure that they will follow. The connection between the present nuclear shape and the comet's origin should be investigated, but convincing or unique results may be difficult to obtain. The nuclear mass now is estimated at roughly one-half the original mass. The shape now, which involves evolution from an unknown original shape, reflects structure and processes in material already lost.

J. A. M. McDonnell (*Unit for Space Sciences, University of Kent at Canterbury, U.K.*) Concerning the shape of Comet Halley's nucleus, and the suggestion that it might have been modified by impact processes, I would like to remind the meeting that by all current evidence Halley has certainly lost a mass equal to its present mass since injection, based even on a 20 000 year existence near its present orbit. Therefore its present shape – be it a potato or a peanut – does have to be ascribed to the ablation by solar radiation perhaps reflecting inhomogeneity of internal structure. At each perihelion passage it loses an average of some 0.5–1 m over the whole surface – perhaps 5 m in active regions – and therefore impact erosion, which corresponds to a rate of 10^{-6} mm a^{-1} at 1 AU as derived from lunar data, is a negligible force in determining its morphology now.

M. K. Wallis (*Department of Applied Mathematics and Astronomy, University College, Cardiff, U.K.*). From Dr Brandt's conception of the jet-emitting regions, I estimate roughly five craters of 1 km radius and depth. At an erosion rate of 10 m per orbit, such craters must be pretty stable. Yet he conceives the surface between them as consisting of 1 cm deep crust over ice, which is rather unstable under sublimation (Shul'man 1972). Are these two scales for evolution of the nucleus surface not contradictory?

Reference

Shul'man 1972 In *Motion, evolution of orbits and origins of comets* (IAU Symp. no. 45), p. 271.

J. C. Brandt. We are all in the position of attempting to synthesize vast quantities of new data into a coherent picture. The roughly 1 cm crust thickness refers to the active (gas and dust-emitting) regions; the crust between the active regions is expected to be thicker. This clarification may resolve the contradiction in time scales.

J. A. M. McDonnell (*Unit for Space Science, University of Kent at Canterbury, U.K.*). How can the 2.2 μm observation of the two water states be inferred as indicative of the interior temperature of Halley, when the major cross section of Comet Halley resides in the surrounding dust coma?

J. C. Brandt. The water observations at 2.65 μm refer to vapour that has already sublimated and left the nucleus. Because alteration of the spin state of the water molecules takes a very long time, the water vapour retains the signature of the interior temperature through the *ortho*:*para* ratio.

Phil. Trans. R. Soc. Lond. A **323**, 447 (1987)

Printed in Great Britain

General discussion

D. McNally

University of London Observatory, Mill Hill Park, London, U.K.

A possible cometary absorption line

Twenty spectra were obtained of τ' Ari during the 2 h period centred on closest approach $(2'.7')$ to the nucleus of Halley's Comet on 19 November 1986. The star was thus occulted by the coma of Halley's Comet. Two further spectra were obtained of τ' Ari on 22 November 1986 when star and comet were well separated. A test spectrum of HD 26571 was taken 20 minutes before the observations of τ' Ari on 19 November. The spectra were taken at *ca.* 17 Å mm^{-1}† with the Intermediate Dispersion Spectrograph with CCD detector of the Isaac Newton Telescope at the La Palma Observatory. The spectral range included the diffuse interstellar line at 5780 Å and the NaD lines. (The observations were carried out as a service observation by Dr R. Terlevitch asisted by Mr I. A. Crawford.)

The object of these observations was to attempt to establish whether or not the carriers of the diffuse interstellar absorption were present in the material of the coma of Halley's Comet. It was found that no cometary contribution to the diffuse interstellar line at 5780 Å could be detected placing an upper limit, an order of magnitude less than in interstellar space, on the abundance of such carriers if present in cometary material.

However, in all the spectra of τ' Ari of 19 November an additional absorption line was found at 5820.8 ± 0.5 Å (rest wavelength if of cometary origin). This line did not appear in the spectrum of HD 26571 or in τ' Ari of 22 November. Flat fields were used to correct for rogue pixels in the CCD detector. There is therefore a strong presumption that this absorption line must arise in the cometary coma. Nevertheless some small doubt must remain about reality because, although a similar set of observations by G. H. Herbig of the Lick Observatory found a line at the same wavelength, it had less than half the strength found by us. The line is also contaminated by stellar S II and Ne I. Having divided the spectra on and off the comet, we found a residual equivalent width of 45×10^{-13} m for the suspected cometary absorption line.

Unfortunately no identification of the line has so far been possible. The nearest match in wavelength is a line of Ni IV at 5820.7 Å. However, Ni IV is an unlikely state of ionization in the coma of a comet. Various other possibilities have been sought but excluded on poor wavelength match, unlikely abundance or absence of other lines from the same species within the observed wavelength range. From the width of the line, an unresolved molecular band or a blend of atomic lines might be expected.

† Å $= 10^{-1}$ nm $= 10^{-10}$ m.